Computational and Structural Approaches to Drug Discovery
Ligand–Protein Interactions

RSC Biomolecular Sciences

This Series is devoted to coverage of the interface between the chemical and biological sciences, especially structural biology, chemical biology, bio- and chemo-informatics, drug discovery and development, chemical enzymology and biophysical chemistry.

Ideal as reference and state-of-the-art guides at the graduate and post-graduate level.

Titles in the Series:

Biophysical and Structural Aspects of Bioenergetics
Edited by Mårten Wikström, *University of Helsinki, Finland*
Computational and Structural Approaches to Drug Discovery: Ligand-Protein Interactions
Edited by Robert M Stroud and Janet Finer-Moore, *University of California in San Francisco, San Francisco, CA, USA*
Exploiting Chemical Diversity for Drug Discovery
Edited by Paul A. Bartlett, *Department of Chemistry, University of California, Berkeley* and Michael Entzeroth, *S*Bio Pte Ltd, Singapore*
Protein–Carbohydrate Interactions in Infectious Disease
Edited by Carole A. Bewley, *National Institutes of Health, Bethesda, Maryland, USA*
Quadruplex Nucleic Acids
Edited by Stephen Neidle, *The School of Pharmacy, University of London, London, UK* and Shankar Balasubramanian, *Department of Chemistry, University of Cambridge, Cambridge, UK*
Sequence-specific DNA Binding Agents
Edited by Michael Waring, *Department of Pharmacology, University of Cambridge, Cambridge, UK*
Structural Biology of Membrane Proteins
Edited by Reinhard Grisshammer and Susan K. Buchanan, *Laboratory of Molecular Biology, National Institutes of Health, Bethesda, Maryland, USA*
Structure-based Drug Discovery: An Overview
Edited by Roderick E. Hubbard, *University of York, UK and Vernalis (R&D) Ltd, Cambridge, UK*

Visit our website on www.rsc.org/biomolecularsciences

For further information please contact:
Sales and Customer Care, Royal Society of Chemistry, Thomas Graham House, Science Park, Milton Road, Cambridge, CB4 0WF, UK
Telephone: +44 (0)1223 432360, Fax: +44 (0)1223 426017, Email: sales@rsc.org

Computational and Structural Approaches to Drug Discovery

Ligand–Protein Interactions

Edited by

Robert M Stroud and Janet Finer-Moore

University of California in San Francisco, San Francisco, CA, USA

RSC Publishing

ISBN: 978-0-85404-365-1

A catalogue record for this book is available from the British Library

Published by The Royal Society of Chemistry,
Thomas Graham House, Science Park, Milton Road,
Cambridge CB4 0WF, UK

Registered Charity Number 207890

For further information see our web site at www.rsc.org

Preface

The idea that disease could be cured by inhibiting a specific protein target, which was articulated by Ehrlich as early as 1909,[1] is at the heart of the computational drug-design methods discussed in this book. Even before any protein structures were known, mechanism-based inhibitors had been developed that very specifically inhibited a chosen enzyme targ.[2,3] In some cases, the structure of a target's binding site could be deduced from the binding surfaces of its ligands.[4] However, scientists appreciated that a general, high-affinity inhibitor would need to be highly complimentary to the shape and electrostatic properties of a target's active site, and designing such inhibitors would require atomic resolution protein structures. Thus modern methods of rational, target-directed drug design did not emerge until the late 1970's and early 1980's, when protein crystal structures could be routinely determined, and increasingly sophisticated computer graphics software allowed visual analysis of protein–ligand interactions.[5–7]

Searching the biomedical literature in the PubMed database with the search phrases '"drug design" AND (structure OR comput*)' roughly gauges growth of this field: 12 papers were found for years 1971–1975, 323 were found for 1986–1990, and 5126 were found for 2001–2005 (see Figure 1). More impressive than the growth in number of published papers, perhaps, is the speed with which computational methods of drug design moved from being academic exercises to actually producing drugs that entered clinical trials. The history of the development of these methods, an assessment of their successes and failures, and prospects for improvement are all topics covered in this volume.

The book is divided into five sections. Section 1 includes two chapters that provide overviews of the drug-discovery field, written from the points of view of structural chemists and a medicinal chemist, respectively. Chapter 1 is an assessment of the reasons why structural and computational approaches have failed to live up to their potential for producing "designer drugs", and the sorts of technological developments required to reliably predict compound specificity and affinity from structure. Novel methods for targeting unconventional sites in proteins are cited as some of the most promising areas for progress in drug discovery.

In Chapter 2 a scientist with over 40 years of experience in medicinal and combinatorial chemistry traces the development of drug-discovery methods

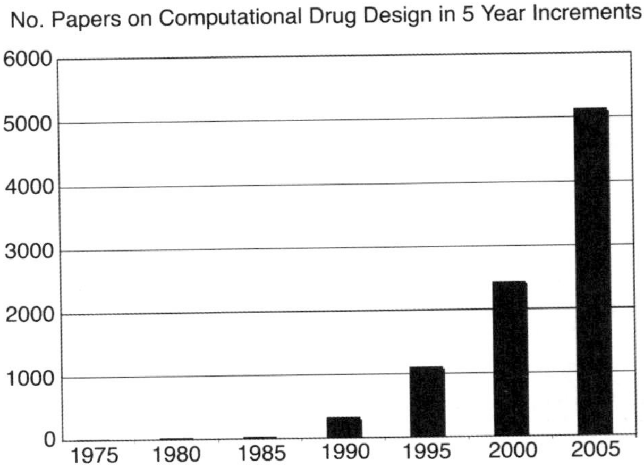

Figure 1 The growth of impact of computational approaches to drug design is indicated by the sum on the ordinate, over each five-year interval ending in the year displayed along the abscissa. The growth follows an exponential growth that essentially doubles every five years.

from the 1960's to the present. This chapter shows how new technological breakthroughs, such as the ability to clone and express proteins, increases in computer power, or progress in protein-structure determination, have led to novel approaches for rapidly identifying new drug targets and drug leads. But the first applications of these approaches have been disappointing in terms of introducing new drugs to the clinic. Initial optimism has been tempered with the realization that complex biological issues have to be considered even in the early stages of drug discovery. Nevertheless, each of the methods has validity and, after refinements that addressed early failures, has served to facilitate drug discovery.

The themes in Section 1 are explored in greater detail in the next four Sections. Subsequent chapters in the book include in-depth discussions and examples of several of the computational methods used in four popular approaches to drug discovery: structure-based design, docking, high-throughput screening, and fragment-based design. Section 2 includes several chapters that are illustrations, descriptions, or evaluations of structure-based drug design methods. One of the earliest and most successful applications of structure-based drug design is described in Chapter 3. In the early 1990's, high-affinity, specific inhibitors of human purine nucleoside phosphorylase were designed using high-resolution crystal structures. These efforts culminated in a transition-state analog inhibitor that entered clinical trials for treatment of cutaneous T-cell lymphoma.

Balancing the success story described in Chapter 3 is the warning in Chapter 4 that not all crystal structures extracted from the Protein Data Bank are reliable enough for structure-based design, and some may actually lead the chemist astray. Water structure in particular can be inaccurate in low-resolution X-ray crystal structures, or in nuclear magnetic resonance structures, and a well-defined water structure is critical for calculating binding free energies. Fortunately, as discussed in Chapter 5, computational simulations can be used to both predict the positions of bound waters in protein cavities and to determine the solvent density around a protein. Chapter 5 also includes a discussion of how reorganization of waters upon ligand binding contributes to the binding free energy and how well-ordered waters in the active site may enhance affinity for a ligand or the plasticity of the binding site. Useful guidelines are presented for when to include explicit waters in a target protein structure during docking.

Knowledge-based methods of predicting binding modes are reviewed in Chapter 6. An explosion in the number of publicly available structures of ligand-bound macromolecules has provided an opportunity to statistically analyze binding preferences of atom types or of chemical groups. Incorporation of these preferences into "binding propensity surfaces" has been a useful way of identifying binding hot spots in proteins. Knowledge-based scoring functions have also been used for virtual screening.

Finally, a novel strategy for designing robust drugs, not subject to drug resistance, is discussed in Chapter 7. This strategy grew out of analysis of numerous crystal structures of substrate-bound and inhibitor-bound HIV-protease.

Section 3 focuses on docking algorithms and their use in drug discovery. Chapter 8 is a critical assessment of the success rates of docking methods in binding-mode prediction, virtual screening, and prediction of potencies. The bottom line of this analysis is that docking algorithms are now quite successful in predicting binding modes, but identification of the most promising hits often fails because of inadequate scoring functions. Protocols for unbiased evaluations of docking algorithms and scoring functions are proposed, and address a critical need in the field.

The various roles of docking in drug discovery are illustrated in Chapter 9. Here, a scenario is presented for employing docking methods during each stage of drug discovery, starting with target identification and ending with lead-compound optimization. The appropriate protocols and scoring functions for each stage are described.

Docking algorithms are becoming increasingly sophisticated. Many recent applications have taken into account flexibility in both ligands and receptors. Chapter 10 is a discussion of the types of protein conformational changes that most often occur upon ligand binding, and the protocols that take these movements into account during docking and screening. Chapter 11 provides several examples of how using a flexible receptor was critical for correctly identifying inhibitors.

Section 4 includes several chapters on library screening. High-throughput screening of chemical libraries is a widely accepted approach to drug discovery. However, as noted in Chapter 2, early applications of this approach yielded a large number of hits that, for various reasons, could never have been developed into drugs. The importance of constructing chemically diverse libraries whose members have drug-like properties is emphasized throughout Section 4. Chapter 12 is a summary of several methods used to predict whether a compound will have drug-like properties. Computational methods are being developed to predict not only a compound's physicochemical properties, such as solubility and lipophilicity, but also its metabolism and hERG channel blockade activity.

Using libraries of drug-like molecules for high-throughput screening does not guarantee that hits will be useful drug leads. Inhibitors discovered by screening often include a number of promiscuous inhibitors, which by many measures are "drug like", but which actually inhibit by forming large aggregates in the presence of the target protein. The fascinating story of how the phenomenon of inhibition by aggregation was discovered and characterized, and a suggested counter-screen for aggregating inhibitors, are the subjects of Chapter 13.

Virtual (*in silico*) ligand screening is a cost-effective way of extracting a subset of likely inhibitors from a large database of compounds. Some of the ways docking is used for virtual screening are described in Section 3. Chapter 14 contains detailed accounts of how one docking program, Autodock, has been used in an iterative screening approach to screen diverse libraries and rank the hits. The iterative screening strategy is compared to other screening strategies, and new directions in virtual ligand screening are described.

One of the major challenges in *in silico* screening is ranking sets of candidate drugs according to binding affinity. Chapter 15 is a review of how binding free energies can be rigorously calculated from first principles and used to rank compounds. While great progress has been made in calculating the electrostatic components of the binding free energy, entropies of binding and the effects of changes to water structure in the active site are not yet adequately accounted for in binding free-energy calculations.

Section 5 includes three chapters on use of fragment-based methods in drug design. These methods combine high throughput screens of low molecular weight compounds with iterative structure-based drug design. Initial hits from screening are generally nonspecific, weak binders, but they can often be elaborated into specific, high-affinity inhibitors without increasing their size beyond the ideal molecular weight range for drugs. A comprehensive review of the common fragment-based approaches is found in Chapter 16, while Chapter 17 is focused on one specific approach, Tethering. Tethering is a fragment-based method that is especially useful for discovering inhibitors that bind to traditionally difficult target sites on receptors. Libraries of disulfide-containing compounds are screened for members that bind specifically near a cysteine at the target site, and subsequently form a disulfide link to that cysteine. Applications of Tethering to five protein targets are described.

One of the most explored classes of drug targets is the class of protein kinases. Therefore it is fitting to conclude the book with a discussion of drug development against these proteins. Chapter 18 is a summary of successful kinase inhibitors in the clinic today, including their binding modes and inhibition mechanisms. A challenge for treating complex diseases involving kinase-signaling pathways is to optimize kinase inhibitors for inhibiting a specific subset of the kinome. As described in this chapter, fragment-based methods are well suited to discovering selective inhibitors because small compounds can bind to nonconserved pockets, and can also induce conformational changes in order to access nonconserved residues. The chapter ends with a description of a comprehensive approach to kinase drug discovery involving high-throughput fragment-based screening, high-throughput crystallography, bioassays against the whole set of human kinases, and ultimately, bioinformatics to guide the design of inhibitors with the desired inhibition profiles.

Janet S. Finer-Moore and Robert M. Stroud

References

1. P. Ehrlich, *Ber. Dtsch. Chem. Ges.*, 1909, **42**, 17.
2. S. S. Cohen, *Science*, 1979, **205**, 964–71.
3. C. Heidelberger, P. V. Danenberg and R. G. Moran, *Adv. Enzymol. Relat. Areas Mol. Biol.*, 1983, **54**, 58–119.
4. P. Gund, J. D. Andose, J. B. Rhodes and G. M. Smith, *Science*, 1980, **208**, 1425–31.
5. C. R. Beddell, P. J. Goodford, F. E. Norrington, S. Wilkinson and R. Wootton, *Br. J. Pharmacol.*, 1976, **57**, 201–9.
6. J. M. Blaney, E. C. Jorgensen, M. L. Connolly, T. E. Ferrin, R. Langridge, S. J. Oatley, J. M. Burridge and C. C. Blake, *J. Med. Chem.*, 1982, **25**, 785–90.
7. P. J. Goodford, *J. Med. Chem.*, 1985, **28**, 849–57.

Contents

Section 1
Overview

CHAPTER 1

Facing the Wall in Computationally Based Approaches to Drug Discovery

JANET S. FINER-MOORE[a], JEFF BLANEY[b] AND ROBERT M. STROUD[*a]

[a]Department of Biochemistry and Biophysics, University of California in San Francisco, S412 Genentech Hall, 600 16th Street, San Francisco, California, 94158-2517, USA
[b]Structural GenomiX, 10505 Roselle Street, San Diego, California, 92121, USA

1.1 The Promise, and the Problem

It has been 36 years since the first protein structure was determined and the promise that structure could guide drug discovery was born.[1,2] Yet today even the ability to design a molecule that will target a nominated site and bind there still remains a tantalizing and intellectually enticing prospect. Most good lead compounds fail for reasons to do with lack of efficacy, toxicity, or interference with metabolic pathways. These properties, too, are ripe for computational evaluation before synthesis rather than after. These important areas are being addressed by computational approaches. But the real challenge for drug design is in the intellectual process of appreciating what is actually coded within macromolecular interactions of the target with small ligands, and the nature of specificity for metabolic enzymes that degrade the compound. This code is at the heart of biology, as it is of chemistry.

Let us first recognize the impact that just one, rather incremental, computational approach could have on the process. If it were possible, we might be able to take the crystal structure of a good lead compound and predict where introducing a specific substituent would increase affinity of the compound with even a 10% chance of moving affinity in the favorable direction. This could provide for successive improvements in affinity of the lead compound with just 10 alternative chemical synthetic modifications at each round, one of which would be an improvement (presuming the 10% chance of success). But there is a wall beyond which the probability of success becomes very low.[3,4] If we could even recognize the cusp at which that will happen ahead of hitting 'the wall', alternative scaffolds could be introduced. This would save enormous resources in chemistry. The same is true for many steps, such as predicting metabolic fate or toxicity of compounds. A small improvement coded in computational screening can reduce downstream efforts by factors of hundreds to thousands. Hence the rational encoding of valid principles, and even empirical wisdom, into the process will remain a key to human health, and one of the highest intellectual challenges of the next decade.

Another area of drug discovery that is hungry for computational assistance is the personalization of medicine. Why is it that some persons have devastating, even life-threatening, responses to Celebrex or Vioxx, while others find them to be the best treatment? Genetic screening is required, but the translation of this into therapeutic strategy is paramount to the salvage of vast potential in the efforts that have already been made to develop potentially good drugs, that could be highly valuable, though for only a defined cohort of the population. If we could anticipate the patients who would respond badly we would save billion dollar markets for present and future drugs. In many ways we are at an exciting frontier in drug discovery, and a frontier where computational biology will need to take the lead.

Lead discovery in the traditional pharmaceutical industry typically involves screening of up to 5 million chemical entities in a few weeks using high-throughput methods, at a cost of \$0.50–\$1.00 for each compound. A screen of that local library could easily cost \$5 000 000. There is consequently real motivation to maximize the return on investment by finding fast and more effective ways to screen.[5] The time and money are compounded by the need to test lead compounds and derivatives for absorption, distribution, metabolism, and excretion (ADME) and toxicity. The spectrum of activity against off-target proteins is something rarely built into even computational screens, and is often left simply to trials on cells, animals, and then humans. With such a wealth of knowledge gathered over decades, why are we not further along in proscribing affinity and favorable properties?

We are in the adolescence of drug discovery. As even empirical rules emerge it is increasingly pertinent that those rules be incorporated into computational analysis, then understood, and then translated into rational rules and algorithms. Ultimately the computer will extract everything we learn in our laboratories, and translate that knowledge into 'wisdom'. Can we blame the industrial leader who shuns the knowledge-based approach for the tried-and-true screening approach to lead discovery? As the knowledge base explodes, the intellectual ability to understand that knowledge and translate it into improved drug discovery has to catch up and demonstrate its contribution to enhanced drug design. Unfortunately much of the knowledge of the binding affinities of series of compounds is sometimes protected, or otherwise unpublished in the knowledge bases of pharmaceutical companies. Another great challenge for the industry and for computational approaches is to release and collate the combined knowledge across the scientific community, for that could help chart the course between scaffolds or elaborations that we should avoid. Some of these areas are already in good hands, though are not yet established, recognized, or used routinely. Like the turn of the tide in an inland sea, the tide often turns long before the reverse flow begins.

The difference between a lead compound and a drug-like compound, or a drug, involves an increase in complexity, and can be encoded in a Venn diagram. This reminds us that optimizing a lead generally develops a compound of modest affinity for its target protein site, to one with drug-like properties, to one with increased affinity and specificity for the target.[6] The requirements for each of these three types of compounds are inherently different. A tool compound for elaboration has different requirements to the drug-like compound, or the final drug. Likewise the context of use for a drug presents different requirements depending on clinical circumstances or the chosen mode of delivery. The classic paper of Lipinski *et al.*[7] sought to address the frustration at Pfizer on finding that many hits were not optimizable. Thus new and separate criteria can now be described for leads, and for drug-like compounds,[6] and these criteria can be phased in during development.

The challenge to make drug discovery a more rational process is balanced by the need to develop new drugs. The theorist enjoys some freedom to evolve and test ideas and produce algorithms for 'docking', 'fragment linking', scoring theoretical libraries, defining algorithms that predict ADME/Toxicity well, *etc.* The real thermometer of how well we can do relies on heavy commitment from the pharmaceutical industry to take the best ideas and algorithms of the moment and test them in a way that pushes them against the real criteria of making drugs. Even the most conservative chemists will readily take advantage of protein structure as a guide, but perhaps not as a driving plan for the next design, in part because of the limitations in its 'interpretation'. And the insights from a crystal structure are chemical and not well coordinated with perspectives of toxicity or other physiological properties. The balance of the 'tide' here reflects both the needs of the professional medicinal chemist to produce compounds, and the need to better generate understanding and rules that can at least bias toward improved properties. Thus the intellectual quest for rationality and the practical need to produce something quantifiable beautifully constrain each other into one of the grand challenges of our generation. There is a beautiful synergy of ideas and objective evaluation. The solution requires a much closer open connection between the vast experience of the pharmaceutical industry and the academic, since industry has less incentive for interpreting failure at any iteration, and academics lack the primary knowledge that is largely inaccessible.

1.2 Current Limitations in Structure-guided Lead Design

The structure of a lead compound bound to a drug target protein is powerful in directing alterations to the lead compound that produce greater affinity,[1] but only up to a point. There seems to be a bounded space within which current schemes prevail, even when one accounts for protein flexibility in some manner.[8] What constitutes the wall beyond which further alteration does not improve affinity, or 'efficiency' (the free energy divided by the number of atoms[9])? Surely if we understood the intermolecular code, we should be able to continue to optimize a lead, at least in affinity for its target, by logical addition of functional groups. In many ways getting past this 'wall' in optimization remains one of the most promising, rewarding, and challenging goals in pharmaceutical chemistry, and in chemical biology.

The process of interpreting the structure of a protein target with a bound fragment, and using that knowledge to design an improvement in binding affinity (arguably one of the easier parts of drug discovery) will be understood and encoded in time. There are some obvious weaknesses in our current analyses. Free energies of associations are not additive because binding entropy is not additive, and the contribution of entropy to the binding free energy is hard to encode, especially the entropy contributions from water molecules that are displaced from drug and site on drug binding to the site. This can be seen, for example, in mutational analysis that removes one, then another interaction. In a case where the crystal structures indicate no significant change in the overall interaction or dynamics, the loss of the first, no matter which one, is small, while the second produces a large effect.[10] Also the hydrophobic effect is a major determinant of binding that is not easily incorporated into the exact 'Newtonian' approach to molecular mechanics. Combination of 'atomic solvation parameters' with the force fields that attempt to describe the 'Newtonian' forces yields improved results.[11] Yet the necessity for empirical solvation parameters also reminds us that molecular mechanics does not yet adequately deal with the solvation and desolvation that occurs when a ligand binds to a macromolecule.

In principle, freeing of an ordered water molecule while an interface is made can yield an entropy-driven advantage of about $\Delta G \approx T\Delta S = RT \ln W$ where W is the number of degrees of freedom generated. Thus $W \approx 6$ for a freed water molecule (three translational and three rotational degrees of freedom) so $\Delta G \approx 0.6 \ln 6 \approx 1.2\,\text{kcal mol}^{-1}$, or a factor of 10 in K_d. Displacement of water contributed to the binding affinity of the human immunodeficiency virus (HIV) protease cyclic urea drugs produced by Merck.[12] However, other water molecules become ordered to make the interface. We are not yet able to pay enough attention to the enormous contribution of water. Free energy perturbation methods provide one of the more exact ways of taking what we know of molecular interactions and testing small alterations to ask what will be the consequence.[13–15]

One limitation to structure-based guidance of drug discovery lies in crystallography itself. Clearly in order to predict molecular behavior we need more than a single set of coordinates that represents the electronic structure of a molecular complex, which is what the usual interpretation of a crystal structure most readily provides. Flexibility is both a key to the ability of the protein to adapt to and bind different chemical moieties, and a basis for relatively poor compatibility of others.[8] Thus it is an essential element in computational approaches. From its inception protein crystallography has struggled to minimize the parameters used to define structure, in order to maintain a good data to parameter ratio, a typical ratio being ~10:1 at a resolution of ~2 Å. The results are usually a median structure, perhaps with occasional alternative conformations in certain positions, and a set of isotropic B factors, which are used to account for thermal vibrations. The parsimonious set of parameters defining the structure is all that can be justified based on the number of observations. With restraints in refinement there are perhaps four to five torsional parameters and eight 'B factors' per amino acid. The B factors really represent the radius of a distribution of trapped states in their librational trajectories that are far from harmonic, or anharmonic. The focus of crystallography has been on developing technology that ensures the quality of the atomic-level median structures. Rather different issues are important for drug development. Drug discovery needs an understanding of the dynamic trajectories of the protein chain, the change in the drug lead on forming a complex with the target, and the change in solvent organization, in order to estimate the various steric, strain, and entropic contributions to binding. Very few of us have interpreted X-ray structures in a way that would allow incorporation of anharmonic motion and of changes in entropy of the side chains and water into structure-guided ligand design.[16–18] This remains a 21st century challenge that is within reach even now. At the very least protein structures used for these purposes should refine a manifold of conformers in the region of any binding site.

One of the ways to appreciate the nature of the prospects and limitations is to track the history of efforts in one highly validated drug target of high impact for human health. These criteria guarantee that the best attempts from all routes available for drug discovery have been stretched and harnessed cooperatively with each other. Therefore they provide tracks to successful drugs, cautionary tales, lessons for how methods can contribute to the process, and a focus on where new improvements are most needed. Thymidylate synthase (TS) is one such example. It has been recognized as a target for antiproliferative anticancer drugs for 50 years, the first structure was determined 20 years ago,[19] and the landscape has been comprehensively mapped ever since. Its mechanism is among the most thoroughly characterized by chemistry, mutational analysis, and crystal structures.[20,21]

1.3 Lessons in Structure-based Drug Design from Thymidylate Synthase

1.3.1 Mechanism-based Inhibitors and Enzyme-catalyzed Therapeutics

In our own laboratory and elsewhere TS has been a rewarding test bed for computational methods of structure-guided drug discovery. TS is essential for DNA replication, especially in transformed cells, and so has been one of the key targets for anticancer drugs. It became one of the first targets for mechanism-based drug design in 1983, with Heidelberger's discovery of 5-fluorouracil.[22] 5-Fluorouracil is converted into 5-fluoro-2′-deoxyuridine-5′-monophosphate (FdUMP), an analog of the TS substrate 2′-deoxyuridine-5′-monophosphate (dUMP), inside cells. Like dUMP, FdUMP forms a covalent ternary complex with the enzyme and cofactor, but is unable to proceed through subsequent catalytic steps. Structure–activity relationships (SARs) for other dUMP analogs indicated that only C5 on the dUMP base could be extensively modified without loss of affinity for TS.[23] The reasons for this became clear when we determined the crystal structure of TS.[19] The preformed dUMP-binding site included side chains or backbone amides within hydrogen-bonding distance of every heteroatom of the pyrimidine ring, and only C5, the position that would become methylated to produce thymidine, was surrounded by enough space to accommodate a bulky substituent.

The structure of TS allowed rational design of a new class of mechanism-based drugs, named 'enzyme-catalyzed therapeutic agents' (ECTAs). These compounds were non-toxic C5-substituted dUMP analogs that were metabolized to release toxic compounds in the rapidly dividing cancer cells via the initial steps of TS catalysis.[24] Cancer cells that had developed resistance to TS inhibitors through overproduction of TS were most susceptible to ECTAs.[25]

1.3.2 Iterative Structure-based Drug Design

In the 1980s technological advances in DNA recombinant biology and in X-ray diffraction detection greatly speeded up the rate at which protein structures could be solved. Several structures of TS in complexes with substrate and cofactor analogs were solved, and the protein became one of the first proving grounds for iterative structure-based drug design.[26–28] Scientists at Agouron Pharmaceuticals employed an iterative structure-based drug-design strategy to design novel antifolates that were more lipophilic, thus better able to diffuse into cells, than the antifolates that had previously entered clinical trials.[29] 5,10-Methylene-5,6,7,8-tetrahydrofolate (CH_2H_4folate) comprises three moieties that occupy different subsites in a large active site cavity. In classical antifolates, such as ZD1694 (Tomudex),[30] one or more moieties are replaced with a new chemical group with similar shape and electronic properties, resulting in an inhibitor that closely resembles the cofactor (Figure 1.1). The Agouron scientists, guided by crystal structures and chemical principles, were able to design inhibitors of human TS with K_i's in the 10–20 nm range that filled the same binding subsites with moieties quite different in structure from those in the cofactor[31] (Figure 1.1).

1.3.3 Docking, Fragments and Optimizability

TS was also an early proving ground for *de novo* drug discovery by molecular docking methods.[32] While the leads that the Agouron team had designed were novel, they still resembled the cofactor in having two aromatic ring systems

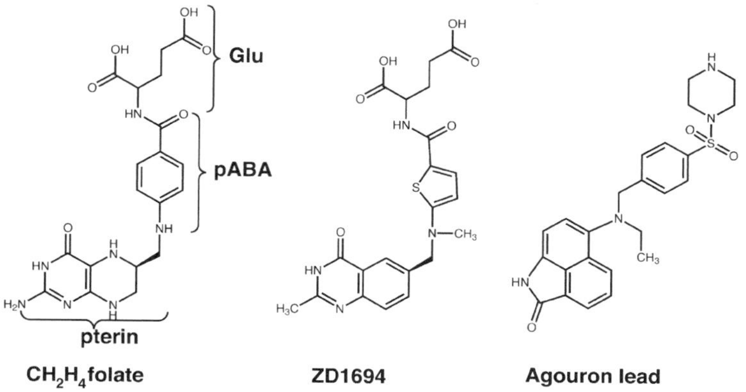

Figure 1.1 Chemical structures of the thymidylate synthase cofactor, 5,10-methylenetetrahydrofolate (CH_2H_4folate) and two antifolate inhibitors. The three moieties of the cofactor are indicated by brackets and labeled. The antifolate ZD1694 (Tomudex) closely resembles the cofactor, while the Agouron compound, which was developed by iterative structure-based design, has two of the three moieties substituted by novel ring systems. The Agouron lead was further elaborated into inhibitors with K_is against human thymidylate synthase in the 10–20 nM range.

connected by a nitrogen-containing linker, and they occupied most of the cofactor-binding site. The goal of the docking screen was to find a TS inhibitor that bore no resemblance to either the substrate or inhibitor. *In silico* screens of the molecules in the Fine Chemical Directory (FCD) using the program DOCK[33,34] identified several molecules that inhibited *Lactobacillus casei* TS with inhibition constants in the high micromolar range. Narrowing the screens to subsets of chemicals that resembled one of the initial low-affinity hits eventually identified phenolphthalein as a 15 μM non-competitive inhibitor of TS.[32] The crystal structure of the TS–phenolphthalein complex confirmed that the inhibitor was bound in the novel site predicted by DOCK. However, the hydrogen bonds phenolphthalein formed with the protein had not been predicted by DOCK because they involved a side chain in a new conformation or were mediated by active site waters not seen in the apo-TS crystal structure. Thus the experiment showed that, in principle, DOCK could be used to identify novel drug leads, but that a rigid receptor target was inadequate for accurately predicting the binding interactions.

The flexibility of the TS active site allows the enzyme to form TS–dUMP–antifolate ternary complexes with a wide variety of antifolates. The protein conformational changes invoked by the binding of an antifolate have sometimes allowed the inhibitor to access non-conserved side chains outside the active site. Binding of the bacterial-specific antifolate *N*,*O*-didansyl-L-tyrosine (DDT) to the *Escherichia coli* TS–dUMP binary complex, for example, induced major rearrangements of both main chain and side chain conformations that exposed several bacterial specific residues to the inhibitor.[35] Interactions of

DDT with these residues were most probably responsible for the inhibitor's species-specificity profile.

DDT had been discovered by in-parallel chemical elaboration of a virtual screening hit to the *L. casei* TS cofactor-binding site, followed by enzyme assays against a panel of TSs from different species.[36] Initially DOCK was used to model the TS–dUMP–DDT complex, but the model was not consistent with SARs. In the DOCK model, the *N*-dansyl moiety of DDT was not positioned correctly because the DDT-induced conformational changes had not been anticipated.[36]

Not surprisingly, the results of virtual screening for TS inhibitors with DOCK are highly dependent on the TS conformation used as the target. Two potent TS inhibitors, the antifolates CB3717 (10-propargyl-5,8-dideazafolate) and BW1843U89 (2-desamino-2-methylbenzoquinazoline) induce different protein conformational changes when they bind to TS–dUMP binary complexes (Figure 1.2). When BW1843U89 was docked to TS fixed in the CB3717-bound conformation, its interaction score was worse than those for the 500 top scoring compounds from the Available Chemicals Directory (ACD). A virtual screen for inhibitors to TS in the CB3717-bound conformation would not have selected BW1843U89 as a likely inhibitor, yet it has a significantly higher affinity (~10 pM) than the ~10 nM CB3717. Similarly, a screen of the ACD using the BW1843U89-bound conformation of TS as the target would have identified over 500 hits with better interaction scores than CB3717 docked to this target.[37] The results of these experiments made it clear that screening against a rigid receptor would, in general, miss a large proportion of potential inhibitors.

Fortunately, the many crystal structures of TS complexes showed that the flexible regions of the TS active site are limited to a few loops and side chains.

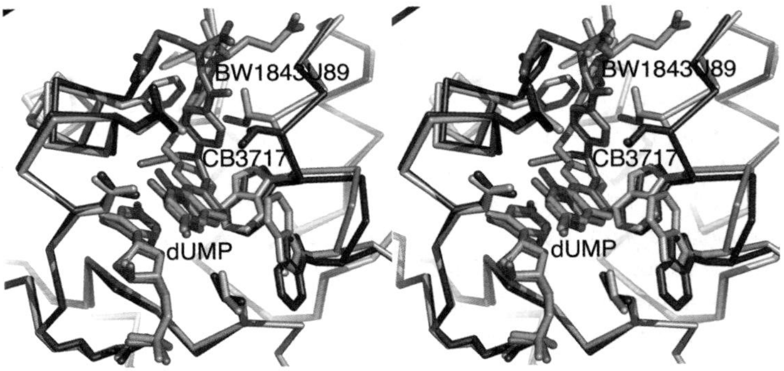

Figure 1.2 Stereo cartoon of the superposition of the active sites of two *E. coli* thymidylate synthase ternary complexes, one containing dUMP and the antifolate CB3717 (gray protein, magenta ligands) and the other containing dUMP and the antifolate BW1843U89 (blue protein, orange ligands). The two antifolates induce different conformational changes to side chains and flexible loops in the active site. The figure was made with PyMOL (Delano Scientific, San Carlos, CA).

It was possible to get significantly more accurate docking and scoring results just by incorporating into the target model a spectrum of side-chain rotamers for a few active-site residues that are seen to change conformation upon complex formation.[37] In a recent test, it was found that *in silico* screening for TS inhibitors with DOCK gave a dramatically better enrichment factor if the compounds were docked against an ensemble of 18 distinct TS conformations, generated by allowing three flexible regions of the TS cofactor-binding site to adopt independently the conformations seen in three crystal structures.[38] In a screen of the ACD using this flexible docking algorithm, 55.7% of known TS inhibitors were in the top 10% of ranked compounds, compared to ~23% when docking was against any single TS conformation.

Another lesson learned from these early attempts at *in silico* drug discovery was that, even though a compound has inhibition kinetics consistent with it being a good drug lead, it in fact may not be a very good starting point for structure-guided drug design. Phenolphthalein bound near a non-conserved domain of *L. casei* TS, suggesting that it could be elaborated into an *L. casei*-specific TS inhibitor. If so, it might be feasible to use TS, which has a highly conserved active site, as a target for antimicrobials. Derivatives were designed that would specifically interact with non-conserved residues and several of these showed improved specificity for *L. casei* TS compared with the human enzyme. However, crystal structures of TS complexed with two of the inhibitors revealed that they bound in different sites and orientations than phenolphthalein[39] (Figure 1.3). Apparently, binding affinity was largely based on hydrophobic effects rather than on specific interactions, so the binding modes of derivatives were not easily predictable. A stable binding mode for a drug lead is a prerequisite to iterative structure-based design.

Therefore one challenge is how to determine whether a given hit binds in a stable mode, *i.e.* whether its close analogs bind to the same site in the same pose (pose = conformation + orientation). This partly determines how 'optimizable' a given hit will be – how far it can be modified before it hits an affinity plateau or wall. We need to be able to design the optimization path for any given hit based on the predicted affinities of its analogs. A recent analysis by Hajduk,[4] partially addresses this issue by fragmenting, or deconstructing, 18 highly optimized inhibitors to identify the minimal fragments. These data place well-defined limits on the ideal size and potency of fragment leads that are being considered for use in fragment-based drug design. Surprisingly the relationship between pK_d and molecular weight is close to being perfectly linear ($R^2 = 0.98$), with a gradient–binding efficiency index (defined as the pK_d divided by the MW in kDa) of ~12, which corresponds to 0.30 kcal mol^{-1} per heavy atom. Kuntz *et al.*,[3] plotting diverse sets of ligands to protein sites, observed an initial rapid increase of 1.5 kcal mol^{-1} per heavy atom for the first four atoms, and a similar gain per atom of 0.3 kcal mol^{-1} beyond that for the range 4–25 atoms, but with no net gains in potency beyond 25 heavy atoms. This represents the affinity 'wall' where further additional atoms no longer increase affinity.

At this limit it was argued that the maximal affinity of ligands may be due to significant shielding of van der Waals and hydrophobic interactions, such that

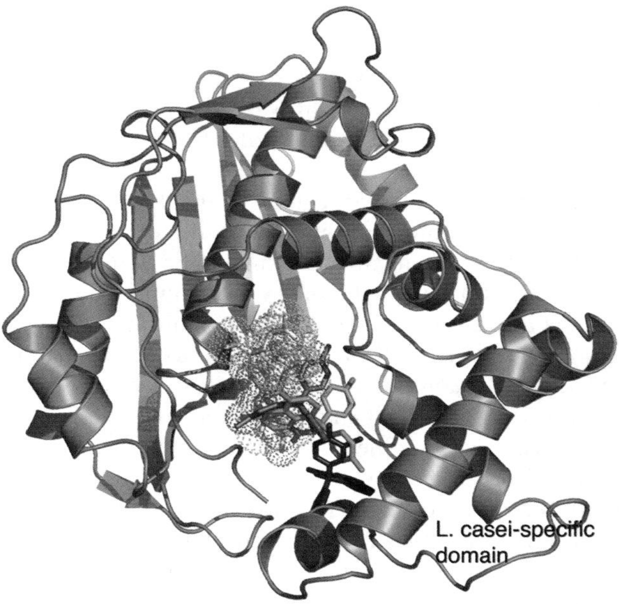

Figure 1.3 Ribbon drawing of one monomer of *L. casei* thymidylate synthase illustrating the binding sites for the substrate and cofactor (shown in gray sticks and dotted spheres), phenolphthalein, an inhibitor discovered by *in silico* screening (shown in purple sticks), and two inhibitors based on the phenolphthalein scaffold, MR20 and α156 (orange and blue sticks, respectively). MR20 and α156 are analogues of phenolphthalein that were designed to contact residues in the *L. casei*-specific domain. However, all three inhibitors bound in different positions and orientations, and thus phenolphthalein proved to be a poor lead compound for optimization. The figure was made with PyMOL (Delano Scientific, San Carlos, CA).

as the number of ligand atoms increases, the average contribution per atom to the binding energy decreases due to shielding by other regions of the molecule. Nevertheless what is abundantly clear is that each protein site has its own size that determines the limiting molecular complexity and size of compound that gains potency there. Thus, for example, compounds of 36 atoms are needed to attain potency for HIV protease. Since the size of a protein cavity or active site is tuned to its biological role, the substrate determines the size and chemical complexity of the target site, which in turn determines the complementary size and complexity of an inhibitor, and so the degree to which potency can increase with optimum additional atoms.

The 'deconstruction analysis' of 18 drug leads from 15 Abbott discovery programs[4] is useful in predicting a relationship between the molecular weight of the fragment and its potency, and the final molecular weight of an elaborated molecule if based on that fragment lead. This useful algorithm suggests, for example, that fragment leads with molecular weights over 250 should have potencies below 30 μM if they are to retain drug-like properties (that is, be compatible with Lipinski's 'rule of 5') at the end of an elaboration. The ability to predict which fragments can be advanced and retain a stable mode is still extremely challenging and not possible yet.

A part of the limitation in the ability to elaborate a lead lies in the Gibbs free energy equation and the logarithmic relationship between binding free energy and affinity. A 10-fold difference in affinity at the final stages of lead optimization can be the difference between success and failure, yet this is only 1.4 kcal mol^{-1}, a very small amount. This amount is the equivalent of a displaced water molecule, or a favorable intramolecular hydrogen bond formation, after either contributor is balanced by the loss of its favorable interaction before complexation. This leads to the apparent paradox that computational methods tend to be less valuable in the 'end game' of lead optimization: the noise intrinsic to the most rigorous computations (*i.e.* free energy perturbation) is about 1.5–2.0 kcal mol^{-1} in the best cases, so predicting relative activities to an accuracy of even 10-fold is rarely possible in practice.

1.4 New Developments in Structure-based Drug-design Methods

1.4.1 Fragment-based Methods

Even the largest libraries of compounds in major pharmaceutical corporations ($\sim 10^7$) are miniscule in comparison with the possible chemical diversity space, estimated to be over 10^{60} possible compounds for molecules based on up to 30 non-hydrogen atoms.[40] A significant reflection is that a collection of just one molecule of each of these 10^{60} would exceed the mass of the earth by 10^{10} fold! Much effort is therefore focused onto ways of screening chemical space by experimental and virtual screening of fragments.[41] Computationally, active sites can be mapped in terms of shape and chemical features and encoded as bit strings that can be used to screen for scaffolds that will bind there.[42] This is a powerful route to focused elaboration aimed at a class of similar targets. Kuntz developed DOCK[43] using a reference database of small molecules to describe binding sites by the bit string of dock hits, and then used these bit strings to compare sites.[44] This procedure can now reproduce crystal structures to within about 2 Å for 70% of test cases for flexible ligands with up to seven rotatable bonds, diminishing with complexity and increasing with fewer rotatable bonds. More realistic models of the receptor would improve this rate.

High-throughput crystallography has been harnessed very efficiently to screen automatically cocktails of compounds from fragment libraries that bind to particular targets that can be crystallized.[45–47] This approach has several advantages since it identifies the soluble fragments that are selectively bound by the target with high (μM or better) or low (mM) affinity, and shows immediately how and where they bind to the protein. Compounds that target the selected 'active site' can readily be identified relative to those that bind at other sites, such as allosteric sites.[47]

The underlying idea behind attempts to discover fragments that bind in a particular target site and to further elaborate these weakly binding small molecules derives from several key concepts. First, recent papers,[6,48] on 'lead-like' compounds noted that there is an inverse relationship between screening hit rate and ligand complexity: smaller molecules provide higher hit rates with a higher probability of discovering novel hits and interactions (see below). Fragment optimization can proceed more efficiently than typical optimization of high-throughput screening hits, since molecular weight can be kept to a minimum while affinity is systematically improved. This results in leads with higher 'ligand efficiency',[9] which has been proposed to be a key factor in clinical success. Likewise, sites that do not have pockets that can bind fragments are less likely to be tractable drug targets.[49]

The higher hit rate for fragments is because they do not simultaneously need to satisfy as many complementary interactions with a site as a larger molecule. This has important consequences for discovering novel fragment hits. Consider a fully elaborated drug-sized molecule which contains a fragment that can bind to a key subsite in an active site, for example, the benzamidine moiety common to many non-peptide serine protease inhibitors which bind in the P1 pocket. Benzamidine binds weakly to trypsin-like serine proteases, but fully elaborated molecules which contain a benzamidine moiety will only inhibit the enzyme if the rest of the molecule is complementary to the rest of the active site. The recent discovery of potent serine protease inhibitors that contain a non-basic P1 moiety is a striking example of this concept. The cost of an 'insult' to the interaction can be lethal. Serine protease enzyme-substrate and enzyme-inhibitor SARs have been extensively studied throughout the past 40 years, beginning with the papers of Schechter and Berger in the 1960s[50,51] until the present, producing a vast literature. Yet the fact that simple haloaromatic moieties at P1 can provide potent inhibition was only discovered recently because of the unusual, non-intuitive nature of this interaction in the P1 pocket and presumably due to the incorrect presentation of haloaromatics group in molecules present in screening libraries. SGX Pharmaceuticals also recently showed that simple haloaromatic fragments bind to the P1 pocket of blood clotting factor VIIa,[47] even though no haloaromatic inhibitors have yet been reported for this particular serine protease (they have only been reported for thrombin, factor Xa, and trypsin).

Finally, the gain in binding free energy by linking two adjacent fragments together in an appropriate manner can be much more than the sum of the binding free energies of the fragments themselves. This fact was well examined in the paper of Jencks in 1981,[52] and now re-evaluated from the drug discovery-by-parts view.[53] Jencks was concerned as to how an enzyme that makes adenosine triphosphate (ATP) from adenosine diphosphate (ADP) and phosphate (Pi) could ever release its product. He postulated that the linked ATP should be bound so much more tightly than ADP + Pi as to become an inhibitor. He tested this notion by the reverse experiment. He fragmented biotin, a known tight-binding ligand of avidin $\Delta G = -17\,\text{kcal mol}^{-1}$, and measured the binding affinities of the two biotin components as $-6\,\text{kcal mol}^{-1}$ and $-5\,\text{kcal mol}^{-1}$. The binding free energy of the linked fragments, that is, of biotin itself, was greater than the sum of binding free energies of the fragments by $\sim 6\,\text{kcal}\,\text{mol}^{-1}$. Biotin achieved higher affinity than the sum of its parts since one component effectively brought the other component to a high local concentration at its site. Put another way, it paid the entropic price for bringing in the 'linked fragment'. That entropic contribution shows up as a gain in favorable free energy since the linkage is fixed in the conjugated pair, to be offset by the lesser entropy cost of fixing the more complex linked molecule in the target site. With the benefit of a more comprehensive database a more recent analysis suggests that the cost of fixing the larger conjugated molecule is actually less than Jencks would have predicted, perhaps only accounting for a factor of 10^3 in affinity. This provides an even stronger case for linking,[53] while pointing out that the apposition of non-overlapping fragments is so rare as to make the situation not general at least.

The fragment method is very popular and has yielded novel drug leads from a variety of approaches. These include biophysical methods for detecting and assigning the binding of a fragment, including nuclear magnetic resonance (NMR), crystallography, *in silico* screens, calorimetry, and functional screens. A landmark paper in 1996 demonstrated that fragments could be screened for binding to a protein target using NMR. Termed 'SAR by NMR' the method defined the idea that 'fragment' sized compounds could be screened for binding to a protein when binding was too weak to ever be detected in a typical high-throughput screen.[54] The location of the site could be determined in more detailed NMR experiments, and the fragment elaborated. Also fragments can often be more easily elaborated and tested than can products of larger library screens.

A particularly innovative approach to fragment discovery is by disulfide-mediated tethering.[55] The approach requires a library of 'fragment' compounds that are synthesized with added thiols at various points around the molecule. The thiol on the compound is covalently linked to a small disulfide-linked cystamine-blocking group in the library. A site on the protein structure is nominated as the target. Residues around the periphery of the site are singly mutated to cysteine and the libraries screened for fragments that bind by forming disulfides. Typically those that can cross-link to the introduced cysteine can be detected at $\sim$20-fold lower concentration than can the fragment alone. The compounds that bind and form disulfide bonds are titrated according to stability in a reducing agent, typically β-mercaptoethanol, that then reflects the strength that the fragment itself has for the site, and their identities are determined by mass spectrometry. The binding mode can be seen by crystallography. The thiol on the fragment hit is removed and the resulting thiol-free fragment is tested against the wild-type protein. The compound is then elaborated to develop selectivity.[56]

There is still much need to understand what is required to maintain ligand efficiency during elaboration.[57] The fragment approach has succeeded best by elaboration of fragments[58] rather than linkage of pairs of fragments. The main reason for this is probably that adjacent fragment binding is very rare, and rarer still with geometry correct for linking. In part, the nature of the linker is key to the properties of the conjugated compound. It is usually the case that where multiple non-overlapping fragments have been observed, out of hundreds of fragment hits, the geometry generally isn't appropriate for the linking of fragments other than by extremely demanding chemistry.[53] Certainly fragment linking cannot be relied upon, as the probability of finding two linkable fragments is seemingly so low.

The SAR-by-NMR group at Abbott has championed the fragment-linking approach, with a few instructive publications. However in no case is it clear that the route to discovery was that the fragments were first discovered and then linked. This would need to be explicitly shown in any such case since reverse engineering of discovered, more complex inhibitors (a nice route to extend into other permutations, granted) does not show that fragment linking is effective. Thus fragment-linking (as opposed to fragment discovery) is theoretically attractive and should be attempted if and when the possibility arises, but it may be so rare in practice as not to merit much concern here.[53]

Finally, it is not always the case that binding of inhibitor recapitulates the binding of inhibitor fragments.[59] Some complex inhibitors would not be discovered by screening libraries containing their fragments. Powerful as fragment-based methods are, they don't replace screening of libraries of larger compounds. Thus fragment-based methods are at present complementary to classical high-throughput screening approaches.

1.4.2 Identifying Drug Target Sites on a Protein

One of the goals of proteomics, and the protein structure initiatives around the world, is to be able to recognize targt sites on proteins that might be effector-binding sites, and that seem like they should bind compounds or proteins as part of their biological role. Many sites often can be detected by 'pocket detection' methods. If the pockets are close to known functional regions the choice of a target site is simplified. However, in many cases there are biological regulatory sites that are not yet recognized. These, too, are sites that we need to consider as drug target sites.

The recent structure of an ammonia channel from the Rh family, AmtB, in complex with a cytoplasmic protein, GlnK, showed that the process of nitrogen assimilation is highly regulated, not only at the transcriptional level, but also at the protein level, by blockade of the ammonia channel by GlnK.[60] There are several known effectors that control the association of GlnK with the ammonia channel. GlnK binds ATP at one site, which augments the interaction that leads to blockade. This site could be seen in the crystal structure of the complex, which had ADP bound at the site. But GlnK also binds 2-oxoglutarate, sensing the health of the carbon pool in the cell, and this dissociates the complex. Thus 2-oxoglutarate could not be seen in the complex, but rather will be seen in the GlnK structure alone. Based on DOCKing analysis there are three likely ligand pockets in GlnK besides the ATP site. In all probability one of these binds 2-oxoglutarate, but the others may also bind different ligands.

Detecting 'pockets' in protein structures was pioneered because, while the largest pocket is often the substrate site, others may well be allosteric sites for cellular components that are not yet recognized. Protein structures often show pockets that look as if they were made to bind ligands! Since almost every protein molecule interacts with at least one, and often several, proteins and ligands, it is no longer surprising to find that pockets on protein surfaces are allosteric effector sites. Some such sites have long been recognized, for example those on phosphorylase.[61,62] Others were discovered from structures before any particular effector was ever recognized. For example an inhibitor targeted to a pocket on caspase-7 directly inhibits the apoptosis-controlling enzyme by converting it to a zymogen-like conformation. The pocket presumably has a function in biology that is not yet recognized, however the discovery of an inhibitor that binds at this site has already established it as a key target for anti-cancer drugs that can modulate apoptosis.[63]

More recent methods of detecting effector-binding sites involve phylogenetic analysis that locates regions on the protein surface with a potential for high binding affinity by similarity with other proteins. The prediction of binding sites is incorporated into drug discovery technologies, and into several docking tools.[64]

This avenue of drug discovery, looking for sites on a protein that seem to be receptive to small-molecule binding, can lead to the detection of genuine sites for as yet unknown regulatory molecules. Alternatively, it may identify *de novo* sites that have no biological role. But what is certain is that there will be many of these cavities, and that proteins will adapt to molecules that bind in those sites. This opens up a new horizon for drug design, even against targets for which active site-targeting drugs or compounds are known already. An example is the discovery of two different non-competitive inhibitory sites on Factor VIIa, discovered by phage display of a peptide library. Subsequent peptide synthesis and optimization yielded low nM inhibitors that acted at these allosteric sites. Knowing that these sites could be inhibitory, Structural GenomiX (SGX) found fragment hits at these sites using fragment-based crystallographic screening.[47]

1.4.3 Targeting Protein–Protein Interactions

A new horizon in drug discovery is the targeting of protein–protein interactions.[65] Protein–protein interfaces are often sterically 'flat' and ~1200–3000 Å^2 in contact area, however the binding contributions are not evenly distributed throughout the surfaces. 'Hot spots' exist, which represent focal points of the interaction.[66] A major theme that needs to be incorporated into computational thinking is the concept of 'hot spots'. These are sites that are malleable in the protein, and so can adapt to a protein partner. Protein–protein interaction can be redesigned to achieve large (*e.g.* ~300-fold) selectivity gains.[67]

Because protein–protein interacting sites seem to be flexible prior to complexation, and seem to change on forming the complex, they provide a manifold of new target sites for drug design and discovery. In the case of IL-2–IL-2 receptor, the receptor and a discovered small molecule do bind to the same hot spot on IL-2, however they trap very different conformations of IL-2.[68] This suggests that precise structural mimics of receptors are not required for high-affinity binding of small molecules. More importantly, there are multiple solutions to tight binding at shared and adaptive hot spots. A novel approach to designing inhibitors that target protein–protein interfaces is to use chemical templates that are capable of mimicking protein molecules as scaffolds that can be adapted to modulate a particular interaction by adding substituents.[69]

Given the wealth of interacting substrates and partners within a protein system, there are a wealth of sites that can be targeted to modulate the activity of the system. 'Hot spots' encode the adaptation required to bind other partners. These can be discovered using molecular mechanics, and targeted using flexible docking.[70] Nevertheless, most small molecule modulators found so far have not been found this way, but rather by serendipity coupled with structural analysis of discovered effectors. Such discoveries are important in demonstrating that these effector sites exist. They probably exist for genuine biological modulation inside the cell. Their importance now is in showing that these sites are much more common than we thought in the pre-proteomic world.

1.4.4 Computational Docking to Nominated Sites

New developments in docking and scoring algorithms are beginning to make virtual screening a more effective way of discovering drugs that target a nominated site. A combinatorial approach to library design based on four-point pharmacophores provides a way of selecting compounds that match the target site.[42,43] Docking can be combined with molecular dynamics (MD) simulations to more accurately dock small molecules into protein receptors. More focused, accurate MD simulations are applied when a few selected ligand candidates remain. MD simulations can be used:

(i) during the preparation of the protein receptor before docking, to optimize its structure and account for protein flexibility;
(ii) for the refinement of docked complexes, to include solvent effects and account for induced fit;
(iii) to calculate binding free energies, to provide an accurate ranking of the potential ligands; and
(iv) in the latest developments, during the docking process itself to find the binding site and correctly dock the ligand *a priori*.[71]

Accurate docking of compounds into nominated sites can be accomplished in many cases. However tests of current methods seem to reinforce the idea that accurate docking of a compound to a site is not sufficient for calculating its affinity. Improved treatment of electrostatics, and assessments of changes in water structure, of induced conformation changes in the protein, and of binding entropy are needed to rank sets of docked inhibitors. While many scoring functions have been proposed, all remain challenged to define ranking among real ligands. In many cases performance is good enough to yield high enrichment for real versus false ligands, but not to rank the hits accurately.[72] Effective ranking, when it works, will actually be a tremendous assist in drug discovery. If one could adequately rank derivatives of a known lead compound, one could save the vast efforts of chemists to synthesize derivatives that would not advance. In principle, computational methods can achieve this, but there is great need for the vision that can develop an effective scoring scheme from empirical results and theoretical considerations.

For certain specialized applications docking and scoring can be highly accurate. For example, when inhibitors of BACE1 were docked to the target site using an algorithm that incorporates flexibility in both the protein and the inhibitor, the mean error in predicted binding affinity of 50 compounds was just 1.2 kcal mol^{-1}. 80 compounds were well ranked according to affinity.[73] Familiarity with a target enables tailoring the screening method to great advantage. An evaluation of 10 docking programs and 37 scoring functions conducted against eight proteins of seven protein types was made.[74] It showed that, while the docking programs were able to generate ligand conformations similar to crystal structures of protein–ligand complexes structures for at least one of the targets, scoring functions were less successful at distinguishing the crystallographic conformation from the set of docked poses. Docking programs identified active compounds from a pharmaceutically relevant pool of decoy compounds; however, no single program performed well for all of the targets. None of the docking programs or scoring functions made a useful prediction of ligand binding affinity.

Proteomics research has led to more accurate homology modeling of potential protein targets. These models are a big step back from crystal structures of the target. Nevertheless, a recent application to models of Aurora kinases A and B led to a new series of compounds with selectivity for these kinases versus others.[58] Another good approach is targeted towards proteins with common ligands or substrates. The kinases have been a prominent palette for anti-cancer drugs where selectivity, or lack of it, often can be used to great advantage. Another application to purine-binding proteins shows that $\sim$13% of the human genome encode such proteins. Structure-based approaches have defined new classes of compounds that are leads.[75]

1.5 Conclusion

Computation is ultimately a powerful primary engine for drug discovery, depending totally on chemistry, structures of drug–target complexes, and the vast experience in the pharmaceutical industry. At present medicinal chemists, crystallographers or NMR spectroscopists, and computational biologists can all benefit greatly from strengthening the tenuous synchrony that exists between us. Crystallographers need to provide the asynchronous trajectories within the target sites. The docking team and molecular mechanicists need to recognize the manifold of structures represented, and we need to decode the rules for

ADME/Toxicity and translate these into the medicinal chemists rules of triage. These fields are among the major great challenges for human health control in the foreseeable future. They promise exponential growth (see the Preface) and deserve to attract the best scientific minds at this particular time when the opportunities of synergy between these disciplines offer so much to human kind.

References

1. T. L. Blundell, H. Jhoti and C. Abell, *Nat. Rev. Drug Discovery*, 2002, **1**, 45.
2. G. Scapin, *Curr. Pharm Des.*, 2006, **12**, 2087.
3. I. D. Kuntz, K. Chen, K. A. Sharp and P. A. Kollman, *Proc. Natl. Acad. Sci. U. S. A.*, 1999, **96**, 9997.
4. P. J. Hajduk, *J. Med. Chem.*, 2006, **49**, 6972.
5. J. W. Davies, M. Glick and J. L. Jenkins, *Curr. Opin. Chem. Biol.*, 2006, **10**, 343.
6. T. I. Oprea, A. M. Davis, S. J. Teague and P. D. Leeson, *J. Chem. Inf. Comput. Sci.*, 2001, **41**, 1308.
7. C. A. Lipinski, F. Lombardo, B. W. Dominy and P. J. Feeney, *Adv. Drug Delivery Rev.*, 1997, **23**, 3.
8. S. J. Teague, *Nat. Rev. Drug Discovery*, 2003, **2**, 527.
9. A. L. Hopkins, C. R. Groom and A. Alex, *Drug Discovery Today*, 2004, **9**, 430.
10. R. J. Morse, S. Kawase, D. V. Santi, J. Finer-Moore and R. M. Stroud, *Biochemistry*, 2000, **39**, 1011.
11. C. A. Schiffer, J. W. Caldwell, R. M. Stroud and P. A. Kollman, *Protein Sci.*, 1992, **1**, 396.
12. P. Y. Lam, P. K. Jadhav, C. J. Eyermann, C. N. Hodge, Y. Ru, L. T. Bacheler, J. L. Meek, M. J. Otto, M. M. Rayner, Y. N. Wong, C. -H. Chang, P. C. Weber, D. A. Jackson, T. R. Sharpe and S. K. Erickson-Viitanen, *Science*, 1994, **263**, 380.
13. J. Carlsson and J. Aqvist, *Phys. Chem. Chem. Phys.*, 2006, **8**, 5385.
14. V. B. Luzhkov, M. Almlof, M. Nervall and J. Aqvist, *Biochemistry*, 2006, **45**, 10807.
15. R. Mahajan, D. Kranzlmuller, J. Volkert, U. H. Hansmann and S. Hofinger, *Phys. Chem. Chem. Phys.*, 2006, **8**, 5515.
16. C. A. Schiffer, J. W. Caldwell, P. A. Kollman and R. M. Stroud, *Proteins*, 1990, **8**, 30.
17. C. A. Schiffer, P. Gros and W. F. van Gunsteren, *Acta. Crystallogr. Sect. D: Biol. Crystallogr.*, 1995, **51**, 85.
18. R. M. Daniel, R. V. Dunn, J. L. Finney and J. C. Smith, *Annu. Rev. Biophys. Biomol. Struct.*, 2003, **32**, 69.
19. L. W. Hardy, J. S. Finer-Moore, W. R. Montfort, M. O. Jones, D. V. Santi and R. M. Stroud, *Science*, 1987, **235**, 448.

20. C. W. Carreras and D. V. Santi, *Annu. Rev. Biochem.*, 1995, **64**, 721.
21. J. S. Finer-Moore, D. V. Santi and R. M. Stroud, *Biochemistry*, 2003, **42**, 248.
22. C. Heidelberger, P. V. Danenberg and R. G. Moran, *Adv. Enzymol. Relat. Areas Mol. Biol.*, 1983, **54**, 58.
23. D. V. Santi and P. V. Danenberg, in *Folates in Pyrimidine Nucleotide Biosynthesis*, ed. R. L. Blakley and S. J. Benkovic, New York, 1984.
24. D. B. Lackey, M. P. Groziak, M. Sergeeva, M. Beryt, C. Boyer, R. M. Stroud, P. Sayre, J. W. Park, P. Johnston, D. Slamon, H. M. Shepard and M. Pegram, *Biochem. Pharmacol.*, 2001, **61**, 179.
25. Q. Li, C. Boyer, J. Y. Lee and H. M. Shepard, *Mol. Pharmacol.*, 2001, **59**, 446.
26. D. A. Matthews, K. Appelt, S. J. Oatley and N. H. Xuong, *J. Mol. Biol.*, 1990, **214**, 923.
27. W. R. Montfort, K. M. Perry, E. B. Fauman, J. S. Finer-Moore, G. F. Maley, L. Hardy, F. Maley and R. M. Stroud, *Biochemistry*, 1990, **29**, 6964.
28. K. M. Perry, E. B. Fauman, J. S. Finer-Moore, W. R. Montfort, G. F. Maley, F. Maley and R. M. Stroud, *Proteins*, 1990, **8**, 315.
29. K. Appelt, R. J. Bacquet, C. A. Bartlett, C. L. Booth, S. T. Freer, M. A. Fuhry, M. R. Gehring, S. M. Herrmann, E. F. Howland, C. A. Janson, T. R. Jones, C. C. Ken, V. Kathardekar, K. K. Lewis, G. P. Marzoni, D. A. Matthews, C. Mohr, E. W. Moomaw, C. A. Morse, S. J. Oatley, R. C. Ogden, M. R. Reddy, S. H. Reich, W. S. Schoettlin, W. W. Smith, M. D. Varney, J. E. Villafranca, R. W. Ward, S. Webber, S. E. Webber, K. M. Welsh and J. White, *J. Med. Chem.*, 1991, **34**, 1925.
30. A. L. Jackman, D. I. Jodrell, W. Gibson and T. C. Stephens, *Adv. Exp. Med. Biol.*, 1991, **309A**, 19.
31. M. D. Varney, G. P. Marzoni, C. L. Palmer, J. G. Deal, S. Webber, K. M. Welsh, R. J. Bacquet, C. A. Bartlett, C. A. Morse, C. L. Booth, S. M. Herrmann, E. F. Howland, R. W. Ward and J. White, *J. Med. Chem.*, 1992, **35**, 663.
32. B. K. Shoichet, R. M. Stroud, D. V. Santi, I. D. Kuntz and K. M. Perry, *Science*, 1993, **259**, 1445.
33. R. L. DesJarlais, R. P. Sheridan, G. L. Seibel, J. S. Dixon, I. D. Kuntz and R. Venkataraghavan, *J. Med. Chem.*, 1988, **31**, 722.
34. I. D. Kuntz, J. M. Blaney, S. J. Oatley, R. Langridge and T. E. Ferrin, *J. Mol. Biol.*, 1982, **161**, 269.
35. T. A. Fritz, D. Tondi, J. S. Finer-Moore, M. P. Costi and R. M. Stroud, *Chem. Biol.*, 2001, **8**, 981.
36. D. Tondi, U. Slomczynska, M. P. Costi, D. M. Watterson, S. Ghelli and B. K. Shoichet, *Chem. Biol.*, 1999, **6**, 319.
37. A. C. Anderson, R. H. O'Neil, T. S. Surti and R. M. Stroud, *Chem. Biol.*, 2001, **8**, 445.
38. B. Q. Wei, L. H. Weaver, A. M. Ferrari, B. W. Matthews and B. K. Shoichet, *J. Mol. Biol.*, 2004, **337**, 1161.

39. T. J. Stout, D. Tondi, M. Rinaldi, D. Barlocco, P. Pecorari, D. V. Santi, I. D. Kuntz, R. M. Stroud, B. K. Shoichet and M. P. Costi, *Biochemistry*, 1999, **38**, 1607.
40. R. S. Bohacek, C. McMartin and W. C. Guida, *Med. Res. Rev.*, 1996, **16**, 3.
41. R. S. Bohacek and C. McMartin, *Curr. Opin. Chem. Biol.*, 1997, **1**, 157.
42. J. E. Eksterowicz, E. Evensen, C. Lemmen, G. P. Brady, J. K. Lanctot, E. K. Bradley, E. Saiah, L. A. Robinson, P. D. Grootenhuis and J. M. Blaney, *J. Mol. Graph. Model.*, 2002, **20**, 469.
43. D. T. Moustakas, P. T. Lang, S. Pegg, E. Pettersen, I. D. Kuntz, N. Brooijmans and R. C. Rizzo, *J. Comput. Aided Mol. Des.*, 2006, **20**, 601.
44. T. J. Ewing, S. Makino, A. G. Skillman and I. D. Kuntz, *J. Comput. Aided Mol. Des.*, 2001, **15**, 411.
45. I. Tickle, A. Sharff, M. Vinkovic, J. Yon and H. Jhoti, *Chem. Soc. Rev.*, 2004, **33**, 558.
46. W. T. Mooij, M. J. Hartshorn, I. J. Tickle, A. J. Sharff, M. L. Verdonk and H. Jhoti, *Chem. Med. Chem.*, 2006, **1**, 827.
47. J. Blaney, V. Nienaber and S. K. Burley, in *Fragment-Based Lead Discovery and Optimization Using X-Ray Crystallography Computational Chemistry, and High-Throughput Organic Synthesis*, ed. W. Jahnke and D. Erlanson, Wiley, 2006.
48. M. M. Hann, A. R. Leach and G. Harper, *J. Chem. Inf. Comput. Sci.*, 2001, **41**, 856.
49. J. Fejzo, C. A. Lepre, J. W. Peng, G. W. Bemis, Ajay, M. A. Murcko and J. M. Moore, *Chem. Biol.*, 1999, **6**, 755.
50. I. Schechter and A. Berger, *Biochem. Biophys. Res. Commun.*, 1968, **32**, 898.
51. I. Schechter and A. Berger, *Biochem. Biophys. Res. Commun.*, 1967, **27**, 157.
52. W. P. Jencks, *Proc. Natl. Acad. Sci. U. S. A.*, 1981, **78**, 4046.
53. C. W. Murray and M. L. Verdonk, *J. Comput. Aided Mol. Des.*, 2002, **16**, 741.
54. S. B. Shuker, P. J. Hajduk, R. P. Meadows and S. W. Fesik, *Science*, 1996, **274**, 1531.
55. D. A. Erlanson, A. C. Braisted, D. R. Raphael, M. Randal, R. M. Stroud, E. M. Gordon and J. A. Wells, *Proc. Natl. Acad. Sci. U. S. A.*, 2000, **97**, 9367.
56. D. A. Erlanson, J. A. Wells and A. C. Braisted, *Annu. Rev. Biophys. Biomol. Struct.*, 2004, **33**, 199.
57. D. A. Erlanson, *Curr. Opin. Biotechnol.*, 2006.
58. S. L. Warner, S. Bashyam, H. Vankayalapati, D. J. Bearss, H. Han, D. D. Von Hoff and L. H. Hurley, *Mol. Cancer Ther.*, 2006, **5**, 1764.
59. K. Babaoglu and B. K. Shoichet, *Nat. Chem. Biol.*, 2006, **2**, 720.
60. F. Gruswitz, J. O'Connell and R. M. Stroud, *Proc. Natl. Acad. Sci. U. S. A.*, 2007, **104**, 42.
61. L. N. Johnson and M. O'Reilly, *Curr. Opin. Struct. Biol.*, 1996, **6**, 762.
62. J. L. Buchbinder, V. L. Rath and R. J. Fletterick, *Annu. Rev. Biophys. Biomol. Struct.*, 2001, **30**, 191.

63. J. A. Hardy, J. Lam, J. T. Nguyen, T. O'Brien and J. A. Wells, *Proc. Natl. Acad. Sci. U. S. A.*, 2004, **101**, 12461.
64. A. T. Laurie and R. M. Jackson, *Curr. Protein Pept. Sci.*, 2006, **7**, 395.
65. C. D. Thanos, W. L. DeLano and J. A. Wells, *Proc. Natl. Acad. Sci. U. S. A.*, 2006, **103**, 15422.
66. C. D. Thanos, M. Randal and J. A. Wells, *J. Am. Chem. Soc.*, 2003, **125**, 15280.
67. L. A. Joachimiak, T. Kortemme, B. L. Stoddard and D. Baker, *J. Mol. Biol.*, 2006, **361**, 195.
68. M. R. Arkin, M. Randal, W. L. DeLano, J. Hyde, T. N. Luong, J. D. Oslob, D. R. Raphael, L. Taylor, J. Wang, R. S. McDowell, J. A. Wells and A. C. Braisted, *Proc. Natl. Acad. Sci. U. S. A.*, 2003, **100**, 1603.
69. Y. Che, B. R. Brooks and G. R. Marshall, *J. Comput. Aided Mol. Des.*, 2006, **20**, 109.
70. D. Gonzalez-Ruiz and H. Gohlke, *Curr. Med. Chem.*, 2006, **13**, 2607.
71. H. Alonso, A. A. Bliznyuk and J. E. Gready, *Med. Res. Rev.*, 2006, **26**, 531.
72. A. Jain, *Curr. Protein Pept. Sci.*, 2006, **7**, 407.
73. N. Moitessier, E. Therrien and S. Hanessian, *J. Med. Chem.*, 2006, **49**, 5885.
74. G. L. Warren, C. W. Andrews, A. M. Capelli, B. Clarke, J. LaLonde, M. H. Lambert, M. Lindvall, N. Nevins, S. F. Semus, S. Senger, G. Tedesco, I. D. Wall, J. M. Woolven, C. E. Peishoff and M. S. Head, *J. Med. Chem.*, 2006, **49**, 5912.
75. M. Knapp, C. Bellamacina, J. M. Murray and D. E. Bussiere, *Curr. Top. Med. Chem.*, 2006, **6**, 1129.

CHAPTER 2
The Changing Landscape in Drug Discovery

HUGO KUBINYI

D-67256 Weisenheim am Sand, Germany

2.1 Introduction

Looking back on 40 years of research in medicinal chemistry, later in molecular modeling and combinatorial chemistry, I see a significant change in the science and art of drug discovery. When I started my industrial career in 1966, in the search for new drugs, my colleagues and I synthesized a few compounds per week, at most. The biggest problem was to convince the pharmacologists to investigate them in due time, because they were still busy with the compounds delivered the weeks before. The chemical structures of our candidates resulted from working hypotheses that were often based on poor evidence. Once the compounds showed some activity in animals, we hoped that this would also be the case in humans. This situation, as well as books on drug discovery from that time, like Frank Clarke's *How modern medicines are discovered*,[1] Alfred Burger's *A Guide to the Chemical Basis of Drug Design*,[2] and Walter Sneader's *Drug Discovery: The Evolution of Modern Medicines*,[3] sound like stories from an ancient time, long ago. Nevertheless, they tell us how medicinal chemistry, biology, and pharmacology successfully(!) worked together to discover new medicines.

In an excellent review, Ralph Hirschmann characterized the decades between 1950 and 1990 as "Medicinal chemistry in the golden age of biology".[4] Indeed, this time was the golden age of drug discovery. New research results in biochemistry and biology paved the way from neurotransmitters and hormones to more-or-less selective agonists and antagonists of G protein-coupled receptors (GPCRs) and nuclear receptors. Within this relatively short period, Paul Janssen and his company were able to introduce about 80 new drugs into human therapy, many of which are still highly valuable therapeutics.[5] Whereas this unique yield may be (correctly) interpreted as the success of a genius, one should also consider that his research techniques were at the frontier of science. He used a toolbox of biologically interesting substructures (today we would call them "privileged fragments") and assembled them in different combinations, long before combinatorial chemistry or fragment-based design appeared on the scene. Nonlinear mapping methods were applied to investigate the relationships between animal and *in vitro* test models, and to characterize the biological activity profiles of the compounds.[6] But the main ingredient for success, as in most other companies, was a deep understanding of the underlying structure–activity relationships (SARs) and creative intuition in medicinal chemistry.

Nowadays genomics, proteomics, combinatorial chemistry, high-throughput screening (HTS), structure-based and computer-aided design, and virtual screening have completely changed the strategies of the drug discovery process. Without focussing on experimental techniques, the most important computational approaches in drug discovery are discussed in the following sections, which evaluate their strengths and limitations.

2.2 QSAR – Understanding Without Prediction

The very first computer-aided approach in drug design developed in the early 1960s, when Corwin Hansch started the QSAR (quantitative structure–activity relationships) discipline.[7,8] He considered drug action to result from two independent processes, that is:

(i) transport of the drug from the site of application to the site of action, and
(ii) non-covalent interactions of the drug with its binding site at a receptor.

As neither very polar nor very lipophilic compounds have a good chance to permeate several lipid and aqueous phases, he formulated a nonlinear lipophilicity relationship for the transport. Then, following a proposal by his postdoc Toshio Fujita, he combined lipophilicity terms and electronic parameters, and later molar refractivity and steric terms also, in a linear free-energy related (LFER) model to describe the ligand–receptor interaction. His third contribution was the definition of a lipophilicity parameter π, in the same manner as Louis Hammett had defined the electronic σ parameter 30 years before. Whereas Hansch and his group, as well as many others, were able to derive thousands of QSAR models for all kinds of biological activities, this approach was not much accepted by medicinal chemists. The very same happened to 3D QSAR methods, like comparative molecular field analysis (CoMFA), which were introduced about 20 years later.[9–12]

In principle, 3D QSAR is more powerful than classic QSAR, because:

(i) 3D structures are considered instead of only 2D structures;
(ii) more heterogeneous sets of compounds can be included than in classic QSAR;
(iii) molecular fields are calculated instead of just substituent constants;
(iv) contour maps show the effect of certain properties in certain regions.

However, these advantages are associated with several problems:

(i) most often neither the bioactive conformation nor the binding mode of the molecules of the data set are known;
(ii) a model-based superposition of the molecules always remains hypothetical;
(iii) minor displacements of the box around the molecules generate different and irreproducible results, because of the artificial cut-offs of the Lennard-Jones and Coulomb potentials – this problem can be avoided in the CoMSIA (comparative molecular similarity index analysis) modification[13];
(iv) variable selection produces fragmented contour maps that are difficult to interpret – instead of single variables, regions should be selected.

Thus, two reasons may be responsible for the lack of acceptance of QSAR and 3D QSAR by medicinal chemists: first, a detailed knowledge of statistics and much practical experience are needed to apply these methods in a proper manner and, second, even "good" models, with sufficient internal predictivity, are often poor in test-set prediction.[14–17] In particular, the best-fitting QSAR models, which result from variable selection and are validated by all reasonable statistical criteria, including cross-validation and y scrambling, are externally less predictive than models with inferior fit, an observation that has been called the "Kubinyi paradox".[18,19] However, it's no paradox, but results from the fact that these best-fitting models include variables that fit the error in the data, whereas some other models do not. The variation in test-set prediction results from the distribution of data with major experimental error (not necessarily outliers): if they are included in the test set, external predictivity is poor; if they are included in the training set, fit is poor, but external predictivity may be much better.[17]

So, what remains from QSAR for the medicinal chemist? The best answer has been given by Robin Ganellin, one of the leading medicinal chemists of our time, when he was asked by Steve Carney, "Has there been a single development that, in your opinion, has moved the field of medicinal chemistry ahead more than any other?" and Ganellin responded, "I would go back to the 1960s to the work of Corwin Hansch on the importance of lipophilicity. . . . I think that changed the way of thinking in medicinal chemistry . . . . I think that the application of physical organic chemical approaches to structure–activity analysis have been very important".[20] There is nothing more to say. Today, ligand–receptor interactions are considered in terms of hydrophobic interactions, polarizability, and ionic and neutral hydrogen bonds. The influence of lipophilicity, as well as of the dissociation and ionization of acids and bases, on transport and distribution is well understood. Medicinal chemists, who did not care about the pK_a values of their acids or bases, are now well aware of the risks that arise from those values being too far away from 7, the neutral pH value.

2.3 Gene Technology – from Mice to Humans

Gene technology created a lot of hype and it still creates a lot of fear. However, many of the anticipated benefits have not resulted. In a critical analysis, Glassman presents a long list of technologies for which the "initial hype of healthcare innovations did not live up to expectations",[21] including immunotherapy for cancer, stem cell technology, antisense technology, pharmacogenomics, genomics-based target identification, and gene therapy. He argues that the targets of several blockbuster drugs [statins, proton-pump inhibitors, leukotriene antagonists, selective serotonin reuptake inhibitors (SSRIs), taxol, angiotensin-converting enzyme (ACE) inhibitors, antihistamines, and kinase inhibitors] would not have been discovered by a systematic search for mutated "disease" genes. However, he admits that these targets might have been discovered through knockout technology.

Gene therapy has not yet been a success and we do not know whether it will be applicable in the future. Also, genomics-based target identification has not delivered to the expected extent. On the other hand, gene technology contributes to the production of human proteins for substitution therapy [insulin, growth hormone, erythropoietin (EPO), *etc.*] and to a better understanding of the function of enzymes and receptors. By far, the most important application of gene technology is in drug research. After the identification of a potential target, by any technology, the deoxyribonucleic acid (DNA) or messenger RNA (m-RNA) sequence of the corresponding gene directly provides the protein sequence. In many cases, even the folding and the function of a protein can be derived from its sequence. Larger amounts of the protein are produced in bacteria, insect cells, or higher organism cells, which enables the development of (high-throughput) screening models and, if more material becomes available, also a 3D structure elucidation by protein crystallography or multidimensional nuclear magnetic resonance (NMR) methods. Going back 25 years, this was the situation: with certain exceptions, we could only test in animals or with organs or other material from animals, which too often produced misleading results. Now we screen and develop our potential drug candidates with human (!) proteins – this is most probably the biggest achievement of gene technology for the benefit of mankind.

Before the sequence of the human genome became known, there were estimates of about 100 000 or even more human genes. This number had immediately to be corrected to about 30 000–35 000; recent estimates are closer to 20 000–25 000 than to these larger numbers. In the year 2000, Jürgen Drews counted 483 targets of current therapies and he speculated that, in total, there might be about 5000–10 000 drug targets, starting from an estimate of about 1000 "disease" genes and 5–10 proteins linked to such a disease gene.[22] Hopkins and Groom arrived at much smaller numbers in their estimate of the "druggable genome". Starting from a number of 30 000 genes in the human genome, they assumed that about 10% are disease-modifying genes and about 10% are druggable genes (*i.e.* genes for which the corresponding proteins can be modulated by a small molecule). As both subsets do not completely overlap,

600–1500 drug targets were estimated.[23] This estimate has to be criticized, because the term "druggable genome" is highly misleading. The relatively small number of genes within our genome codes for a few hundred thousands different proteins, because of alternative splicing and post-translational modifications. Many more potential drug targets result from protein-complex formation – a few protein chains form a multitude of different receptors and ion channels [*e.g.* the integrins, heterodimeric GPCRs and nuclear receptors, γ-aminobutyric acid (GABA) and nicotinic acetylcholine receptors, *etc.*].[24] Without speculating about numbers of druggable, disease-relevant proteins, we can conclude that there are many more potential targets than anticipated by Hopkins and Groom; the number might even surpass Drews' estimate. Indeed, the situation is more complex – several drugs, especially central nervous system (CNS) drugs, do not act against only one target, they modulate several targets at the same time.

Thus, two questions are associated with target-based screening. The first is whether high target selectivity is a desirable or unfavorable property? In certain cases it might be imperative to have such a high target selectivity [*e.g.* for an human immunodeficiency virus (HIV) protease inhibitor], but even here activity against a multitude of protease mutants of resistant strains is highly desirable. In other cases, it might be good to have a defined but broader selectivity against several related targets [*e.g.* metalloprotease inhibitors, kinase inhibitors (newer investigations show that even so-called "selective" inhibitors, like imatinib, show a broader spectrum of inhibitory activities against several kinases[25] than originally anticipated), and especially CNS-active drugs, *e.g.* the atypical neuroleptic olanzapine, which has nanomolar affinities at more than a dozen different GPCRs and the 5-hydroxytryptamine ($5HT_3$) ion channel]. Nobody will ever know whether this promiscuous binding behavior shows the right pattern or whether activities at a certain receptor should be higher or lower; but even if we knew, how could we design a compound that has exactly this slightly modified binding pattern?

The second question is, "Do we lose too many potential drugs by target-based screening?" The unexpected prodrug sulfamidochrysoidine would not have been discovered in cell culture; omeprazole acts only in acid-producing cells, after acid-catalyzed rearrangement; aciclovir is monophosphorylated only by a viral thymidine kinase, thus it works only in virus-infected cells. For this purpose, chemical biology, which aims to discover new leads by searching for phenotypical changes in cells or small animals, is a step in the right direction.

2.4 Combinatorial Library Design – Driven by Medicinal Chemistry

Combinatorial chemistry really had a poor start. In a bid for numbers, huge libraries were prepared as more-or-less undefined mixtures of compounds, driven by chemical accessibility. Biological activities, if discovered, often disappeared after deconvolution (*i.e.* the preparation of pure, single compounds that should be present in the mixture). Even when chemists realized some of the problems of such libraries, they searched for potential solutions in the wrong direction. In the 1990s, similarity and diversity were a big issue in chemoinformatics, despite the fact that the similarity of compounds can only be defined from a chemical point of view, never from a biological perspective.[26] But even from a chemical point of view, how similar or dissimilar are benzene and cyclohexane? Looking at aromaticity – no similarity at all; looking at lipophilicity and many other properties – very similar.

In 1998, Stuart Schreiber impressed organic chemists with the stereoselective synthesis of a library of 2.18 million "natural product-like compounds", starting from shikimic acid via a tricyclic intermediate, which allowed specific chemical reactions in many different directions.[27] In a misunderstanding of chemical diversity and in an over-optimistic consideration of its potential for producing biologically active compounds, this approach was called "diversity-oriented organic synthesis" (DOS).[28] In a later retrospective, Schreiber had to admit that "the field of DOS has not yet come close to reaching its goals . . . even a qualitative analysis of the members . . . reveals that they are disappointingly similar. Of even greater concern is that the selection of compounds has so far been guided only by the organic chemist's knowledge of candidate reactions, creativity in planning DOS pathways, and intuition about the properties likely to yield effective modulators. Retrospective analyses of these compounds show that they tend to cluster in discrete regions of multidimensional descriptor space. Although algorithms exist to identify subsets of actual or virtual compounds that best distribute in chemical space in a defined way . . . these are of little value to the planning of DOS".[29] This is exactly the dilemma of chemistry-driven combinatorial chemistry.

Recently Lipinski and Hopkins presented a cartoon which shows the chemical space as a box with embedded regions that stand for bioavailable compounds [the absorption, distribution, metabolism, and excretion (ADME) space], GPCR ligands, kinase inhibitors, protease inhibitors (in this cartoon they do not overlap with the ADME region – a sad experience in the search for bioavailable thrombin inhibitors).[30] If one understands chemical space as being of almost infinite size and the "bioactivity regions" just as very tiny spots, like the stars within our universe, we can more easily understand and accept the almost complete failure of chemistry-driven combinatorial libraries with respect to new biologically active compounds. The situation is even worse – biological activity space seems not to be evenly distributed in chemical space. There are groups of islands with higher density, so-called "privileged structures",[31] which definitely have a higher chance to produce biologically active molecules than others. Such privileged structures (*e.g.* the benzodiazepines, steroids, phenethylamines, diphenylmethanes, diphenylamines, and tricyclics, to mention only a few) are also called chemical masterkeys,[32] following the "lock and key" principle of Emil Fischer. Searching for "new" chemistry increases the risk of ending up with biologically inactive molecules because it avoids the privileged "activity islands".

Camille Wermuth, another great medicinal chemist of our time, proposed the "selective optimization of side activities" (SOSA) approach,[33,34] which starts from any side activity of a certain drug and aims to optimize this activity and to generate a new selectivity in this direction. There are several examples from the past where such side activities have been discovered and used to create new drugs, *e.g.* the development of diuretic and antidiabetic sulfonamides from the antibacterial sulfonamides (reviewed by Wermuth[33,34] and Kubinyi[35]).

In 1999, Roger Lahana argued that not a single new lead resulted from HTS and combinatorial chemistry.[36] Whereas this statement was and is wrong, his other conclusion, "when trying to find a needle in a haystack, the best strategy might not be to increase the size of the haystack" is absolutely correct. Another truism was formulated by Ashton and Moloney,[37] "Combinatorial chemistry has certainly failed to meet early expectations. Does this mean the technology has failed? Or does the problem lie in the manner in which the technology has been applied?"

In 1997 Chris Lipinski had already observed that some properties of the Pfizer in-house compounds, especially molecular weight and lipophilicity, developed in a wrong direction. As a consequence of increasing molecular weight and increasing lipophilicity, many screening hits could not be profiled as potential leads. Investigating a collection of drugs and drug candidates, he realized that only a minor percentage of these compounds had a molecular weight >500, a lipophilicity (expressed as log P) >5, more than five hydrogen bond donors, and more than 10 N+O atoms (as a rough estimate of the number of hydrogen bond acceptors). From this observation he defined his now famous "Rule of Five" (also called the Lipinski rule of 5 and Pfizer rule of 5) that low permeability of a molecule is to be expected if more than one of the following rules is violated: molecular weight <500, log $P<5$, no more than five hydrogen bond donors, and no more than 10 hydrogen bond acceptors.[38] Originally intended only as a warning flag for the Pfizer chemists, the rule was immediately accepted by the scientific community. It helped to clear screening collections from inappropriate compounds and to avoid the synthesis of meaningless combinatorial libraries. Nowadays, application of the Lipinski rule is mandatory in compound acquisition and in almost every library design.

Sometimes the Lipinski rule is misunderstood in the sense that it could define a drug-like character of the compounds. This is not the case – it defines drug-like properties with respect to bioavailability, but not drug-like structures; most of the Available Chemicals Directory (ACD) compounds pass the Lipinski filter, but they are by no means drug-like with respect to their structures. In a rare coincidence, two groups at Vertex and BASF independently developed almost identical neural net filters to characterize drug-likeness with respect to chemical structures.[39,40] Both groups used chemical descriptors, training sets from drug collections and from the ACD, and two different versions of supervised neural nets. To some surprise, the trained nets are able to differentiate between drugs and chemicals with a precision of about 75–80%, even if in different runs complete sets of drugs (*e.g.* CNS drugs, cardiovascular drugs, hormones) were eliminated from the training sets. Whereas a failure rate of 20–25% is acceptable for the evaluation of libraries, it is not suitable to accept or discard a certain compound. Thus, such tools should be used only to enrich or rank libraries and compound collections, and not for individual compounds.

In addition to drug-like properties and drug-likeness, lead structure properties have also been defined, starting from the observation that in recent years the optimization of leads produced most often (much) larger and (much) more lipophilic analogues. Different recommendations were given, most of which restricted molecular weight and lipophilicity to relatively low values.[41–43] However, an independent investigation of 470 lead–drug pairs showed that molecular weight increased, on the average, from the lead to the final drug only by 38 mass units.[44] It was early combinatorial chemistry that misled chemists simply to decorate lead structures with additional rings and other large substituents. Medicinal chemists of the past demonstrated how to modify lead structures in a more intelligent manner, often producing smaller analogues with higher activity or selectivity (*e.g.* several major analgesics, derived from morphine).

How to apply combinatorial chemistry or, better, automated parallel synthesis in medicinal chemistry? Its contribution to lead finding is relatively poor, because of the unfavorable ratio of chemical *vs.* biological activity space, as discussed above. If lead discovery libraries are to be designed:

(i) they should create real diversity by producing many small libraries with different, nonplanar scaffolds (*e.g.* natural products), instead of just one huge library with diverse decoration, and
(ii) the libraries should be checked for their drug-like or lead-like properties and their drug-like character.

The biggest potential of parallel automated synthesis is in chemogenomics and in the early steps of lead optimization. Chemogenomics aims to discover selective ligands of a certain target within a family of proteins or to shift biological activity and/or selectivity from one target to a related one. This is achieved by testing chemically related compounds in classes of evolutionary related targets (GPCRs, integrins, nuclear hormone receptors, aspartyl, metallo-, serine and cysteine proteases, kinases, phosphatases, ion channels, *etc.*).[45] Following this strategy, it is mandatory to synthesize and test a large number of analogues around a lead structure to find directions for further optimization, in any direction. In lead optimization, one should cover the chemical space around the current lead as completely as possible, in order not to lose any interesting candidates and to obtain a solid intellectual property position. Whenever improved candidate molecules are observed (*e.g.* with higher affinity, selectivity, bioavailability, and/or therapeutic range), the process can be repeated around the new structure, if chemically feasible. In the very last steps of lead structure optimization, dedicated syntheses will be necessary – then classic medicinal chemistry is back again.

2.5 Docking and Scoring – Solved and Unsolved Problems

Molecular modeling also had a difficult start. Early limitations in calculation power meant molecules were considered *in vacuo*. However, humans are aqueous systems and all drug targets are surrounded, even "filled", with water. Conformations of molecules were not understood as populations of several to many low-energy geometries – only the minimum energy conformation of a molecule was considered. Slowly chemists and modelers realized that ligand conformations *in vacuo*, in aqueous solution, in the crystal, and at the binding site of a protein may be very different. Whereas it is true that a ligand of a protein will not bind in a high-energy conformation of the ligand and/or the protein, the net free energy of binding results from the balance of entropy gain and entropy loss, as well as enthalpy gain and enthalpy loss. This includes also minor distortions of the ligand and/or the protein, which are to be compensated by other, favorable effects.

Why mention entropy first? The role of entropy is less well understood than the influence of enthalpy and it is most often underestimated. A ligand and its binding site at the surface or in a cavity of a protein are completely covered by water molecules. Some of them are relatively happy, despite the fact that they are more-or-less immobilized (unfavorable entropy), because they form hydrogen bonds to polar groups of either the ligand or the protein (favorable enthalpy). Some other molecules do not feel well, because they are loosely ordered at nonpolar surfaces; there is no favorable enthalpic interaction, only an unfavorable entropic contribution. When the ligand and the protein form a complex, (almost) all water molecules at the interacting surfaces have to be stripped. The water molecules formerly ordered at the hydrophobic surfaces are now happy; they can freely move in the aqueous medium surrounding the complex – this is the driving force of hydrophobic interaction. It is important to know that a perfect fit of hydrophobic residues into their hydrophobic cavities contributes most to ligand affinity; partially filled pockets ("*horror vacui*") or trapped water molecules in such nonpolar surroundings are highly unfavorable.

However, there is never a free lunch, and so also not in ligand binding: if too many hydrophobic groups are present, the solubility of the ligand decreases beyond a level that is acceptable for a drug molecule. The enthalpy terms of the stripping of water molecules, being hydrogen-bonded either to the ligand or to the protein surface, are partially or completely compensated by the entropy gain of their release and the interaction enthalpy of the new hydrogen bond between the ligand and the protein. Instead of increasing affinity, this sometimes results in an unfavorable contribution to binding affinity. In general, hydrogen bonds are important for recognition and for the orientation of a ligand to its binding site, but their affinity contributions are hard to predict.[46] The contributions of hydrogen bonds of the ligand to easily accessible polar groups at the surface of a protein are often overestimated in their affinity-enhancing effect.

In addition to all these effects, there is the unfavorable entropy of freezing the translational and rotational degrees of freedom of the ligand, as well as freezing the internal rotational degrees of freedom of the ligand and the binding site, and the unfavorable enthalpy terms of some distortions of the ligand and/or the binding site.

In his early studies on structure-based design, Peter Goodford developed the computer program GRID, which rolls chemical probes around the surface of a protein to discover regions where certain chemical functionalities should provide favorable interactions.[47] Out of the many interesting applications of the program GRID,[48] probably the most exciting one is the computer-aided design of the neuraminidase inhibitor zanamivir.[49] When Mark von Itzstein applied the program to investigate the 3D structure of neuraminidase, it uncovered a pocket for a positively charged group, close to a hydroxyl group of a weakly active lead structure. With only a few chemical modifications, *e.g.* by exchanging this hydroxyl group with a guanidinium group, the *in vitro* activity of the ligand could be increased by four orders of magnitude – a world record in computer-aided design.[49]

Docking programs go a step further. They use 3D structures of potential ligands to automatically position them into the known 3D structure of the binding region of a target protein. The very first version of the program DOCK[50] considered only geometric complementarity, without searching for potential interactions. LUDI, a hybrid of a docking and *de novo* program,[51] defines interaction sites within the binding site and searches for molecules which have the corresponding functionalities exactly in these positions. In a further step, other small molecules or groups may be attached to such ligands. A simple scoring function was developed to rank the results according to their quality of geometric fit and interaction energies.[52]

Many different programs for rigid and flexible docking and for *de novo* design have been developed in the meantime (for recent reviews see Schneider and Böhm,[53] Schulz-Gasch and Stahl,[54] and Warren *et al.*[55]), most of which generate reasonable poses for the potential ligands. The problem lies in ranking the results – which pose is "better" than the others and how can the ligands be ranked according to their estimated affinities? As different docking programs and scoring functions are reviewed in detail elsewhere,[54–60] only the inherent problems of scoring functions are discussed here.

An obvious problem in the calibration and validation of general scoring functions, *i.e.* scoring functions that will be applicable to any ligand–protein complex, is the quality of the biological data. These stem from different laboratories and correspondingly differ in test conditions, precision, and reliability. If one considers the difficulties to reproduce K_i or IC_{50} values (concentration required for 50% inhibition) from one laboratory to another one, standard deviations of one log unit are a reasonable error estimate.

The next problem comes from the protein 3D structures. Even if they are absolutely correct, with respect to electron density interpretation, they often lack a careful inspection and orientation of the hydrogen-bonding groups.

Thus, it might happen that incorrectly oriented hydroxyl groups and asparagine, glutamine, threonine, or histidine side chains, even at a distant site, are responsible for incorrect orientations of amino acid side chains within the binding site. pH shifts may influence the protonation state of histidine and lysine.

Whereas the Protein Database[61] provides many hundreds to thousands of 3D structures of ligand–protein complexes (depending on the desired resolution), it does not contain too many complexes with low, *i.e.* millimolar, ligand affinities. Although such information would be important to study unfavorable interactions, it is under-represented in the calibration of scoring functions. Entropic effects, which are discussed in detail above, are only roughly considered, *e.g.* by the area of the interacting hydrophobic surfaces. Internal rotational energies of ligands and the entropy loss on freezing the bioactive conformation are only considered by a constant term per rotatable bond. Unfavorable geometries of the ligand or the binding site are not considered at all.

Scoring functions do not consider the molecular electrostatic potential of the protein and the dipole moment of the ligand. Inserted water molecules have to be placed "by hand" – whether their replacement is favorable or not can be estimated by the program GRID.[47] Although some docking programs are able to use various conformations of a binding site, the scoring functions do not consider the residual flexibility of the ligand–protein complex. Large ligands have lower affinities than expected from the sum of favorable interactions,[62] an effect that has so far been considered only in some docking studies (*e.g.* Huang *et al.*[63] and Krämer *et al.*[64]).

The importance of desolvation in ligand binding was discussed about 20 years ago, in much detail. In a series of thermolysin inhibitors, a hydrogen bond of the ligands from an –NH– group to a backbone carbonyl oxygen of the protein.[65–68] Replacement of the –NH– group by –O– reduced affinities by three orders of magnitude, which is to be expected because of the lacking hydrogen bond and an electrostatic repulsion between the two oxygen atoms. Replacement of the –NH– group by a $-CH_2-$ group retained affinity,[68] an effect which had already been predicted from modeling two years earlier.[69] The $-CH_2-$ group cannot form a hydrogen bond in the complex, but there is also no negative effect from desolvation, as in the –NH– and –O– analogues.

This is a long list of problems and it is far from complete. Very recently, the group of Brian Shoichet discussed decoys in docking and scoring,[70] *i.e.* molecules with favorable rankings from several scoring functions but without binding affinity. They realized three problems:

(i) scoring functions may tolerate ligands that are too large;
(ii) scoring functions with the hard 12-6 van der Waals potential may miss potential ligands because of steric conflicts;
(iii) scoring functions do not consider the desolvation of the ligands in an adequate manner and therefore overestimate the affinity of polar compounds.

Scoring functions will have to consider such details, otherwise they will continue to fail. In a most comprehensive comparison, GlaxoSmithKline modeling groups at three different locations cooperated to evaluate 10 docking programs and 37 scoring functions against eight therapeutically interesting proteins of seven protein types. In the publication of their results[55] they arrive at the conclusion " . . . no single program performed well for all of the targets. For prediction of compound affinity, none of the docking programs or scoring functions made a useful prediction of ligand binding affinity". Graves *et al.* proposed the use of typical docking decoys as test cases in to further improve scoring functions.[70]

2.6 Virtual Screening – the Road to Success

Virtual screening covers a series of computer techniques, from simple filtering[71] and pharmacophore searches to docking and scoring.[72–77] The title of this section has a double meaning: all virtual screening techniques have to be applied in a proper manner to be successful – if these recommendations are followed, they will be successful in the search for new leads.

Virtual screening starts from a database of real compounds or from a virtual database, in which the chemical structures exist only in the computer. Even at this stage a careful preparation of the database is necessary. Besides the elimination of duplicates and counterions, compounds with undesired functionalities (reactive compounds, organometallics, *etc.*) should be eliminated. If not assigned in a unique manner, all configurations, as well as all enantiomers and diastereomers of chiral compounds, have to be generated. Carboxylic acids should be deprotonated, and amidines and guanidines should be protonated. All other acids (activated sulfonamides, phenols, *etc.*) and bases (amines, nitrogen-containing heterocycles) should be generated, in parallel, in the neutral and ionized forms. Compounds existing as tautomers should be generated in the correct form or as several different tautomers, *e.g.* by the program Agent.[78] 3D structures must be generated if 3D searches or docking are to be performed. If different ligand conformations are not considered "on the fly", multiple low-energy conformations have to be generated. Pharmacophoric features have to be defined in a correct manner, avoiding the attribution of acceptor properties to oxygen atoms with low electron density (*e.g.* the oxygen atom that is attached to the carbonyl group of esters, oxygen atoms in five-membered aromatic heterocycles, *etc.*).[24] All these processes are an absolute must, otherwise pharmacophore searches and docking will fail.

The next steps are options, but they are highly recommended. According to the needs of the user, different filters can be applied to narrow down the originally large size of the database. The filters can set molecular weight ranges, ranges for lipophilicity, and upper numbers for hydrogen bond donors and acceptors (*e.g.* by using the Lipinski rule), but also for polar surface area, number of rotatable bonds, number of rings, maximum number of halogens (more fluorine atoms than chlorine, bromine or iodine atoms may be accepted), *etc*. Such filters should be handled with the appropriate flexibility. The next step is the elimination of certain groups that should not be present (*e.g.* multiple amide groups, to eliminate peptides) or the inclusion of groups that should be present ("warhead" groups, like activated sulfonamides in carbonic anhydrase inhibitors, zinc-complexing groups in metalloprotease inhibitors, *etc*.).

Additional options are filters for lead-likeness, neural nets to prioritize compounds according to their drug-likeness or potential cytotoxicity, pharmacophore models for CYP 450 inhibition, hERG channel inhibition, and other antitarget activities. Too many filters should not be applied at the same time, because they all have a certain error range – thus, using too many might eliminate too many interesting candidate structures.

These preliminary steps are followed by pharmacophore searches. It is appropriate to search first for the presence of the desired pharmacophore groups and only then perform the more time-consuming step of a topological or 3D pharmacophore search. The pharmacophore can be derived from the 3D structures of active and inactive ligands, in a classic manner (*e.g.* by using the "active analogue approach") or by appropriate software. If 3D structures of a ligand–protein complex are available, the new program LigandScout is an attractive option to generate a pharmacophore.[79] As an alternative to classic 3D pharmacophore searches, the much faster FTree program may be used,[80,81] which showed the best performance in a recent comparison of different virtual screening protocols.[82] Even higher enrichment factors of active analogues can be achieved by using the newly developed multiple ligand-based MTree approach.[83]

The best candidates from the pharmacophore FTree or MTree searches should be flexibly docked into the experimental 3D structure of the target (beware, 3D structures of unliganded proteins may differ significantly from complex structures; if dimers or oligomers of the protein are the biologically active form, their 3D structures have to be used instead of the monomer 3D structure).

Experience shows that several different scoring functions should be applied to evaluate the docking results – which one will have the better performance for a certain target cannot be predicted *a priori*. Some investigators "spike" the database with known actives to find out which scoring function produces the best results. Of utmost importance is a visual inspection of the results for unreasonable ligand geometries, unreasonable binding modes, potential van der Waals clashes, polar interactions at the protein surface, *etc*.

The docking results may be clustered according to their chemical similarity. Only some compounds within a cluster may be picked for biological testing and groups of compounds with already known scaffolds may be eliminated.

A comprehensive review of such virtual screening procedures showed that in almost all cases interesting lead structures were observed, with micromolar to subnanomolar affinities to their target.[77] Even homology models of soluble proteins and GPCRs yielded good docking results, showing the enormous potential of virtual screening in the search for new leads of all potential targets.

2.7 Fragment-based and Combinatorial Design – A New Challenge

In 1975 Green had dissected biotin, the femtomolar ligand of avidin and streptavidin. Elimination of the sulfur atom reduced affinity to some extent. Dissecting this desthiobiotin further, into 4-methylimidazolin-2-one and caproic acid, produced millimolar to submillimolar ligands.[84] A similar effect was obtained by Kati *et al.*, by dissecting a subpicomolar transition state inhibitor of adenosine deaminase.[85] Seemingly, nobody considered to go the opposite way, *i.e.* to design a high-affinity ligand by combining low-affinity fragments.

Only in 1996 did Stephen Fesik develop the SAR-by-NMR (structure–activity relationships by nuclear magnetic resonance) method. This is an experimental approach in which, first, a low-affinity binder for a certain pocket of a binding site is searched for by NMR measurement, then this site is saturated with the ligand and another ligand is sought for an adjacent binding pocket. In the last step, the two ligands have to be linked in a relaxed conformation, to end up with a high-affinity ligand.[86–89] Other experimental techniques followed, based on protein crystallography, NMR, and mass spectrometry (for a comprehensive review, see Erlanson *et al.*[90]). Schuffenhauer *et al.* recently described the design of a dedicated library for fragment-based screening.[91]

The computer program RECAP has been used to dissect drug molecules into pieces that can easily be re-assembled by typical organic reactions.[92] Such fragments can be used for scaffold hopping (*i.e.* to generate a virtual library of new drug-like molecules) by connecting the pieces in a combinatorial manner, and to compare the similarity of the resulting structures with a lead structure.[93] In their search for cyclin-dependent kinase 4 (CDK4) inhibitors, Honma *et al.* first constructed a homology model, starting from the 3D structure of activated CDK2, and then performed a *de novo* design of ligands in the binding site, using the programs LEGEND and SEEDS.[94] Grzybowski *et al.* started from *p*-carboxamidophenylsulfonamide, a submicromolar ligand of carbonic anhydrase.[95,96] Within the binding site of this protein they generated 100 000 different N-substituents of the carboxamido group from a limited number of very small groups, by using the program CombiSMoG (combinatorial small molecular generator). The highest-scoring molecule showed 30 pM affinity. Krier *et al.* propose a scaffold-linker functional (SCF) group approach to convert active ligands into high-affinity analogues. Starting from the phosphodiesterases 4 (PDE4) inhibitor zardaverine, they designed a virtual combinatorial library that combined zardaverine and a few close analogues with different carbon-chain linkers and different functional groups, *e.g.* amines and aromatic rings. The results of this relatively small library were analogues with 40-fold to 900-fold improved inhibitory activity.[97]

These few examples show that the use of computer programs for fragment-based and combinatorial ligand design is just starting (for a recent review see Schneider and Fechner[98]). Programs for incremental flexible docking, like

FlexX,[99,100] could, in principle, be directly used for such a purpose. Instead of the original bits and pieces of the ligand, thousands of alternative structural elements could be used, creating a virtual multitude of potential ligands. These potential ligands need not even be constructed in the computer, and only the best intermediate results of the incremental design would be forwarded to the next step, to assemble the next partial structure (a corresponding tool FlexNovo[101] is in development). However, the problem of ranking the intermediate results is now even much more difficult than that for docking only the original ligand. Further improvements in the scoring functions will be necessary to apply this appealing approach as a routine technique. In addition, the computer programs should include simple rules (*e.g.* the RECAP reactions[92]) for the chemical accessibility of the potential ligands.

2.8 Summary and Conclusions

Within the past few decades the strategies of drug design have changed significantly. Whereas chemistry, biological activity hypotheses, and animal experiments dominated drug research, especially in its "golden age", in the 1960s and 1970s, many new technologies have developed over the past 20 years.[102] A vast number of new drugs was expected to result from combinatorial chemistry and HTS. In the meantime, most groups learned that this is not the case; the yield of new drug candidates was relatively poor and the number of new chemical entities (NCEs) is steadily declining.[103] It is now evident that chemistry-driven syntheses are most often a waste of resources.

Genomics, proteomics, and pharmacogenomics support the discovery of new targets for human therapy. Target validation is performed with genetically modified animals or with the new small interfering RNA (siRNA) technology. System biology and orthogonal ligand–receptor pairs help us to understand the effect of a modulator of a certain protein, long before such a compound is discovered. However, two problems remain: first, will the target be "druggable" (*i.e.* can it be modulated) and will such a modulator be discovered with reasonable effort and within reasonable time? Several protein–protein interactions seem to be not druggable, at least so far. Second, will the modulator of the new target at the very end, after years and years of research, preclinical profiling, and clinical testing, be suited as an efficient and safe drug in human therapy?

Once a target is identified, its 3D structure can be elucidated by structural biology or, at least in many cases, be modeled from the 3D structures of related proteins. With the ongoing progress in protein crystallography and multidimensional NMR studies, the 3D structures of many important proteins, especially enzymes, have been elucidated at atomic resolution. This information enables the structure-based design of therapeutically useful enzyme inhibitors, many of them still in preclinical or clinical development. Whereas structure-based design can be regarded as the predominant strategy of the past decade, several computer-assisted methods were developed more recently. If thousands of candidates and even larger structural databases are to be tested as to whether there are suitable ligands of a certain binding site, this can no longer be performed by hand. The design process has to be automated, *i.e.* investigated with the help of the computer. Virtual screening selects compounds or libraries that are either lead-like, drug-like, have a good potential of oral bioavailability, or are similar to a lead, by sets of rules, neural nets, similarity analyses, pharmacophore analyses, or docking and scoring.

Correspondingly, the identification of lead structures of a druggable target, by HTS, by structure-based design, or by computer-aided approaches, is now more-or-less a routine approach, as well as their optimization with respect to target affinity. Successful applications of virtual screening, 3D structure-based design, and docking demonstrate the value of these techniques in the selection and rational design of high-affinity protein ligands. Whereas the application of all modern technologies in this step is highly desirable, the risk increases that the accumulated experience in medicinal chemistry[104] becomes increasingly forgotten. Chemists of our time, especially if they lack medicinal chemistry know-how, tend to "decorate" their lead structures, instead of taking the more difficult route of systematic chemical variation, including the formation of new rings (rigidization of a bioactive conformation), replacing ring or chain atoms, *etc.*

The time from selecting a new target to discovering a series of promising leads and optimizing them to nanomolar ligands is now much shorter than in the past. However, this is often accompanied by a neglect of favorable ADME properties. High affinity to a disease-relevant target is only a necessary property of a drug candidate, not a sufficient one. In addition, a drug must have the right selectivity, it must be orally bioavailable, should have favorable pharmacokinetics and metabolism, and should lack serious side effects. The desired degree of selectivity cannot be defined in an absolute manner. In some cases high selectivity is mandatory, in other cases (*e.g.* for kinase inhibitors) a certain lack of selectivity might be tolerable (*e.g.* imatinib, Gleevec™), whereas in the case of CNS-active drugs a high degree of promiscuity might be better than a one-target selectivity.

What are the reasons for the so-called "productivity gap" in pharmaceutical industry, *i.e.* the situation that research costs steadily increase, but output is declining[103]? There is no unique answer and there are no simple reasons. One possible explanation is the already relatively high standard in the symptomatic treatment of "simple", acute diseases. Poor ADME properties are often cited as the most common reason for failure in clinical development, creating a demand for ADME prediction tools; however, this conclusion is based on old data[105] and is not even generally supported by these data.[24,106] ADME became an issue in the attempt to optimize large, lipophilic hits from early combinatorial chemistry and HTS; it was never a major reason for the failure in clinical development, neither in the early period of 1964–1985 (7% attrition rate due to ADME, excluding anti-infectives),[24,106] nor in the years 1992–2002 (11% attrition rate due to ADME).[107] Bioavailability problems can now be minimized at a very early stage, *e.g.* by applying the Lipinski rule.[38]

In our time, the two most common reasons for clinical failure are lack of efficacy and/or toxic side effects; they account for about 75% of all terminated clinical studies[106,107]; both effects cannot be completely separated, because often just the therapeutic window is small – low doses are without sufficient efficacy, but high doses cause toxic side effects. In human therapy, chronic diseases, especially cell-degenerative diseases, are much harder to prevent or treat than acute disorders: often the disease is already too far advanced before it can be diagnosed, *e.g.* in cancer, Alzheimer's disease, or Parkinson's disease. For some progressive diseases it would be desirable to treat healthy people to prevent the development of certain pathological conditions. For this purpose, the drug must be totally free of any side effects, even after treatment over years, which seems to be an impossible task.

Our current expectations on the efficacy and safety of a drug are much, much higher than they were decades ago. Correspondingly, restrictions by the health agencies have increased from decade to decade. It is questionable whether, *e.g.* acetylsalicylic acid (aspirin™) or corticosteroids, would nowadays be approved as "safe" drugs, despite the fact that long-term application shows that they are well tolerated and effective, if applied in the right manner. Rare side effects, which cannot be uncovered in some thousands of animals or by treating a few thousand patients under controlled conditions, end the use of otherwise successful drugs, *e.g.* cerivastatin (Lipobay™) or rofecoxib (Vioxx™).[107]

So, the question is not, "Why aren't we more successful with all these modern technologies?" The question must be, "Where would we stand without genomics, molecular biology, combinatorial chemistry, HTS, structure-based and computer-aided approaches, and virtual screening?" The answer is that the situation would be much worse. Indeed, in the long process of drug discovery and development, from target discovery to launching the new drug, the phase of lead discovery and optimization is nowadays a very fast and most effective one.

How to increase success in drug research? We should merge the know-how of classic medicinal chemistry with the new technologies and follow the recommendations of George de Stevens, who formulated the following as long as 20 years ago. "The (drug) discovery process is at times slow, somewhat tedious, always exciting and requiring patience, tenacity, objectivity and above all intellectual integrity. Therefore, scientists, to be innovative, must work in a corporate environment in which the management not only recognizes these factors but makes every effort to let their importance be known to the scientists. The people in research don't have a need to be loved but they do need to feel that they are understood and supported and not to be manipulated according to short-term business cycles. . . . Drug discoveries are made by scientists practicing good science. By and large these discoveries are usually made in a company with an enlightened management which encourages its scientists with freedom of action, freedom to think widely and to challenge dogma, and freedom in risk-taking. Moreover, important drug discoveries are not made by committees but by individual scientists working closely together, sharing ideas, testing hypotheses, looking for new solutions to difficult problems, accepting negative results and learning from these results so that the next group of compounds synthesized and tested will open the door to new and improved therapy".[108] This is more true than ever!

[December 2005]

References

1. F. H. Clarke, *How Modern Medicines are Discovered*, Futura Publishing, Mount Kisko, NY, 1973.
2. A. Burger, *A Guide to the Chemical Basis of Drug Design*, John Wiley & Sons, New York, 1983.
3. W. Sneader, *Drug Discovery: The Evolution of Modern Medicines*, John Wiley & Sons, New York, 1985.
4. R. Hirschmann, *Angew. Chem., Int. Ed.*, 1991, **30**, 1278–1301; *Angew. Chem.* 1991, **103**, 1305–1330.
5. T. A. Galemmo, Jr., F. E. Janssens, P. J. Lewi and B. E. Maryanoff, *J. Med. Chem.*, 2005, **48**, 1686.
6. P. J. Lewi, *Arzneim.-Forsch. (Drug Res.)*, 1976, **26**, 1295–1300.
7. C. Hansch and T. Fujita, *J. Am. Chem. Soc.*, 1964, **86**, 1616–1626.
8. H. Kubinyi, *QSAR: Hansch Analysis and Related Approaches*, Volume 1 of *Methods and Principles in Medicinal Chemistry*, R. Mannhold, P. Krogsgaard-Larsen and H. Timmerman, Eds, VCH, Weinheim, 1993.
9. R. D. Cramer III, D. E. Patterson and J. D. Bunce, *J. Am. Chem. Soc.*, 1988, **110**, 5959–5967.
10. H. Kubinyi, *3D QSAR in Drug Design. Theory, Methods and Applications*,ESCOM Science Publishers BV, Leiden, 1993.
11. H. Kubinyi, G. Folkers and Y. C. Martin, *3D QSAR in Drug Design, Volume 2, Ligand–Protein Complexes and Molecular Similarity*, Kluwer/ESCOM, Dordrecht, 1998; also published as in *Persp. Drug Disc. Des.*, 1998, **9–11**, 1–416.
12. H. Kubinyi, G. Folkers and Y. C. Martin, *3D QSAR in Drug Design, Volume 3, Recent Advances*, Kluwer/ESCOM, Dordrecht, 1998; also published as *Persp. Drug Disc. Des*, 1998, **12–14**, 1–352.
13. G. Klebe, U. Abraham and T. Mietzner, *J. Med. Chem.*, 1994, **37**, 4130–4146.
14. H. Kubinyi, F. A. Hamprecht and T. Mietzner, *J. Med. Chem.*, 1998, **41**, 2553–2564.
15. A. Golbraikh and A. Tropsha, *J. Mol. Graphics Model.*, 2002, **20**, 269–276.
16. A. Doweyko, *J. Comput. Aided Mol. Des.*, 2004, **18**, 587–596.
17. H. Kubinyi, in *QSAR & Molecular Modelling in Rational Design of Bioactive Molecules (Proceedings of the 15th European Symposium on QSAR & Molecular Modelling, Istanbul, Turkey, 2004)*, E. Aki Sener and I. Yalcin, Eds., CADDD Society, Ankara, Turkey, 2006, pp. 30–33.
18. J. H. van Drie, *Curr. Pharm. Des.*, 2003, **9**, 1649–1664.

19. J. H. van Drie, in *Computational Medicinal Chemistry for Drug Discovery*, ed. P. Bultinck, H. de Winter, W. Langenaeker and J. P. Tollenaere, Marcel Dekker, New York, 2004, pp. 437–460.
20. S. L. Carney, *Drug Discovery Today*, 2004, **9**, 158–160.
21. R. H. Glassman and A. Y. Sun, *Nat. Rev. Drug Discovery*, 2004, **3**, 177–183.
22. J. Drews, *Science*, 2000, **287**, 1960–1964.
23. A. L. Hopkins and C. R. Groom, *Nature Rev. Drug Discovery*, 2002, **1**, 727–730.
24. H. Kubinyi, *Nat. Rev. Drug Discovery*, 2003, **2**, 665–668.
25. M. A. Fabian, W. H. Biggs, D. K. Treiber, C. E. Atteridge, M. D. Azimioara, M. G. Benedetti, T. A. Carter, P. Ciceri, P. T. Edeen, M. Floyd, J. M. Ford, M. Galvin, J. L. Gerlach, R. M. Grotzfeld, S. Herrgard, D. E. Insko, M. A. Insko, A. G. Lai, J. M. Lelias, S. A. Mehta, Z. V. Milanov, A. M. Velasco, L. M. Wodicka, H. K. Patel, P. P. Zarrinkar and D. J. Lockhart, *Nat. Biotechnol.*, 2005, **23**, 329–336.
26. H. Kubinyi, in *3D QSAR in Drug Design*, Volume 2, *Ligand–Protein Complexes and Molecular Similarity*, ed. H. Kubinyi, G. Folkers, and Y. C. Martin, Kluwer/ESCOM, Dordrecht, 1998, pp. 225–252; also published as *Persp. Drug Disc. Des.*, 1998, **9–11**, 225–252.
27. D. S. Tan, M. A. Foley, M. D. Shair and S. L. Schreiber, *J. Am. Chem. Soc.*, 1998, **120**, 8565–8566.
28. S. L. Schreiber, *Science*, 2000, **287**, 1964–1969.
29. S. L. Schreiber, *Chem. Eng. News*, March 3rd, 2003, 51–61.
30. C. Lipinski and A. Hopkins, *Nature*, 2004, **432**, 855–861.
31. B. E. Evans, K. E. Rittle, M. G. Bock, R. M. DiPardo, R. M. Freidinger, W. L. Whitter, G. F. Lundell, D. F. Veber, P. S. Anderson, R. S. L. Chang, V. J. Lotti, D. J. Cerino, T. B. Chen, P. J. Kling, K. A. Kunkel, J. P. Springer and J. Hirshfield, *J. Med. Chem.*, 1988, **31**, 2235–2246.
32. G. Müller, *Drug Discovery Today*, 2003, **8**, 681–691.
33. C. G. Wermuth, *J. Med. Chem.*, 2004, **47**, 1303–1314.
34. C. G. Wermuth, *Drug Discovery. Today*, 2006, **11**, 160–164.
35. H. Kubinyi, in *Chemogenomics in Drug Discovery. A Medicinal Chemistry Perspective*, ed. H. Kubinyi and G. Müller. Volume 22 of *Methods and Principles in Medicinal Chemistry*, ed. R. Mannhold, H. Kubinyi, and G. Folkers, Wiley-VCH, Weinheim, 2004, pp. 43–67.
36. R. Lahana, *Drug Discovery Today*, 1999, **4**, 447–448.
37. M. Ashton and B. Moloney, *Curr. Drug Discovery*, 2003(8), 9–11.
38. C. A. Lipinski, F. Lombardo, B. W. Dominy and P. J. Feeney, *Adv. Drug Deliv. Rev.*, 1997, **23**, 3–25.
39. Ajay, W. P. Walters and M. A. Murcko, *J. Med. Chem.*, 1998, **41**, 3314–3324.
40. J. Sadowski and H. Kubinyi, *J. Med. Chem.*, 1998, **41**, 3325–3329.
41. S. J. Teague, A. M. Davis, P. D. Leeson, and T. Oprea, *Angew. Chem. Int. Ed.*, 1999, 38, 3743–3748 *Angew. Chem.*, 1999, 111, 3962–3967.

42. T. I. Oprea, A. M. Davis, S. J. Teague and P. D. Leeson, *J. Chem. Inf. Comput. Sci.*, 2001, **41**, 1308–1316.
43. J. R. Proudfoot, *Bioorg. Med. Chem. Lett.*, 2002, **12**, 1647–1650.
44. M. M. Hann, A. R. Leach and G. Harper, *J. Chem. Inf. Comput. Sci.*, 2001, **41**, 856–864.
45. H. Kubinyi and G. Müller, *Chemogenomics in Drug Discovery. A Medicinal Chemistry Perspective*, Volume 22 of *Methods and Principles in Medicinal Chemistry*, ed. R. Mannhold, H. Kubinyi, and G. Folkers, Wiley-VCH, Weinheim, 2004.
46. H. Kubinyi, in *Pharmacokinetic Optimization in Drug Research. Biological Physicochemical and Computational Strategies*, ed. B. Testa, H. van de Waterbeemd, G. Folkers and R. Guy, Helvetica Chimica Acta and Wiley-VCH, Zurich, 2001, pp. 513–524.
47. P. J. Goodford, *J. Med. Chem.*, 1985, **28**, 849–857.
48. G. Cruciani, *Molecular Interaction Fields. Applications in Drug Discovery and ADME Prediction*, Volume 27 of *Methods and Principles in Medicinal Chemistry*, ed. R. Mannhold, H. Kubinyi and G. Folkers, Wiley-VCH, Weinheim, 2005.
49. M. von Itzstein, W.-Y. Wu, G. B. Kok, M. S. Pegg, J. C. Dyason, B. Jin, T. V. Phan, M. L. Smythe, H. F. White, S. W. Oliver, P. M. Colman, J. N. Varghese, D. M. Ryan, J. M. Woods, R. C. Bethell, V. J. Hotham, J. M. Cameron and C. R. Penn, *Nature*, 1993, **363**, 418–423.
50. E. C. Meng, B. Shoichet and I. D. Kuntz, *J. Comput. Chem.*, 1992, **13**, 505–524.
51. H.-J. Böhm, *J. Comput. Aided Mol. Des.*, 1992, **6**, 61–78.
52. H.-J. Böhm, *J. Comput. Aided Mol. Des.*, 1994, **8**, 243–256.
53. G. Schneider and H.-J. Böhm, *Drug Discovery Today*, 2002, **7**, 64–70.
54. T. Schulz-Gasch and M. Stahl, *J. Mol. Model.*, 2003, **9**, 47–57.
55. G. L. Warren, C. W. Andrews, A.-M. Capelli, B. Clarke, J. LaLonde, M. H. Lambert, M. Lindvall, N. Nevins, S. F. Semus, S. Senger, G. Tedesco, I. D. Wall, J. M. Woolven, C. E. Peishoff and M. S. Head, *J. Med. Chem.*, 2006, **49**, 5912–5931.
56. I. Muegge and I. Enyedy, in *Computational Medicinal Chemistry and Drug Discovery*, ed. P. Bultinck, H. de Winter, W. Langenaeker and J. P. Tollenaere, Marcel Dekker, New York, 2004, pp. 405–436.
57. E. M. Krovat, T. Steindl and T. Langer, *Curr. Comp. Aided Drug Des.*, 2005, **1**, 93–102.
58. R. Wang, Y. Lu and S. Wang, *J. Med. Chem.*, 2003, **46**, 2287–2303.
59. P. S. Charifson, J. J. Corkery, M. A. Murcko and W. P. Walters, *J. Med. Chem.*, 1999, **42**, 5100–5109.
60. M. Kontoyianni, G. S. Sokol and L. M. McClellan, *J. Comput. Chem.*, 2005, **26**, 11–22.
61. H. M. Berman, J. Westbrook, Z. Feng, G. Gilliland, T. N. Bhat, H. Weissig, I. N. Shindyalov and P. E. Bourne, *Nucleic Acids Res.*, 2000, **28**, 235–242.

62. I. D. Kuntz, K. Chen, K. A. Sharp and P. A. Kollman, *Proc. Natl. Acad. Sci. U.S.A.*, 1999, **96**, 9997–10002.
63. N. Huang, A. Nagarsekar, G. J. Xia, J. Hayashi and A. D. MacKerell Jr, *J. Med. Chem.*, 2004, **47**, 3502–3511.
64. O. Krämer, I. Hazemann, A. D. Podjarny and G. Klebe, *Proteins: Struct. Funct. Genet.*, 2004, **55**, 814–823.
65. P. A. Bartlett and C. K. Marlowe, *Science*, 1987, **235**, 569–571.
66. D. E. Tronrud, H. M. Holden and B. W. Matthews, *Science*, 1987, **235**, 571–574.
67. P. A. Bash, U. C. Singh, F. K. Brown, R. Langridge and P. A. Kollman, *Science*, 1987, **235**, 574–576.
68. B. P. Morgan, J. M. Scholtz, M. D. Ballinger, I. D. Zipkin and P. A. Bartlett, *J. Am. Chem. Soc.*, 1991, **113**, 297–307.
69. K. M. Mertz and P. K. Kollman, *J. Am. Chem. Soc.*, 1989, **111**, 5649–5658.
70. A. P. Graves, R. Brenk and B. J. Shoichet, *J. Med. Chem.*, 2005, **48**, 3714–3728.
71. J. A. Lumley, *QSAR Comb. Sci.* 24, 1066–1075 (2005).
72. B. K. Shoichet, *Nature*, 2004, **432**, 862–865.
73. T. Langer and G. Wolber, *Pure Appl. Chem.*, 2004, **76**, 991–996.
74. A. C. Anderson and D. L. Wright, *Curr. Comp.-Aided Drug Des.*, 2005, **1**, 103–127.
75. J. Alvarez and B. Shoichet, *Virtual Screening in Drug Discovery*, CRC Press, Taylor & Francis Group, Boca Raton, 2005.
76. G. Klebe, *Drug Discovery Today*, 2006, **11**, 580–594.
77. H. Kubinyi, in *Computer Applications in Pharmaceutical Research and Development*, ed. S. Ekins, John Wiley & Sons, Hoboken, 2006, pp. 377–424.
78. P. Pospisil, P. Ballmer, L. Scapozza and G. Folkers, *J. Recept. Signal Transduct. Res.*, 2003, **23**, 361–371.
79. G. Wolber and T. Langer, *J. Chem. Inf. Model.*, 2005, **45**, 160–169.
80. M. Rarey and J. S. Dixon, *J. Comput. Aided Mol. Des.*, 1998, **12**, 471–490.
81. M. Rarey and M. Stahl, *J. Comput. Aided Mol. Des.*, 2001, **15**, 497–520.
82. A. Evers, G. Hessler, H. Matter and T. Klabunde, *J. Med. Chem.*, 2005, **48**, 5448–5465.
83. G. Hessler, M. Zimmermann, H. Matter, A. Evers, T. Naumann, T. Lengauer and M. Rarey, *J. Med. Chem.*, 2005, **48**, 6575–6584.
84. N. M. Green, *Adv. Protein Chem.*, 1975, **29**, 85–133.
85. W. M. Kati, S. A. Acheson and R. Wolfenden, *Biochemistry*, 1992, **31**, 7356–7366.
86. S. B. Shuker, P. J. Hajduk, R. P. Meadows and S. W. Fesik, *Science*, 1996, **274**, 1531–1534.
87. P. J. Hajduk, R. P. Meadows and S. W. Fesik, *Science*, 1997, **278**, 497–499.
88. P. J. Hajduk, G. Sheppard, D. G. Nettesheim, E. T. Olejniczak, S. B. Shuker, R. P. Meadows, D. H. Steinman, G. M. Carrera Jr, P. A.

Marcotte, J. Severin, K. Walter, H. Smith, E. Gubbins, R. Simmer, T. F. Holzman, D. W. Morgan, S. K. Davidsen, J. B. Summers and S. W. Fesik, *J. Am. Chem. Soc.*, 1997, **119**, 5818–5827.
89. E. T. Olejniczak, P. J. Hajduk, P. A. Marcotte, D. G. Nettesheim, R. P. Meadows, R. Edalji, T. F. Holzman and S. W. Fesik, *J. Am. Chem. Soc.*, 1997, **119**, 5828–5832.
90. D. A. Erlanson, R. S. McDowell and T. O'Brien, *J. Med. Chem.*, 2004, **47**, 3463–3482.
91. A. Schuffenhauer, S. Ruedisser, A. L. Marzinzik, W. Jahnke, M. Blommers, P. Selzer and E. Jacoby, *Curr. Top. Med. Chem.*, 2005, **5**, 751–762.
92. X. Q. Lewell, D. B. Judd, S. P. Watson and M. M. Hann, *J. Chem. Inf. Comput. Sci.*, 1998, **38**, 511–522.
93. G. Schneider, O. Clément-Chomienne, L. Hilfiger, P. Schneider, S. Kirsch, H.-J. Boehm and W. Neidhart, *Angew. Chem. Int. Ed.*, 2000, 39, 4130–4133; *Angew. Chem.*, 2000, **112**, 4305–4309.
94. T. Honma, K. Hayashi, T. Aoyama, N. Hashimoto, T. Machida, K. Fukasawa, T. Iwama, C. Ikeura, M. Ikuta, I. Suzuki-Takahashi, Y. Iwasawa, T. Hayama, S. Nishimura and H. Morishima, *J. Med. Chem.*, 2001, **44**, 4615–4627.
95. B. A. Grzybowski, A. V. Ishchenko, C. Y. Kim, G. Topalov, R. Chapman, D. W. Christianson, G. M. Whitesides and E. I. Shakhnovich, *Proc. Natl. Acad. Sci. U.S.A.*, 2002, **99**, 1270–1273.
96. B. A. Grzybowski, A. V. Ishchenko, J. Shimada and E. I. Shakhnovich, *Acc. Chem. Res.*, 2002, **35**, 261–269.
97. M. Krier, J. X. de Araújo-Júnior, M. Schmitt, J. Duranton, H. Justiano-Basaran, C. Lugnier, J. J. Bourguignon and D. Rognan, *J. Med. Chem.*, 2005, **48**, 3816–3822.
98. G. Schneider and U. Fechner, *Nature Rev. Drug Discovery*, 2005, **4**, 649–663.
99. M. Rarey, B. Kramer, T. Lengauer and G. Klebe, *J. Mol. Biol.*, 1996, **261**, 470–489.
100. B. Kramer, M. Rarey and T. Lengauer, *Proteins Struct. Funct. Genet.*, 1999, **37**, 228–241.
101. M. Rarey, *BIOforum*, 2005, **10**, 56–59.
102. H. Kubinyi, *Technologies*, Vol. 3 of *Comprehensive Medicinal Chemistry II*, ed. D. Triggle and J. Taylor, Elsevier Science, Oxford, 2007.
103. J. A. DiMasi, R. W. Hansen and H. G. Grabowski, *J. Health Econ.*, 2003, **22**, 151–185.
104. C. G. Wermuth, *The Practice of Medicinal Chemistry*, Elsevier–Academic Press, New York, 2003.
105. R. A. Prentis, Y. Lis and S. R. Walker, *Br. J. Clin. Pharmacol.*, 1988, **25**, 387–396.
106. T. Kennedy, *Drug Discovery Today*, 1997, **2**, 436–444.
107. D. Schuster, C. Laggner and T. Langer, *Curr. Pharm. Des.*, 2005, **11**, 3545–3559.
108. G. de Stevens, *Prog. Drug Res.*, 1986, **30**, 189–203.

Section 2
Structure-Based Design

CHAPTER 3
Purine Nucleoside Phosphorylase

YANG ZHANG AND STEVEN E. EALICK*

*Department of Chemistry and Chemical Biology, Cornell University, Ithaca, NY 14853, USA

3.1 Introduction

Purine nucleoside phosphorylase (PNP) catalyzes the reversible phosphorolysis of purine ribonucleosides and 2′-deoxypurine ribonucleosides to the free base, and ribose 1-phosphate or 2′-deoxyribose 1-phosphate, respectively.[1–3] PNPs are found in most prokaryotic and eukaryotic organisms where the enzyme functions in the purine salvage pathway (Figure 3.1). PNPs can be divided into two families based on their quaternary structure.[4] Trimeric PNPs have a monomeric molecular weight of approximately 31 kDa and are usually specific for 6-oxopurine nucleosides. Hexameric PNPs have a monomeric molecular

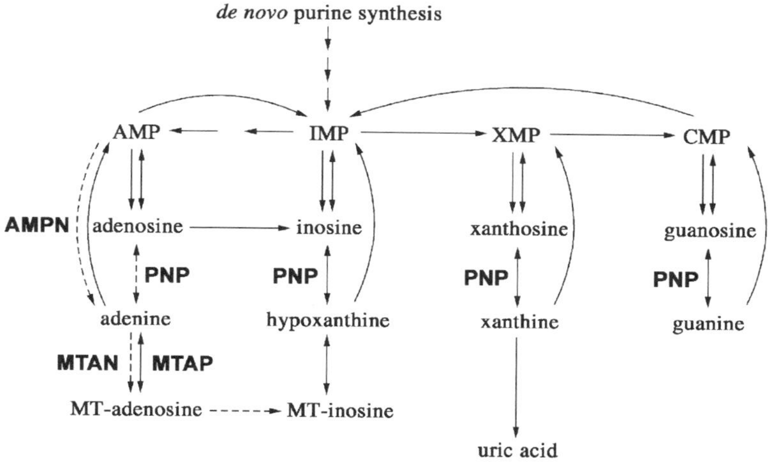

Figure 3.1 Purine salvage pathway. Enzymes utilizing the PNP fold are labeled. Reactions unique to the prokaryotes are indicated as dashed lines.

weight of approximately 25 kDa and often accept both 6-oxo- and 6-amino-purine nucleosides. PNPs have been studied extensively by X-ray crystallography and coordinates for several examples from each family are available in the Protein Data Bank (PDB) (Table 3.1). Although structures of PNPs have been determined from 14 organisms, many of these PNPs have not been characterized biochemically. The PNP fold is also shared by methylthioadenosine phosphorylases (MTAPs),[5] uridine phosphorylases (UPs),[6] adenosine monophosphate nucleosidases (AMPN),[7] and methylthioadenosine/*S*-adenosylhomocysteine nucleosidase (MTAN).[8]

Table 3.1 PNP structures deposited in the Protein Data Bank.

Enzyme	*PDB ID*
Trimeric PNP	
Homo sapiens PNP	1M73,[69] 1PF7,[70] 1PWY,[41] (1RCT, 1V3Q),[71] 1RFG,[72] 1RR6[c], 1RSZ[c], 1RT9[c], (1ULA, 1ULB),[23] 1V2H,[73] 1V41[c], 1V45[c]
Bos taurus PNP	(1A9O, 1A9P, 1A9Q, 1A9R, 1A9S, 1A9T, 1PBN),[25] 1B8N,[52] 1FXU,[74] 1LVU,[75] 1V48,[76] 1VFN,[77] (2AI1, 2AI2, 2AI3),[76] 3PNP[c], 4PNP[c]
Schistosoma mansoni PNP	(1TCU, 1TCV, 1TD1)[78]
Escherichia coli PNP-2[a]	(1YQQ, 1YQU, 1YR3)[18]
Thermotoga maritima PNP	1VMK[c]
Cellulomonas sp. PNP	(1C3X , 1QE5)[79]
Mycobacterium tuberculosis PNP	(1G2O, 1I80),[80] 1N3I[81]
Hexameric PNP	
E. coli PNP	1A69,[82] 1ECP,[19] 1K9S,[26] (1OTX, 1OTY, 1OU4, 1OUM, 1OV6, 1OVG),[55] (1PK7, 1PK9, 1PKE, 1PR0 1PR1, 1PR2, 1PR4, 1PR5, 1PR6, 1PW7)[83]
Vibrio cholerae PNP	(1VHJ, 1VHW)[84]
Trichomoniasis vaginalis PNP	(1Z33, 1Z34, 1Z35, 1Z36, 1Z37, 1Z38, 1Z39)[53]
Bacillus anthracis PNP	1XE3[85]
Thermus thermophilus PNP	(1ODI, 1ODJ, 1ODK, 1ODL)[86]
Sulfolobus solfataricus PNP[b]	(1JDS, 1JDT, 1JDU, 1JDV, 1JDZ, 1JE0, 1JE1, 1JP7, 1JPV)[20]
Plasmodium falciparum PNP	(1NW4, 1Q1G),[87] 1SQ6[c]

[a] This enzyme functions as a homohexamer; however, it can be viewed as dimer of trimers, each of which resembles the trimeric PNP structure.
[b] Also referred to as MTAP; substrate preference is Ino > Guo > Ado > MTA.
[c] PDB entry only; not published.

PNPs are of biomedical interest for several reasons.[9,10] First, in humans the absence of PNP is associated with severe T-cell immune deficiency, while B-cell function is unaffected.[11] The T-cell suppression was first identified in patients with PNP mutations and recent studies show that the cell death is mediated by apoptosis.[12–13] This profile suggests that PNP inhibitors might be useful in the treatment of T-cell proliferative diseases, such as T-cell leukemia or T-cell lymphoma, or to suppress the T-cell response in T-cell mediated autoimmune diseases, such as rheumatoid arthritis, psoriasis, or lupus. Second, the differences in substrate specificities between PNPs from humans and microorganisms might be exploited in designing antimicrobial agents.[14] In *Plasmodium falciparum*, PNP cleaves methylthioinosine in a pathway that is the sole source of purine nucleosides for this organism.[15] Therefore PNP inhibitors might be useful in the treatment of malaria and perhaps other parasitic diseases. Finally, the difference in PNP specificities has been exploited in a gene therapy approach to the treatment of cancers.[16] Efficient tumor-cell killing has been achieved in tumors transfected with a prodrug-specific PNP. Tumor cells activate the prodrug by glycosidic bond cleavage, thus releasing a cytotoxic agent while normal cells are unaffected.

The considerable biomedical interest in both human and microbial PNPs has resulted in a significant effort being invested in inhibitor design. These efforts have resulted in improving the earliest PNP inhibitors, with K_i values in the μM range, to the current best inhibitors of PNP, with K_i values in the pM range. Several of these inhibitors have progressed to clinical trials with encouraging results (see below). This review focuses on the structures of the PNP family enzymes and on the use of structural information in the application of molecular design to PNP and related enzymes.

3.2 Three-dimensional Structures of PNPs

The structure of human PNP was first[17] reported in 1990 and since then almost 90 structures of PNPs from various organisms, with and without bound ligands, have been published or deposited in the PDB (Table 3.1). The collection of structures shows that PNPs can be divided into two groups, trimeric and hexameric, based on quaternary structure (Figure 3.2), but all PNPs share a common α/β monomer fold and active site location (Figure 3.3). The typical PNP monomer fold consists of a central eight-stranded mixed β-sheet and a smaller five-stranded mixed β-sheet. The two sheets pack together to form a distorted β-barrel. The core β-structure is flanked on either side by seven α-helices. Each monomer contains one active site located in a cleft near the β-barrel.

Trimeric PNPs are found in both prokaryotes and eukaryotes and typically contain about 300 amino acids per monomer. Three monomers pack together such that the active site is located roughly at the equator of the molecule, near the monomer–monomer interface.[17]A Dali comparison of the trimeric PNPs starting with human PNP is shown in Table 3.2. Bovine PNP is the most closely related structure, with 87% identity. As crystals of bovine PNP diffract to much

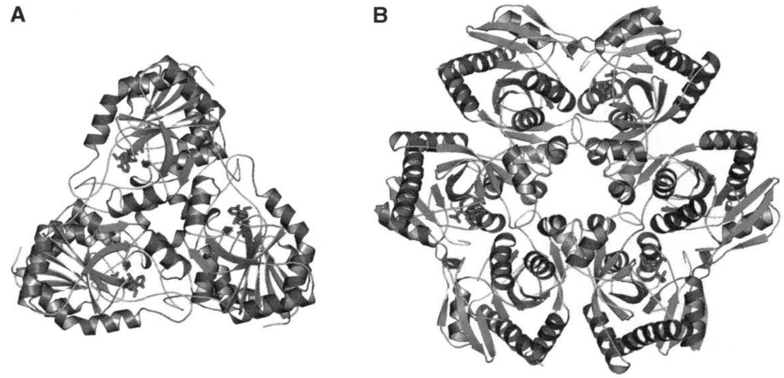

Figure 3.2 Quaternary structures of *H. sapiens* PNP (A) and *E. coli* PNP (B) color coded by secondary structural elements. Ligands at the substrate-binding sites are in stick representation and colored in red.

higher resolution than crystals of human PNP, many of the structural data used for inhibitor design were derived from the bovine enzyme. *Escherichia coli* PNP-2, the XapA gene product, is reported to be a hexamer in which two trimers, each closely resembling the human PNP trimer, pack face to face.[18] The close structural similarity of the trimeric PNPs is also reflected in the high conservation of active site residues.

Hexameric PNPs thus far have been found only in prokaryotes and typically contain about 240 amino acids per monomer. In addition to hexameric PNPs, many prokaryotes also express a trimeric PNP, but the two types of PNPs have different substrate specificities. The packing of monomers in the hexameric PNPs is unrelated to the arrangement in trimeric PNPs. Hexameric PNPs can be thought of as a trimer of dimers, resulting in an overall donut shape.[19] Each dimer contains two complete active sites located near the monomer–monomer interface within the dimer. Hexamer formation results in three active sites on the top of the donut alternating with three active sites at the bottom of the donut. A Dali comparison of the hexameric PNPs starting with *E. coli* PNP is shown in Table 3.2. The fold is highly conserved, as are the active site residues. The most distant family members are *Sulfolobus solfataricus* PNP, which also catalyzes the phosphorolysis of methylthioadenosine (MTA), and *P. falciparum* PNP, which prefers 6-oxopurine substrates, including methylthioinosine.[15]

The existence of two main types of quaternary structure utilizing the same monomeric fold is an unusual feature of the PNP family. The structure of the monomer core fold is highly conserved. Most of the phosphate-binding residues are found in the N-terminal half of the monomer, while most of the nucleoside-binding residues are found in the C-terminal half. Interestingly, the N-terminal half of the molecule shows less sequence similarity than the C-terminal half, even though the nucleoside-binding domain is associated with a greater diversity of substrate-binding specificities. Comparing residues in the phosphate-binding

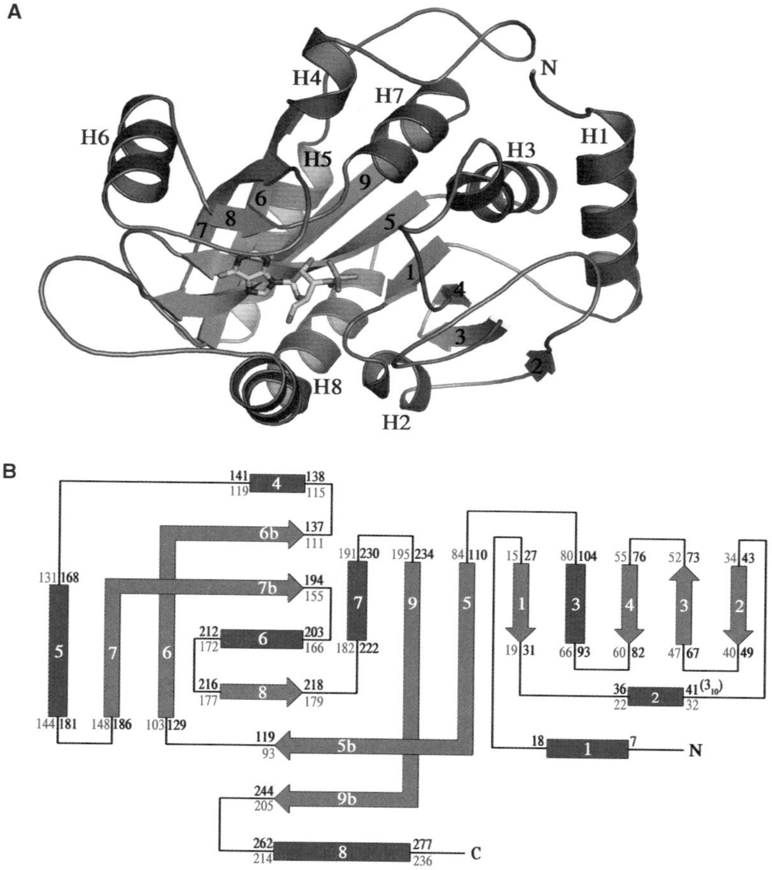

Figure 3.3 PNP monomer fold. (A) The human PNP monomer is represented by a ribbon diagram and color coded by secondary structural elements. The fold has a central β-barrel, formed by two β-sheets, flanked by seven α-helices and a 3_{10} helix. The ligands (inosine and sulfate) bound at the active site are shown in stick representation. (B) The topology diagram of the PNP monomer is labeled with residue numbers – those in black correspond to human PNP and those in red to *E. coli* PNP. Helix 1 (H1) is not present in the structure of *E. coli* PNP. Helix 2 (H2) is a 3_{10} helix in the human PNP structure and an α-helix in the *E. coli* PNP structure.

site shows little similarity between the trimeric and hexameric PNPs, although both evolved to bind phosphate. Significant diversity is found in the loops that mediate the subunit–subunit contacts responsible for trimer and hexamer formation. The subunit–subunit contacts in the trimer are formed by the β4–H3 loop, the C-terminal end of β6, one face of H4, the H4–H5 loop, the β7–H6 loop, one face of H6, and the β9–H8 loop. The subunit–subunit contacts

Table 3.2 Dali comparison of PNPs.

Protein	*PDB*	*Z* [a]	*rmsd* [b]	*LALI* [c]	*LSEQ2* [d]	*IDE* [e]
Starting with human PNP	**1M73**					
Bos taurus PNP	1B8O	43.1	1.2	275	280	87
Schistosoma mansoni PNP	1TCU	39.2	1.4	276	282	45
Escherichia coli PNP-2	1YQQ	35.2	1.9	267	273	40
Thermotoga maritima PNP	1VMK	34.7	1.3	245	249	45
Cellulomonas spp. PNP	1C3X	34.0	2.0	249	266	34
Mycobacterium tuberculosis PNP	1G2O	32.9	2.0	258	262	36
Starting with *E. coli* PNP	**1ECP**					
Vibrio cholerae PNP	1VHJ	42.9	0.4	234	234	79
Trichomoniasis vaginalis PNP	1Z33	39.2	1.1	235	235	59
Bacillus anthracis PNP	1XE3	37.3	1.1	232	233	56
Thermus thermophilus PNP	1ODI	35.6	1.8	234	234	35
Sulfolobus solfataricus PNP	1JDS	34.0	1.3	224	226	32
Plasmodium falciparum PNP	1NW4	32.8	1.8	233	243	27

[a] Z, strength of structural similarity in standard deviations above expected.
[b] rmsd, root mean square deviation.
[c] LALI, length of aligned residues.
[d] LSEQ2, length of the complete sequence.
[e] IDE, percentage of identical residues.

in the hexamer are formed by the N-terminal loop, the β1–H2 loop, the β2–β3 loop, one face of H3, the C-terminal end of β6, H4, H5, the β6–H4 loop, the H4–H5 loop, β7, the β7–H6 loop, one face of H6, and the C-terminal end of H7. Many of the contact regions are positioned differently between the trimer and hexamer; and when similarly positioned loops are utilized for both trimer and hexamer formation, the loops are often dramatically different in size and amino acid conservation. One can therefore imagine a core domain containing most of the active site that can form either trimers or hexamers depending on the size and type of loops attached to the core.

3.3 Related Enzymes of the PNP Family

Several other enzymes that also cleave the N-glycosyl bond have the PNP family fold. They include MTAP, UP, AMPN, and MTAN, although the latter two enzymes utilize water rather than phosphate as the nucleophile for cleaving the glycosidic bond (Table 3.3). Considerable effort has been devoted to the design of MTAP, UP, and MTAN inhibitors, and many of the principles that apply to the design of PNP inhibitors also apply to the design of inhibitors for other enzymes with a PNP fold. Enzymes specific for MTA usually belong to the trimeric PNP family,[5] although hexameric PNPs that cleave MTA in addition to Ado, Gua, and Ino have also been reported.[20] Prokaryotic UP is hexameric,[6] but there is no reported structure for a eukaryotic UP. AMPN

Table 3.3 Other enzymes utilizing the PNP family fold.

Enzyme	*PDB ID*	*Biological Unit*
Homo sapiens MTAP	(1CB0, 1CG6),[5] 1K27,[56] (1SD1, 1SD2)[88]	trimer
Sulfolobus tokodaii MTAP	1V4N	trimer
Escherichia coli UP	1K3F6,[6] 1LX7,[61] (1RXC, 1RXS, 1RXU, 1RXY, 1T0U),[22] (1TGV, 1TGY, 1U1C, 1U1D, 1U1E, 1U1F, 1U1G)[63]	hexamer
Salmonella typhimurium UP	1RYZ, 1SJ9[62]	hexamer
E. coli AMPN	(1T8R, 1T8S, 1T8W, 1T8Y)[7]	hexamer
Bacteroides thetaiotaomicron AMPN	1YBF	hexamer
E. coli MTAN	1JYS,[8] (1NC1, 1NC3),[89] (1Y6Q, 1Y6R),[66] (1Z5N, 1Z5O, 1Z5P)[90]	dimer

belongs to the hexameric PNP family and often contains an extra domain thought to play a regulatory role.[7] MTAN forms a dimer that is unrelated to the dimeric substructure of the hexameric PNP family.[8]

Figure 3.4 shows a phylogenic tree for the PNP family enzymes for which structures have been reported. The analysis shows a clear division between the hexameric and trimeric family members, with dimeric MTAN falling within the hexameric branch. The trimeric enzymes further subdivide into separate branches representing the PNPs and MTAPs. Interestingly, *E. coli* PNP-2, which is a dimer of trimers, does not separate from the trimeric enzymes. Both of the hydrolytic nucleosidases fall outside of the main cluster of hexameric PNPs. It is also notable that *P. falciparum* PNP, which cleaves methylthioinosine, and *S. solfataricus* PNP, which cleaves MTA, show separate branching. Interestingly, *P. falciparum* PNP is more closely related to the hexameric UPs than it is to the hexameric PNPs. Furthermore, *P. falciparum* PNP cleaves uridine at low levels,[21] even though it lacks the active site insertion characteristic of other UPs and also lacks the uridine-binding residues (R168, Q166, R223).[22] Knowledge of differences in active site residues, substrate specificity, and quaternary structure offers strategies for the design of specific inhibitors for related enzymes having the PNP fold.

3.4 PNP Active Sites

The active sites of trimeric PNPs are formed from eight different peptide regions (Figure 3.5).[23] The active site residues are highly conserved among the trimeric PNPs, but show little conservation compared to the hexameric PNPs.[19] Trimeric PNPs that have been both structurally and biochemically characterized are specific for 6-oxopurine nucleoside substrates. The active site can be divided into three subsites. In human PNP, the purine-binding site is formed primarily by Phe200, Glu201, Asn243, and Phe159 from an adjacent monomer.

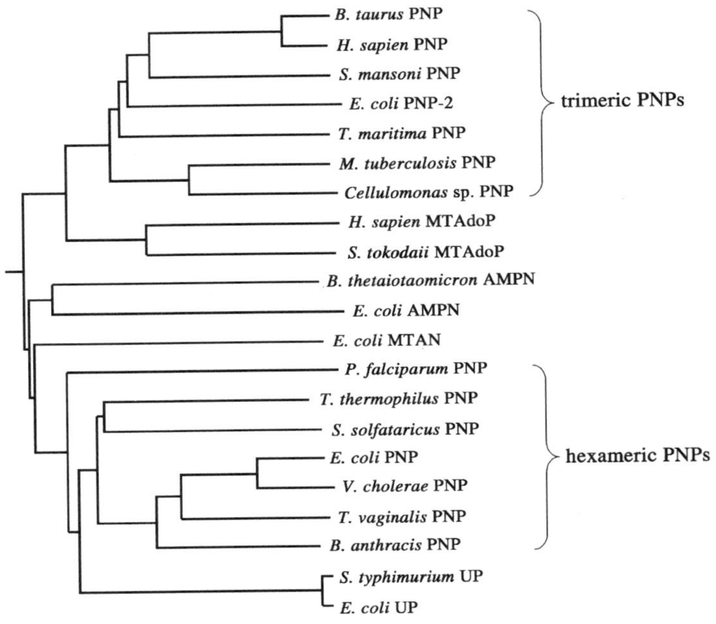

Figure 3.4 Phylogenic tree for the PNP family enzymes with known structures. Hexameric and trimeric family members clearly separate into different branches, with dimeric MTAN falling within the hexameric branch.

Glu201 hydrogen bonds to the purine N1 atom, while Phe200 packs against the purine base. Asn243 hydrogen bonds to the purine N7 atom and is thought to be catalytically important by stabilizing the negative charge that accumulates on the purine during glycosidic bond cleavage.[24] The ribose-binding site is formed by residues Tyr88, Met219, and His257. Tyr88 hydrogen bonds to the O3′ atom of the ribose, while Met219 packs against the hydrophobic face of the ribose. His257 is away from the active site in the structure of unliganded PNP. Atom O3′ also hydrogen bonds to a phosphate oxygen atom and helps position the phosphate for nucleophilic attack at C1′. The phosphate-binding site is formed from Ser33, Arg84, His86, and Ser220, with a total of eight hydrogen bonds. His64 shows variation among the available PNP structures and in some structures forms an additional hydrogen bond to phosphate. It has been proposed that the negatively charged phosphate ion uses substrate-assisted catalysis to stabilize the positive charge of the oxycarbenium ion intermediate, which forms after glycosidic bond cleavage.[24]

Comparison of the available trimeric PNP structures with different substrates, products, or analogs suggests that the primary conformational change upon ligand binding occurs at His257 and the surrounding residues, with a smaller conformational change occurring at His64.[25] The phosphate is rigidly held and shows little movement during glycosidic bond cleavage. The ribosyl

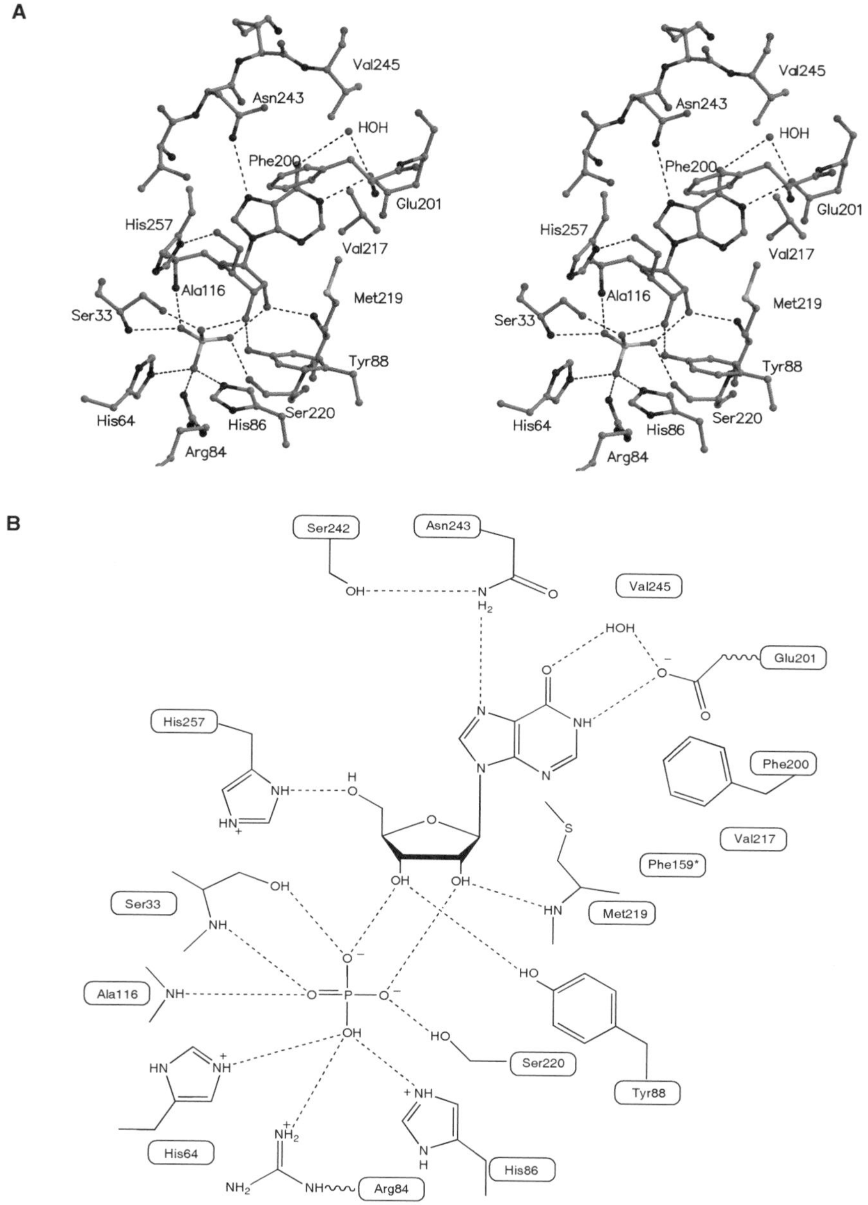

Figure 3.5 Stereo diagram (A) and schematic diagram (B) of the active site of bovine PNP complexed with inosine and phosphate. The active site can be divided into three subsites. The purine-binding site is primarily formed by Phe200, Glu201, Asn204, Val217, and Phe159* (residues from an adjacent subunit are indicated with an asterisk). The ribose-binding site is formed by Tyr88, Met219, and His257. The phosphate-binding site is formed by Ser33, His64, His86, Arg84, Ala116, and Ser220.

group, which undergoes inversion at C1′ during catalysis, shows a flipping of C1′ by about 1.5 Å as the glycosidic bond breaks and the ribose 1-phosphate forms. The remainder of the ribosyl group shows only small changes with the most significant changes occurring closest to C1′. No changes in hydrogen bonding to the ribosyl group are observed between the substrate and product complexes. The purine base shows a slight tilt (15–20°) away from C1′ after glycosidic bond cleavage.

The active sites of hexameric PNPs are structurally analogous to those of trimeric PNPs, but show some important differences (Figure 3.6).[19] First, there is very little conservation of the active site residues (only the position corresponding to Met219 of human PNP is strictly conserved). Second, because of the differences in quaternary structure, hexameric PNPs utilize the N-terminal region and an arginine residue of the adjacent monomer to complete the phosphate- and ribose-binding sites, respectively. This is in contrast to the trimeric PNPs, which utilize a phenylalanine (Phe159 for human PNP) or tyrosine from an adjacent monomer to complete the purine-binding site. The difference in quaternary structures results in an active site for hexameric PNPs that is more exposed than the active site of trimeric PNPs. Third, the active site shows only small conformational changes between unliganded and complexed forms. In *E. coli* PNP the purine-binding site is formed primarily by Phe159 and Asp204. Phe159 packs against the purine base, while Asp204 is analogous to Asn243 in human PNP.[26] It is proposed that Asp204 is protonated during catalysis and donates a hydrogen bond to the purine N7 atom to stabilize the negative charge on the purine base.[19] The ribose binding also utilizes a water molecule to donate a hydrogen bond to the purine N1 position in the case of adenosine or to accept a hydrogen bond in the case of inosine. The ribose-binding site is formed by Glu181 and Met180 from one monomer and His4 of the adjacent monomer. Glu181 forms hydrogen bonds with both O2′ and O3′ of the ribosyl group, His4 hydrogen bonds to the O5′ position, and Met180 packs against the hydrophobic face of the sugar. As is the case for trimeric PNPs, a hydrogen bond is formed between O3′ and an oxygen atom of the attacking phosphate. The phosphate-binding site is formed by Gly20, Arg24, Arg87, and Ser90 of one monomer and Arg43 of the adjacent (twofold related) monomer, resulting in a total of nine hydrogen bonds. Comparison of the available trimeric and hexameric PNP structures suggests that, even though the two types of active sites utilize different amino acid residues, the geometry of the substrate and product complexes is similar and that the two types of PNPs utilize similar catalytic strategies for glycosidic bond cleavage. One notable difference is that trimeric PNPs are usually both more specific and more efficient than their hexameric counterparts.[14]

3.5 Human PNP Inhibitors

Much of the effort to develop PNP inhibitors has been driven by pharmaceutical companies interested in the T-cell specific immunosuppression associated

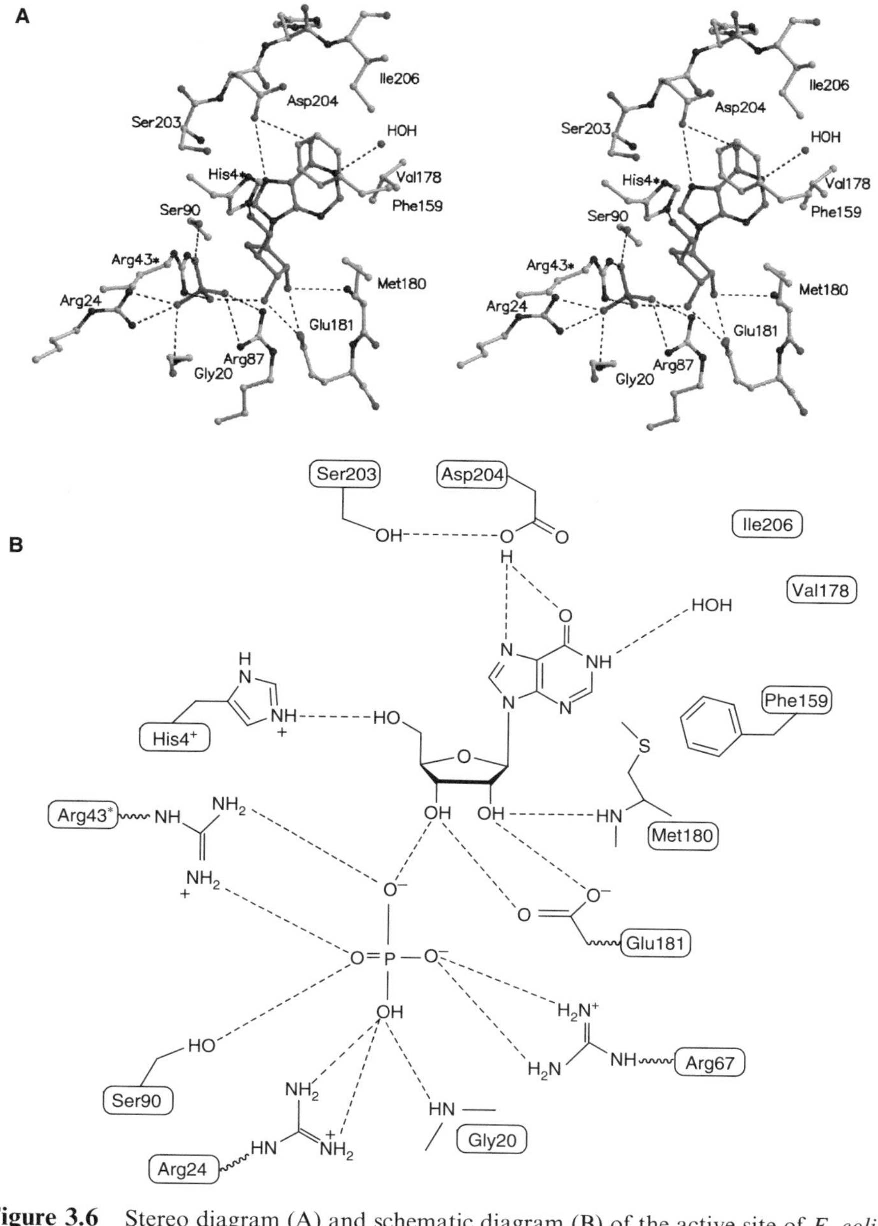

Figure 3.6 Stereo diagram (A) and schematic diagram (B) of the active site of *E. coli* PNP complexed with inosine and sulfate. The purine base binding site is formed by Phe159, Val178, and Asp204. Asp204 is proposed to be protonated and donate a hydrogen bond to the N7 atom of the purine base. The ribose-binding site is formed by Met180, Glu181, and His4* (residues from an adjacent subunit are indicated with an asterisk). The phosphate-binding site is formed by Gly20, Arg24, Arg87, Ser90, and Arg43*.

with the absence of PNP activity. Warner Lambert maintained an active program in the 1980's and both Burroughs Welcome and Merrell Dow developed multisubstrate inhibitors in the late 1980's and 1990's. A partnership between BioCryst and CIBA-Geigy in the late 1980's and early 1990's focused on a structure-based approach. In recent years BioCryst has been the main player in the PNP inhibitor field. Inhibitors of human PNP have evolved through four generations (Figure 3.7). The first-generation inhibitors consisted of substrate analogs resulting from simple modification of purines or purine nucleosides.[27–30] Second-generation inhibitors represent multisubstrate analogs, which simultaneously occupy the purine-, ribose-, and phosphate-binding subsites of the PNP active site.[31,32] Third-generation PNP inhibitors were optimized using iterative structure-based drug design and resulted in a phase III clinical trial.[23,33–36] The fourth-generation PNP inhibitors were designed as transition state analogs and are currently in clinical trials.[37]

The representative human PNP inhibitors shown in Figure 3.7 were designed using a variety of methods, including modification of substrates, *de novo* computational methods, and mimicking the expected transition state. The earliest PNP inhibitor sufficiently potent to be considered as a drug lead was 8-aminoguanosine (**1**) ($K_i \approx 1\,\mu M$). Even with weak inhibition, this drug lead showed selective *in vivo* toxicity towards T-cells when combined with low concentrations of 2′-deoxyguanosine.[28] Other first-generation inhibitors included 5′-iodo-9-deazainosine (**2**)[29] and a promising series of 9-substituted purine analogs.[27] These and other first-generation inhibitors were designed using the traditional medicinal chemistry approach of substrate modification and showed sub-micromolar inhibition. One compound, 8-amino-9-(2-thienylmethyl)guanine (**3**) (PD119,229) showed promise in animal studies,[38] but the high level of PNP activity in erythrocytes meant it was unsuitable as a drug candidate. The first-generation PNP inhibitors have been described in detail.[10,39,40] Acyclovir diphosphate (**4**) was reported as the first multisubstrate inhibitor of PNP ($K_i \approx 1\,nM$ when assayed at 1 mM phosphate). X-ray crystallographic studies later showed that the purine base and terminal phosphate occupied the expected active-site locations and that the acyclic connector lined the hydrophobic side of the ribose-binding pocket.[41] Despite its tight binding affinity, acyclovir diphosphate is both unlikely to cross the cell membrane and is metabolically unstable, thus limiting its usefulness as a drug lead. However, studies on acyclovir diphosphate, a series of 9-phosphonoalkyl analogs (**5**),[32] and a subsequent series of activated multisubstrate phosphonate analogs (**6**)[42] demonstrated the potential for multisubstrate analogs.

In 1991, the structure of human PNP became available, thus providing insight into PNP inhibitor structure–activity relationships and offering the possibility of structure-based approaches to drug design. Early structural studies showed the basis for the tight binding of analogs such as 8-aminoguanine, 9-deazainosine, and 9-benzylhypoxanthine.[23] Both the 8-amino and 9-deaza analogs take advantage of additional hydrogen bonds; however, the improvements in the hydrogen bonding schemes are mutually exclusive, thus rendering the combination of 8-amino and 9-deaza modifications unfavorable.

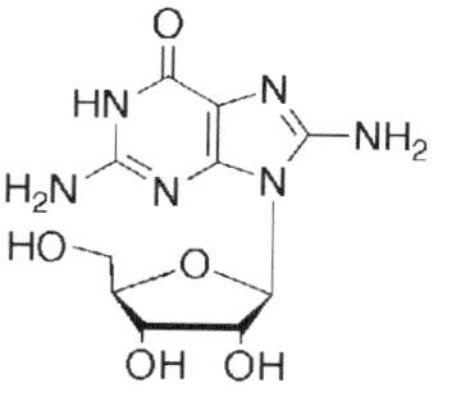

1

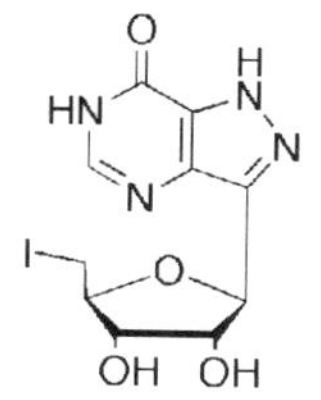

2

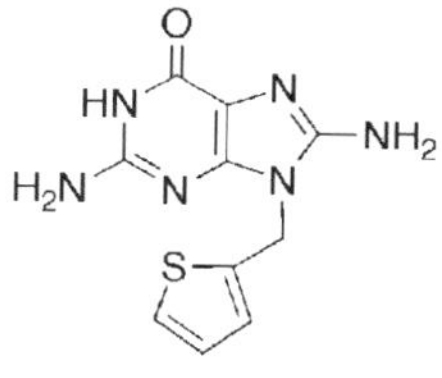

3

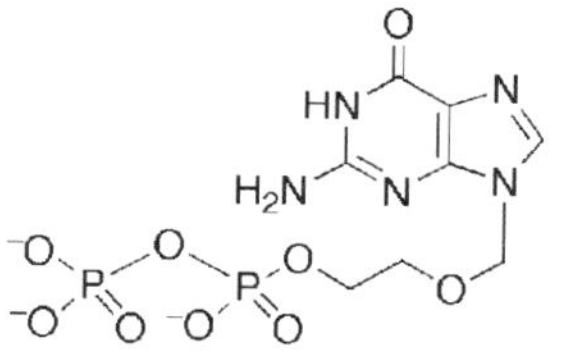

4

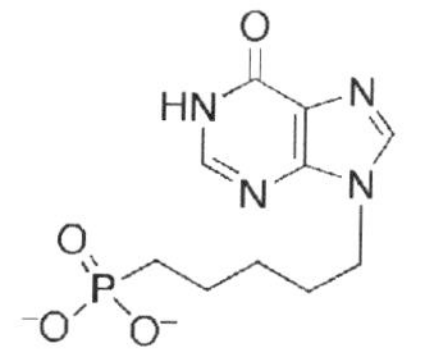

5

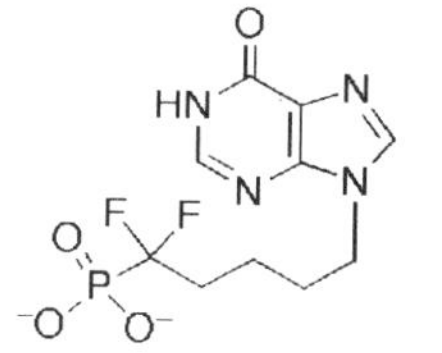

6

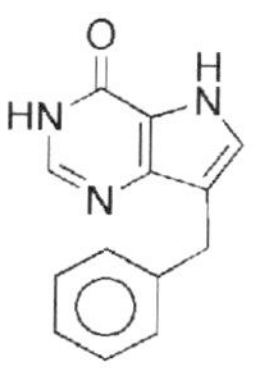

7

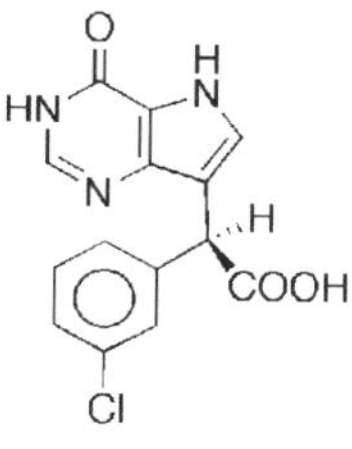

8

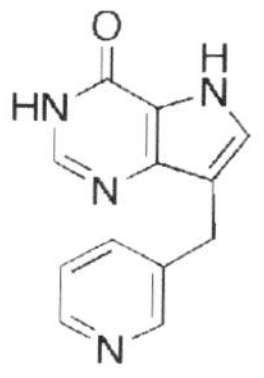

9

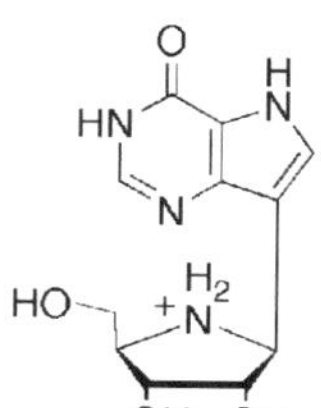

10

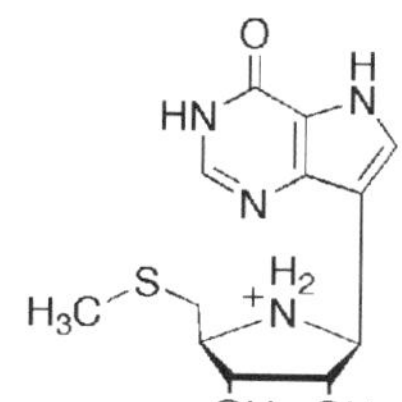

11

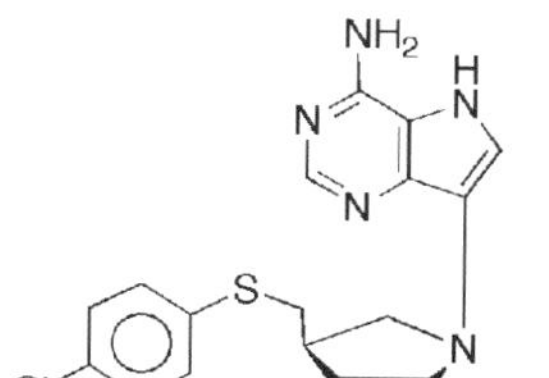

12

Figure 3.7 Representative human PNP inhibitors. Row 1, first generation PNP inhibitors: **1**, 8-aminoguanosine; **2**, 5′-iodo-9-deazainosine; and **3**, 9-(2-thienylmethylguanine). Row 2, second generation PNP inhibitors: **4**, acyclovir diphosphate; **5**, 9-(phosphonopentyl)hypoxanthine; and **6**, 9-(difluorophosphonopentyl)hypoxanthine. Row 3, third generation PNP inhibitors: **7**, 9-benzyl-9-deazahypoxanthine; **8**, [S]-9-3-(2-amino-4-oxo-3*H*,5*H*-pyrrolo[3,2-d]pyrimidin-7-yl)-3-(3-chlorophenyl)propanoic acid; **9**, and 9-(3-pyridyl)-9-deazahypoxanthine. Row 4, fourth generation PNP inhibitors and inhibitors of related enzymes: **10**, immucillin-H; **11**, 5′-deoxy-5′-methylthioimmucillin-H; and **12**, methylthio-DADMe-immucillin-A.

An important breakthrough in the design of PNP inhibitors was the demonstration that hydrophobic 9-substituents packed against the hydrophobic side of the ribose-binding site and increased binding affinity by about two orders of magnitude compared with the unsubstituted parent compound.[23] Combination of either the 8-amino modification or a 9-deaza modification with an aromatic substituent resulted in further improvements. A series of 9-substituted-9-deaza analogs was designed and optimized using iterative structure-based drug design. Starting compounds were generated using Monte Carlo based conformational searching, optimized by molecular mechanics energy minimization, and ranked for synthesis. Synthesized compounds were assayed for activity and in most cases analyzed by X-ray crystallography followed by further cycles of design and analysis. This resulted in the identification of a series of 9-aryl-9-deazahypoxanthine analogs (**7**) with K_i values in the 5–30 nM range. The best compound (**8**) in the series also contains a group with affinity for the phosphate-binding site and has a K_i value of 5 nM. *In vivo* studies identified BCX-34 (**9**) ($K_i = 30$ nM), which has a higher K_i value but better bioavailability, as a drug lead.

The structure of PNP also suggested a novel class of phosphonate multisubstrate analogs in which the phosphonate is attached to O1′ and O2′ through a one- or two-atom bridge.[43] Modeling studies predicted and subsequent crystallographic studies confirmed that the multisubstrate analogs would closely mimic the binding of nucleoside and phosphate in the enzyme–substrate complex. The best compound, which also incorporates a 9-deaza modification, showed an IC_{50} of about 25 nM.

A major breakthrough resulted from the studies of Schramm and coworkers who used the proposed mechanism of human PNP to design transition-state analogs.[37] Mechanistic studies suggested an oxocarbenium intermediate with significant positive charge developing on the 4′-oxygen atom during the transition state.[44,45] Schramm and coworkers designed and synthesized a series of compounds in which the ribosyl group was replaced by the aminoribose analog, which carries a positive charge.[46] The compounds were further enhanced by 9-deaza modification resulting in immucillin-H (I mmH) (**10**) ($K_d \approx 1$ nM), the most potent human PNP inhibitor reported to date.

To date nearly 30 structures of mammalian (human or bovine) PNP complexes with substrates, analogs, or inhibitors have been reported (Table 3.1) and the factors responsible for potent inhibition are well understood.[47–52] Given the extreme potency of ImmH and favorable pharmacological properties, it is unclear whether or not further improvements in PNP inhibition will be beneficial for drug development.

3.6 Other Applications of Molecular Design to PNP

The critical role of PNP in purine nucleoside metabolism and the differences in PNP substrate specificities between prokaryotes and humans mean that the PNPs of microorganisms are obvious targets for antimicrobial agents.

Trichomoniasis is a common sexually transmitted disease caused by the anaerobic protozoan parasite *Trichomonas vaginalis*. *T. vaginalis* lacks purine biosynthetic machinery, so the host provides the parasite's source of purines, which are obtained via the purine salvage pathway. *T. vaginalis* PNP falls in the hexameric PNP family and shows significant sequence identity to *E. coli* PNP.[53] Even though the active site is almost completely conserved, *T. vaginalis* PNP shows a slight preference for adenosine compared with *E. coli* PNP, which shows a slight preference for inosine. As human PNP does not accept 6-aminopurines as substrates, the difference in substrate preferences can be harnessed to design inhibitors. It was recently reported that the nontoxic nucleoside analog 2-fluoro-2′-deoxyadenosine is a "subversive substrate" for *T. vaginalis* PNP, which phosphorolytically cleaves the nucleoside analog to release the highly cytotoxic purine base 2-fluoroadenine. 2-Fluoro-2′-deoxyadenosine has been shown to inhibit the growth of *T. vaginalis* PNP with an apparent IC_{50} of 106 nM.[53] Crystallographic studies showed that 2-fluoroadenine binds to the active site with a slightly distorted geometry, which is likely responsible for the higher inhibition compared with adenine. Further improvements may be achieved by utilizing the structure of the *T. vaginalis* PNP active site.

Another application that takes advantage of PNP's ability to cleave nucleoside analogs with varying substrate specificity is prodrug activation in cancer chemotherapy.[54] Intracellular activation of prodrugs offers the possibility of delivering to tumor cells chemotherapeutic agents that are too toxic for systemic administration. One mechanism for prodrug activation is to introduce into the tumor cell a transfected gene that, when transcribed, produces a protein that activates the prodrug. The nucleoside cleavage reaction catalyzed by PNP can be utilized to activate prodrugs such as 6-methyl-2′-deoxypurine riboside (MePdR). MePdR is relatively non-toxic, while the cleavage product, 6-methylpurine (MeP), is highly cytotoxic. The differences in substrate specificity between human and *E. coli* PNPs can be utilized to activate MePdR to MeP selectively in tumor cells. *E. coli* PNP, but not human PNP, efficiently cleaves MePdR. By transfecting tumor cells with the *E. coli* PNP gene, the prodrug can be selectively activated, thus concentrating the release of cytotoxic MeP to the desired site of action. Parker, Sorscher, and coworkers have successfully demonstrated this concept *in vivo*.[54]

One problem encountered in preclinical studies resulted from low-level cleavage by human PNP and cleavage by intestinal flora. This problem was addressed by using the structures of *E. coli* PNP with and without ligands to design new prodrugs that are not cleaved by either human PNP or *E. coli* PNP and then reengineering *E. coli* PNP to specifically cleave the prodrug.[55] This was accomplished by showing that Me(talo)MePdR (which differs from MePdR by the addition of a methyl group at the 4′-position) is poorly cleaved by both wild-type *E. coli* PNP and human PNP. Modeling studies identified unfavorable steric interactions between the prodrug and the side chain Met64. Mutagenesis of Met64 to valine resulted in a 100-fold increase in the rate of prodrug cleavage. Initial animal studies showed that tumors expressing the M64V mutant were cured using Me(talo)MePdR as the prodrug. The modeling studies were followed by crystallographic studies of the mutant to confirm the structural basis for the substrate specificity of M64V PNP.[55]

3.7 Applications of Molecular Design to Enzymes Related to PNP

Three additional enzymes that have the PNP fold are also targets for drug design. Human MTAP is a member of the trimeric PNP family, with about a 40% identity with human PNP. MTAP functions in the polyamine pathway to recycle MTA as a byproduct resulting from the biosynthesis of spermine and spermidine, and cleavage of MTA by MTAP is the sole source of free adenine in mammals. The polyamine biosynthetic pathway is a validated target for cancer chemotherapy; however, previous efforts have focused largely on the polyamine biosynthetic enzymes, *S*-adenosylmethionine decarboxylase and ornithine decarboxylase. The structure of human MTAP has been determined with and without ligands.[5] The monomer differs from that of human PNP in that it lacks the N-terminal helix, but it has an additional C-terminal helix. Building on their prior success with ImmH as a human PNP inhibitor, Schramm and coworkers synthesized the transition-state analog for MTAP, methylthio-ImmA (**11**) and demonstrated an equilibrium dissociation constant of 1.0 nM.[56] A crystal structure of the MTAP/methylthio-ImmA complex revealed the methylthio group to be in a flexible conformation, suggesting that a bulkier 5′-substituent might result in tighter binding. An ImmA analog with phenylthio replacing the methylthio group also gave a dissociation constant of 1.0 nM. Methylthio-DADMe-ImmA, a pyrrolidine analog with a methylene bridge between the 9-deaza-adenine group and the pyrrolidine, is a slow-onset inhibitor with a dissociation constant of 86 pM. Further studies showed that *p*-Cl-phenylthio-DADMe-ImmA (**12**) binds with a dissociation constant of 10 pM, making it the tightest binding MTAP inhibitor reported to date. This inhibitor exploits both hydrophobic and electrostatic interactions and is sufficiently potent to shut down MTA salvage *in vivo*.

Most organisms have separate nucleoside phosphorylases to cleave uridine and thymidine. UP is a member of the PNP family, while thymidine phosphorylase has a completely unrelated fold.[57] UP functions in the pyrimidine salvage pathway and is critical in balancing the uridine and uracil pools. UP also participates in the metabolism of 5-fluorouracil and prodrugs, such as 2′-deoxy-5-fluorouridine. Since UP levels are elevated in tumor cells, it might be possible for UP inhibitors to potentiate the action of this anticancer drug by preventing back conversion of the prodrug to 5-fluorouracil. However, preclinical studies showed, surprisingly, that combining UP inhibitors with 2′-deoxy-5-fluorouridine was less effective than administering the prodrug alone. At the same time the drug toxicity was decreased, possibly as the result of elevating uridine pools.[58,59] It has also been shown that elevated uridine levels reverse the toxicity caused by azidothymidine in the treatment of acquired immunodeficiency syndrome (AIDS).[60] Despite the potential for UP inhibitors, a structure is not yet available for human or any other mammalian UP. Structures have been determined for native *E. coli* UP[6,61] and *Salmonella typhimurium* UP,[62] and several complexes of *E. coli* UP with substrates and inhibitors have also been reported.[22,63] The X-ray structures showed that *E. coli* UP is a member of the hexameric PNP family. The structure of the *E. coli* UP monomer closely resembles that of the *E. coli* PNP monomer, even though the sequence identity is less than 25%. The main structural difference in the two monomers is a 14 amino acid insertion in UP near the active site, which reduces the active site volume and prevents the binding of purine nucleosides. The phosphate and ribose-binding sites are highly conserved with respect to *E. coli* PNP, while differences in the base-binding site are required to bind uridine and catalyze its cleavage. Specifically, an arginine residue is positioned to hydrogen bond to O4 of the uracil in order to stabilize the negative charge that accumulates during glycosidic bond cleavage. Sequence alignments suggest that human UP has both a similar monomer fold and a similar quaternary structure. The only UP inhibitors thus far structurally characterized are a series of acyclic uridine analogs with either uracil or barbituric acid as the base and with each inhibitor having a bulky, hydrophobic substituent at the pyrimidine 5-position. The K_i of 5-benzyloxybenzylbarbituric acid acyclonucleoside (BBBA) for the human enzyme is 1.1 nM.[63] In the complex of BBBA with *E. coli* UP, the acyclic chain folds such that the terminal hydroxyl group mimics the 3′-hydroxyl group of the natural substrate uridine. The first benzyl group is well ordered, but the second one is not visible in the electron density and presumed to be disordered.[63]

MTAN nucleosidase is a member of the PNP family, but differs from other family members in two important ways. First, MTAN is a dimer and, second, the enzyme utilizes water as the nucleophile for the cleavage reaction rather than phosphate. MTAN is found only in prokaryotes and is required to recycle the byproducts of polyamine biosynthesis and the biosynthesis of autoinducer molecules involved in bacterial quorum-sensing molecules. Disruption of either pathway might be a useful strategy for antimicrobial chemotherapy. Transition-state inhibitors designed for MTAP were analyzed by biochemical, structural, and theoretical methods using *E. coli* MTAN as a target.[64–66] The potent MTAP inhibitor *p*-Cl-phenylthio-DADMe-ImmA (**12**) inhibited MTAN with an equilibrium dissociation constant of 47 fM.[65] Both kinetic isotope effects and theoretical calculations support a highly dissociative mechanism. These studies illustrate the usefulness of transition-state analysis in the design of potent enzyme inhibitors.

3.8 PNP Inhibitors and Clinical Trials

Two of the PNP inhibitors described above have entered clinical trials. 9-(Pyridin-3-yl)-9-deazahypoxanthine (BCX-34; peldesine) reached phase III clinical trials as a topical cream for cutaneous T-cell lymphoma (CTCL).[67] The patient group contained 89 patients with 43 receiving BCX-34 and 46 receiving a placebo. A total of 28% (12/43) of the patients treated with BCX-34 showed a response, but 24% (11/46) of patients who received the placebo also responded. The clinical trial showed that 1% BCX-34 dermal cream was not significantly better than the control as a therapy for patch- and plaque-phase CTCL. A phase I clinical trial in oral formulation for the treatment of AIDS was also initiated (http://www.clinicaltrials.gov/ct/show/NCT00002237). ImmH (BCX-1777; forodesine), a more potent *in vitro* inhibitor than BCX-34, also entered a phase I clinical trial for the treatment of CTLC (http://www.clinicaltrials.gov/ct/gui/show/NCT00098332). Results of a phase I clinical trial using intravenous infusion of BCX-1777 were recently published.[68] Additional information about BCX-34 and BCX-1777 clinical trials may be found at the Biocryst web site.

3.9 Conclusions and Future Directions

PNP and related enzymes offer a number of promising targets for drug design. These enzymes play a critical role in nucleotide metabolism and loss of activity can have profound effects. Human PNP is an early example of structure-based drug design in which inhibition of PNP was improved by 2–3 orders of magnitude and resulted in clinical trials for BCX-34. An additional 2–3 orders of magnitude improvement was achieved by utilizing transition-state chemistry. The current generation of inhibitors includes BCX-1777, which has entered clinical trials. The development of PNP inhibitors evolved from relatively ineffective substrate analogs in the 1980's, to inhibitors designed by computer-aided optimization in the 1990's, to transition-state analogs based on detailed mechanistic understanding at the turn of the 21st century. Clearly, transition-state inhibitors are the big winners *in vitro* for the nucleoside phosphorylase family as these principles have already been extended to other family members, such as MTAP and MTAN. Clinical trials currently underway will provide answers about efficacy.

Given that potent transition-state inhibitors are available for human PNP, *P. falciparum* PNP, MTAP, and MTAN, attention will be focused on other targets. It is likely that transition-state analogs will also inhibit UP and bacterial PNP. Issues of specificity will probably become important in antimicrobial drug design in which, for example, transition-state inhibitors should inactivate bacterial PNP, but not human PNP or human MTAP. The pharmacological profile of BCX-1777 is not yet known and it may be necessary to develop inhibitors that combine the potency of BCX-1777 with improved pharmacological parameters.

In the case of PNP and related enzymes, the success of transition-state inhibitors has been impressive and demonstrates the importance of considering transition-state chemistry in designing inhibitors for other enzyme families.[48]

Note Added in Proof

During the preparation of this chapter, the following structures have been deposited to the protein data bank: 1YRY[91] and (2A0W, 2A0X, 2A0Y, 2OC4,

2OC9, 2ON6)[92] for human PNP, 2AC7 for *Bacillus cereus* PNP, 2B94 for *Plasmodium knowlesi* PNP, 2BSX[93] for *P. falciparum* PNP, 2I4T and 2ISC[94] for *T. vaginalis* PNP, 2A8Y [95] for *S. solfataricus* PNP-2 (also referred to as MTAP-2), 2H8G[96] for *Arabidopsis thaliana* MTAN, and (1Y1Q, 1Y1R, 1Y1S, 1Y1T, 1ZL2) for *S. typhimurium* UP; and papers have been published for previously deposited structures of PNP from *S. mansoni*[97] and *B. anthracis*.[98]

References

1. A. W. Murray, D. C. Elliott and M. R. Atkinson, *Prog. Nucleic Acid Res. Mol. Biol.*, 1970, **10**, 87.
2. A. W. Murray, *Annu. Rev. Biochem.*, 1971, **40**, 811.
3. R. E. Parks Jr. and R. P. Agarwal, in *The Enzymes*, ed. P. D. Boyer, Academic Press, New York, 1972.
4. M. J. Pugmire and S. E. Ealick, *Biochem. J.*, 2002, **361**, 1.
5. T. C. Appleby, M. D. Erion and S. E. Ealick, *Structure*, 1999, **7**, 629.
6. E. Morgunova, A. M. Mikhailov, A. N. Popov, E. V. Blagova, E. A. Smirnova, B. K. Vainshtein, C. Mao, Sh. R. Armstrong, S. E. Ealick, A. A. Komissarov, E. V. Linkova, A. A. Burlakova, A. S. Mironov and V. G. Debabov, *FEBS Lett.*, 1995, **367**, 183.
7. Y. Zhang, S. E. Cottet and S. E. Ealick, *Structure*, 2004, **12**, 1383.
8. J. E. Lee, K. A. Cornell, M. K. Riscoe and P. L. Howell, *Structure*, 2001, **9**, 941.
9. J. A. Montgomery, *Med. Res. Rev.*, 1993, **13**, 209.
10. J. D. Stoeckler, in *Developments in Cancer Chemotherapy*, ed. R. E. Glazer, CRC Press, Boca Raton, 1984.
11. E. R. Giblett, A. J. Ammann, D. W. Wara, R. Sandman and L. K. Diamond, *Lancet*, 1975, **1**, 1010.
12. R. B. Gilbertsen, R. Posmantur, R. Nath and K. K. Wang, *Inflamm. Res.*, 1997, **46** Suppl. 2, S151.
13. E. Arpaia, P. Benveniste, A. Di Cristofano, Y. Gu, I. Dalal, S. Kelly, M. Hershfield, P. P. Pandolfi, C. M. Roifman and A. Cohen, *J. Exp. Med.*, 2000, **191**, 2197.
14. A. Bzowska, E. Kulikowska and D. Shugar, *Z. Naturforsch. C: Biosci.*, 1990, **45**, 59.
15. L. M. Ting, W. Shi, A. Lewandowicz, V. Singh, A. Mwakingwe, M. R. Birck, E. A. Ringia, G. Bench, D. C. Madrid, P. C. Tyler, G. B. Evans, R. H. Furneaux, V. L. Schramm and K. Kim, *J. Biol. Chem.*, 2005 **280**, 9547.
16. W. B. Parker, S. A. King, P. W. Allan, L. L. Bennett Jr., J. A. Secrist, 3rd, J. A. Montgomery, K. S. Gilbert, W. R. Waud, A. H. Wells, G. Y. Gillespie and E. J. Sorscher, *Hum. Gene Ther.*, 1997, **8**, 1637.
17. S. E. Ealick, S. A. Rule, D. C. Carter, T. J. Greenhough, Y. S. Babu, W. J. Cook, J. Habash, J. R. Helliwell, J. D. Stoeckler and C. E. Bugg, *J. Biol. Chem.*, 1990, **265**, 1812.

18. G. Dandanell, R. H. Szczepanowski, B. Kierdaszuk, D. Shugar and M. Bochtler, *J. Mol. Biol.*, 2005, **348**, 113.
19. C. Mao, W. J. Cook, M. Zhou, G. W. Koszalka, T. A. Krenitsky and S. E. Ealick, *Structure*, 1997, **5**, 1373.
20. T. C. Appleby, I. I. Mathews, M. Porcelli, G. Cacciapuoti and S. E. Ealick, *J. Biol. Chem.*, 2001, **276**, 39232.
21. G. A. Kicska, P. C. Tyler, G. B. Evans, R. H. Furneaux, K. Kim and V. L. Schramm, *J. Biol. Chem.*, 2002, **277**, 3219.
22. T. T. Caradoc-Davies, S. M. Cutfield, I. L. Lamont and J. F. Cutfield, *J. Mol. Biol.*, 2004, **337**, 337.
23. S. E. Ealick, Y. S. Babu, C. E. Bugg, M. D. Erion, W. C. Guida, J. A. Montgomery and J. A. Secrist, 3rd, *Proc. Natl. Acad. Sci. U. S. A.*, 1991, **88**, 11540.
24. M. D. Erion, J. D. Stoeckler, W. C. Guida, R. L. Walter and S. E. Ealick, *Biochemistry*, 1997, **36**, 11735.
25. C. Mao, W. J. Cook, M. Zhou, A. A. Federov, S. C. Almo and S. E. Ealick, *Biochemistry*, 1998, **37**, 7135.
26. G. Koellner, A. Bzowska, B. Wielgus-Kutrowska, M. Luic, T. Steiner, W. Saenger and J. Stepinski, *J. Mol. Biol.*, 2002, **315**, 351.
27. D. S. Shewach, J. W. Chern, K. E. Pillote, L. B. Townsend and P. E. Daddona, *Cancer Res.*, 1986, **46**, 519.
28. I. S. Kazmers, B. S. Mitchell, P. E. Dadonna, L. L. Wotring, L. B. Townsend and W. N. Kelley, *Science*, 1981, **214**, 1137.
29. J. D. Stoeckler, C. Cambor, V. Kuhns, S. H. Chu and R. E. Parks Jr., *Biochem. Pharmacol.*, 1982, **31**, 163.
30. J. D. Stoeckler, J. B. Ryden, R. E. Parks Jr., M. Y. Chu, M. I. Lim, W. Y. Ren and R. S. Klein, *Cancer Res.*, 1986, **46**, 1774.
31. J. V. Tuttle and T. A. Krenitsky, *J. Biol. Chem.*, 1984, **259**, 4065.
32. C. E. Nakamura, S. H. Chu, J. D. Stoeckler and R. E. Parks Jr., *Biochem. Pharmacol.*, 1986, **35**, 133.
33. M. D. Erion, S. Niwas, J. D. Rose, S. Ananthan, M. Allen, J. A. Secrist, 3rd, Y. S. Babu, C. E. Bugg, W. C. Guida, S. E. Ealick and J. A. Montgomery, *J. Med. Chem.*, 1993, **36**, 3771.
34. W. C. Guida, R. D. Elliott, H. J. Thomas, J. A. Secrist, 3rd, Y. S. Babu, C. E. Bugg, M. D. Erion, S. E. Ealick and J. A. Montgomery, *J. Med. Chem.*, 1994, **37**, 1109.
35. J. A. Montgomery, S. Niwas, J. D. Rose, J. A. Secrist, 3rd, Y. S. Babu, C. E. Bugg, M. D. Erion, W. C. Guida and S. E. Ealick, *J. Med. Chem.*, 1993, **36**, 55.
36. J. A. Secrist, 3rd, S. Niwas, J. D. Rose, Y. S. Babu, C. E. Bugg, M. D. Erion, W. C. Guida, S. E. Ealick and J. A. Montgomery, *J. Med. Chem.*, 1993, **36**, 1847.
37. V. L. Schramm, *Biochim. Biophys. Acta*, 2002, **1587**, 107.
38. R. B. Gilbertsen, M. E. Scott, M. K. Dong, L. M. Kossarek, M. K. Bennett, D. J. Schrier and J. C. Sircar, *Agents Actions*, 1987, **21**, 272.
39. J. C. Sircar and R. B. Gilbertsen, *Drugs Future*, 1988, **13**, 653.

40. A. Bzowska, E. Kulikowska and D. Shugar, *Pharmacol. Ther.*, 2000, **88**, 349.
41. D. M. dos Santos, F. Canduri, J. H. Pereira, M. Vinicius Bertacine Dias, R. G. Silva, M. A. Mendes, M. S. Palma, L. A. Basso, W. F. de Azevedo Jr. and D. S. Santos, *Biochem. Biophys. Res. Commun.*, 2003, **308**, 553.
42. S. Halazy, A. Ehrhard and C. Danzin, *J. Am. Chem. Soc.*, 1991, **113**, 315.
43. A. V. Toms, W. Wang, Y. Li, B. Ganem and S. E. Ealick, *Acta Crystallogr. D*, 2005, **61**, 1449.
44. P. C. Kline and V. L. Schramm, *Biochemistry*, 1995, **34**, 1153.
45. P. C. Kline and V. L. Schramm, *Biochemistry*, 1993, **32**, 13212.
46. G. A. Kicska, L. Long, H. Horig, C. Fairchild, P. C. Tyler, R. H. Furneaux, V. L. Schramm and H. L. Kaufman, *Proc. Natl. Acad. Sci. U. S. A.*, 2001, **98**, 4593.
47. A. Lewandowicz, E. A. Ringia, L. M. Ting, K. Kim, P. C. Tyler, G. B. Evans, O. V. Zubkova, S. Mee, G. F. Painter, D. H. Lenz, R. H. Furneaux and V. L. Schramm, *J. Biol. Chem.*, 2005, **280**, 30320.
48. V. L. Schramm, *Arch. Biochem. Biophys.*, 2005, **433**, 13.
49. S. Nunez, D. Antoniou, V. L. Schramm and S. D. Schwartz, *J. Am. Chem. Soc.*, 2004, **126**, 15720.
50. A. Lewandowicz and V. L. Schramm, *Biochemistry*, 2004, **43**, 1458.
51. G. B. Evans, R. H. Furneaux, A. Lewandowicz, V. L. Schramm and P. C. Tyler, *J. Med. Chem.*, 2003, **46**, 3412.
52. G. A. Kicska, P. C. Tyler, G. B. Evans, R. H. Furneaux, W. Shi, A. Fedorov, A. Lewandowicz, S. M. Cahill, S. C. Almo and V. L. Schramm, *Biochemistry*, 2002, **41**, 14489.
53. Y. Zhang, W. H. Wang, S. W. Wu, S. E. Ealick and C. C. Wang, *J. Biol. Chem.*, 2005, **280**, 22318.
54. W. B. Parker, P. W. Allan, A. E. Hassan, J. A. Secrist, 3rd, E. J. Sorscher and W. R. Waud, *Cancer Gene Ther.*, 2003, **10**, 23.
55. E. M. Bennett, R. Anand, P. W. Allan, A. E. Hassan, J. S. Hong, D. N. Levasseur, D. T. McPherson, W. B. Parker, J. A. Secrist, 3rd, E. J. Sorscher, T. M. Townes, W. R. Waud and S. E. Ealick, *Chem. Biol.*, 2003, **10**, 1173.
56. V. Singh, W. Shi, G. B. Evans, P. C. Tyler, R. H. Furneaux, S. C. Almo and V. L. Schramm, *Biochemistry*, 2004, **43**, 9.
57. M. R. Walter, W. J. Cook, L. B. Cole, S. A. Short, G. W. Koszalka, T. A. Krenitsky and S. E. Ealick, *J. Biol. Chem.*, 1990, **265**, 14016.
58. O. M. Ashour, O. N. Al Safarjalani, F. N. Naguib, N. M. Goudgaon, R. F. Schinazi and M. H. el Kouni, *Cancer Chemother. Pharmacol.*, 2000, **45**, 351.
59. O. N. Al Safarjalani, X. J. Zhou, R. H. Rais, J. Shi, R. F. Schinazi, F. N. Naguib and M. H. El Kouni, *Cancer Chemother. Pharmacol.*, 2005, **55**, 541.
60. F. N. Naguib, D. L. Levesque, E. C. Wang, R. P. Panzica and M. H. el Kouni, *Biochem. Pharmacol.*, 1993, **46**, 1273.
61. F. T. Burling, R. Kniewel, J. A. Buglino, T. Chadha, A. Beckwith and C. D. Lima, *Acta Crystallogr. D*, 2003, **59**, 73.

62. M. V. Dontsova, Y. A. Savochkina, A. G. Gabdoulkhakov, S. N. Baidakov, A. V. Lyashenko, M. Zolotukhina, L. Errais Lopes, M. B. Garber, E. Y. Morgunova, S. V. Nikonov, A. S. Mironov, S. E. Ealick and A. M. Mikhailov, *Acta Crystallogr. D*, 2004, **60**, 709.
63. W. Bu, E. C. Settembre, M. H. el Kouni and S. E. Ealick, *Acta Crystallogr. D*, 2005, **61**, 863.
64. V. Singh, J. E. Lee, S. Nunez, P. L. Howell and V. L. Schramm, *Biochemistry*, 2005, **44**, 11647.
65. V. Singh, G. B. Evans, D. H. Lenz, J. M. Mason, K. Clinch, S. Mee, G. F. Painter, P. C. Tyler, R. H. Furneaux, J. E. Lee, P. L. Howell and V. L. Schramm, *J. Biol. Chem.*, 2005, **280**, 18265.
66. J. E. Lee, V. Singh, G. B. Evans, P. C. Tyler, R. H. Furneaux, K. A. Cornell, M. K. Riscoe, V. L. Schramm and P. L. Howell, *J. Biol. Chem.*, 2005, **280**, 18274.
67. M. Duvic, E. A. Olsen, G. A. Omura, J. C. Maize, E. C. Vonderheid, C. A. Elmets, J. L. Shupack, M. F. Demierre, T. M. Kuzel and D. Y. Sanders, *J. Am. Acad. Dermatol.*, 2001, **44**, 940.
68. V. Gandhi, J. M. Kilpatrick, W. Plunkett, M. Ayres, L. Harman, M. Du, S. Bantia, J. Davisson, W. G. Wierda, S. Faderl, H. Kantarjian and D. Thomas, *Blood*, 2005, **106**, 4253.
69. W. F. de Azevedo Jr., F. Canduri, D. M. dos Santos, R. G. Silva, J. S. de Oliveira, L. P. de Carvalho, L. A. Basso, M. A. Mendes, M. S. Palma and D. S. Santos, *Biochem. Biophys. Res. Commun.*, 2003, **308**, 545.
70. W. Filgueira de Azevedo Jr., F. Canduri, D. Marangoni dos Santos, J. H. Pereira, M. V. Dias, R. G. Silva, M. A. Mendes, L. A. Basso, M. S. Palma and D. S. Santos, *Biochem. Biophys. Res. Commun.*, 2003, **309**, 917.
71. F. Canduri, D. M. dos Santos, R. G. Silva, M. A. Mendes, L. A. Basso, M. S. Palma, W. F. de Azevedo and D. S. Santos, *Biochem. Biophys. Res. Commun.*, 2004, **313**, 907.
72. F. Canduri, R. G. Silva, D. M. dos Santos, M. S. Palma, L. A. Basso, D. S. Santos and W. F. de Azevedo Jr., *Acta Crystallogr. D*, 2005, **61**, 856.
73. W. F. de Azevedo Jr., F. Canduri, D. M. dos Santos, J. H. Pereira, M. V. Bertacine Dias, R. G. Silva, M. A. Mendes, L. A. Basso, M. S. Palma and D. S. Santos, *Biochem. Biophys. Res. Commun.*, 2003, **312**, 767.
74. M. Luic, G. Koellner, D. Shugar, W. Saenger and A. Bzowska, *Acta Crystallogr. D*, 2001, **57**, 30.
75. A. Bzowska, G. Koellner, B. Wielgus-Kutrowska, A. Stroh, G. Raszewski, A. Holy, T. Steiner and J. Frank, *J. Mol. Biol.*, 2004, **342**, 1015.
76. M. Luic, G. Koellner, T. Yokomatsu, S. Shibuya and A. Bzowska, *Acta Crystallogr. D*, 2004, **60**, 1417.
77. G. Koellner, M. Luic, D. Shugar, W. Saenger and A. Bzowska, *J. Mol. Biol.*, 1997, **265**, 202.
78. H. M. Pereira, A. Cleasby, S. S. Pena, G. G. Franco and R. C. Garratt, *Acta Crystallogr. D*, 2003, **59**, 1096.
79. J. Tebbe, A. Bzowska, B. Wielgus-Kutrowska, W. Schroder, Z. Kazimierczuk, D. Shugar, W. Saenger and G. Koellner, *J. Mol. Biol.*, 1999, **294**, 1239.

80. W. Shi, L. A. Basso, D. S. Santos, P. C. Tyler, R. H. Furneaux, J. S. Blanchard, S. C. Almo and V. L. Schramm, *Biochemistry*, 2001, **40**, 8204.
81. A. Lewandowicz, W. Shi, G. B. Evans, P. C. Tyler, R. H. Furneaux, L. A. Basso, D. S. Santos, S. C. Almo and V. L. Schramm, *Biochemistry*, 2003, **42**, 6057.
82. G. Koellner, A. Stroh, G. Raszewski, A. Holy and A. Bzowska, *Nucleosides Nucleotides Nucleic Acids*, 2003, **22**, 1699.
83. E. M. Bennett, C. Li, P. W. Allan, W. B. Parker and S. E. Ealick, *J. Biol. Chem.*, 2003, **278**, 47110.
84. J. Badger, J. M. Sauder, J. M. Adams, S. Antonysamy, K. Bain, M. G. Bergseid, S. G. Buchanan, M. D. Buchanan, Y. Batiyenko, J. A. Christopher, S. Emtage, A. Eroshkina, I. Feil, E. B. Furlong, K. S. Gajiwala, X. Gao, D. He, J. Hendle, A. Huber, K. Hoda, P. Kearins, C. Kissinger, B. Laubert, H. A. Lewis, J. Lin, K. Loomis, D. Lorimer, G. Louie, M. Maletic, C. D. Marsh, I. Miller, J. Molinari, H. J. Muller-Dieckmann, J. M. Newman, B. W. Noland, B. Pagarigan, F. Park, T. S. Peat, K. W. Post, S. Radojicic, A. Ramos, R. Romero, M. E. Rutter, W. E. Sanderson, K. D. Schwinn, J. Tresser, J. Winhoven, T. A. Wright, L. Wu, J. Xu and T. J. Harris, *Proteins*, 2005, **60**, 787.
85. M. P. Boyle, A. K. Kalliomaa, V. Levdikov, E. Blagova, M. J. Fogg, J. A. Brannigan, K. S. Wilson and A. J. Wilkinson, *Proteins*, 2005, **61**, 674.
86. T. H. Tahirov, E. Inagaki, N. Ohshima, T. Kitao, C. Kuroishi, Y. Ukita, K. Takio, M. Kobayashi, S. Kuramitsu, S. Yokoyama and M. Miyano, *J. Mol. Biol.*, 2004, **337**, 1149.
87. W. Shi, L. M. Ting, G. A. Kicska, A. Lewandowicz, P. C. Tyler, G. B. Evans, R. H. Furneaux, K. Kim, S. C. Almo and V. L. Schramm, *J. Biol. Chem.*, 2004, **279**, 18103.
88. J. E. Lee, E. C. Settembre, K. A. Cornell, M. K. Riscoe, J. R. Sufrin, S. E. Ealick and P. L. Howell, *Biochemistry*, 2004, **43**, 5159.
89. J. E. Lee, K. A. Cornell, M. K. Riscoe and P. L. Howell, *J. Biol. Chem.*, 2003, **278**, 8761.
90. J. E. Lee, G. D. Smith, C. Horvatin, D. J. Huang, K. A. Cornell, M. K. Riscoe and P. L. Howell, *J. Mol. Biol.*, 2005, **352**, 559.
91. R. G. Silva, J. H. Pereira, F. Canduri, W. F. de Azevedo Jr., L. A. Basso and D. S. Santos, *Arch. Biochem. Biophys.*, 2005, **442**, 49–58.
92. A. S. Murkin, M. R. Birck, A. Rinaldo-Matthis, W. Shi, E. A. S. C. A. Taylor and V. L. Schramm, *Biochemistry*, 2007, **46**, 5038–5049.
93. C. Schnick, M. A. Robien, A. M. Brzozowski, E. J. Dodson, G. N. Murshudov, L. Anderson, J. R. Luft, C. Mehlin, W. G. Hol, J. A. Brannigan and A. J. Wilkinson, *Acta Crystallogr.*, 2005, **D61**, 1245–1254.
94. A. Rinaldo-Matthis, C. Wing, M. Ghanem, H. Deng, P. Wu, A. Gupta, P. C. Tyler, G. B. Evans, R. H. Furneaux, S. C. Almo, C. C. Wang and V. L. Schramm, *Biochemistry*, 2007, **46**, 659–668.
95. Y. Zhang, M. Porcelli, G. Cacciapuoti and S. E. Ealick, *J. Mol. Biol.*, 2006, **357**, 252–262.

96. E. Y. Park, S. I. Oh, M. J. Nam, J. S. Shin, K. N. Kim and H. K. Song, *Proteins*, 2006, **65**, 519–523.
97. H. D. Pereira, G. R. Franco, A. Cleasby and R. C. Garratt, *J. Mol. Biol.*, 2005, **353**, 584–599.
98. R. Grenha, V. M. Levdikov, M. J. Fogg, E. V. Blagova, J. A. Brannigan, A. J. Wilkinson and K. S. Wilson, *Acta Crystallogr. F Struct. Biol. Cryst. Commun.*, 2005, **61**, 459–462.

CHAPTER 4

Application and Limitations of X-Ray Crystallographic Data in Structure-Guided Ligand and Drug Design

ANDREW M. DAVIS*[a], SIMON J. TEAGUE[a] AND GERARD J. KLEYWEGT[b]

[a]AstraZeneca R&D Charnwood, Bakewell Road, Loughborough, Leicestershire LE11 5RH, UK
[b]Department of Cell and Molecular Biology, Uppsala University, Biomedical Centre, Box 596, SE-751 24, Uppsala, Sweden

4.1 Introduction

There can be no doubt that drug discovery is an increasingly complex, risky and expensive business. Recent estimates place the cost of delivering a drug to the marketplace at greater than \$800 000 000,[1] with only one in ten compounds that enter clinical development actually making it to the market. Once on the marketplace, many drugs fail to recoup their development costs (as many as 30% according to data from the 1980s[2]), and market withdrawals because of adverse effects add further to the industry's problems. Together with limited patent lifetimes, increasingly complex development programs and the economic pressure of the marketplace, the problems facing the industry are clear. There can be few industries that can match these unfavourable statistics, but then few industries operate in an environment where the underlying rules governing product development are so complex, opaque and fluid. But whatever the underlying difficulties and uncertainties there are very significant economic pressures for the drug-discovery process to be not only faster, but also smarter.[3] Hence the use of protein structural information to guide drug design is an extremely appealing "smart" strategy from the perspectives of both science and business, with the potential to increase speed and quality of drug discovery.

The aim of protein structure-guided design is the optimization of ligand potency, which is usually measured in a simple *in vitro* competitive inhibition or binding assay. However, the aim of all pharmaceutical research projects is the discovery of a potential candidate drug. Here, we highlight the distinction between ligand design and drug design, and illustrate the difference using case histories from studies on human immunodeficiency virus (HIV) protease, neuraminidase and thrombin inhibitors. In most cases the protein structure used in the structure-guided design process has been determined using X-ray crystallography rather than nuclear magnetic resonance (NMR) spectroscopy. The latter technique suffers from severe limitations imposed by constraints on the molecular size of the protein and the requirement for multiple isotopic labeling. Therefore, we highlight some of the pitfalls and limitations in protein-structure determination by X-ray crystallographic methods that might otherwise mislead the unwary user.

4.2 Structure-guided Ligand Design and Drug Design

Structure-guided design is often loosely termed structure-guided drug design or rational drug design. Usually the processes described could more accurately be termed structure-guided ligand design, since the objective is to optimize the potency of a ligand in a simple *in vitro* assay. Drug design requires optimization of many other properties, including dissolution, absorption, metabolic stability, plasma-protein binding, distribution, elimination, toxicological profile, cost of synthesis and pharmaceutical properties. Structure-guided design has already contributed to the discovery of a number of very important drugs, such as the peptidomimetic HIV-protease inhibitors, carbonic anhydrase inhibitor dorzolamide, neuraminidase inhibitors zanamivir and oseltamivir, and recently also the first drug that targets thrombin, ximelagatran.[4]

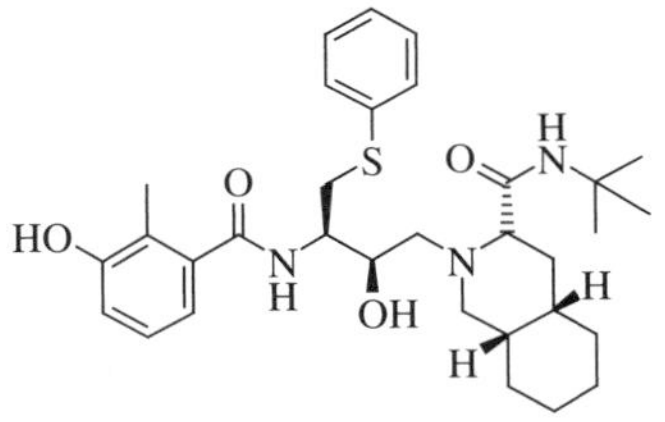

1

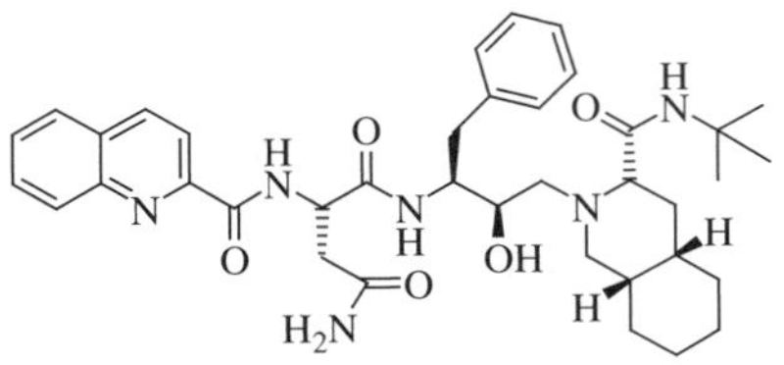

2

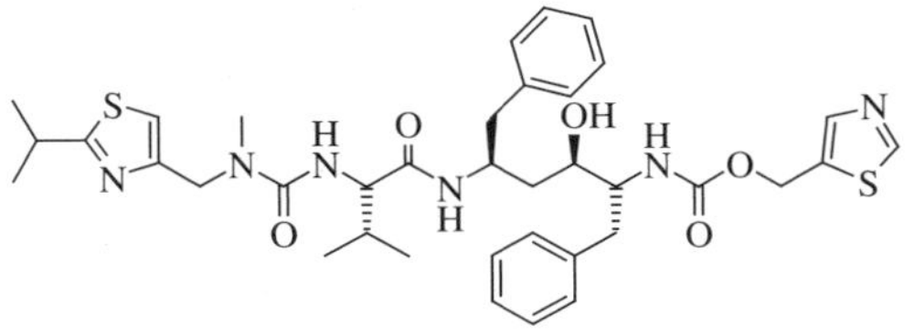

3

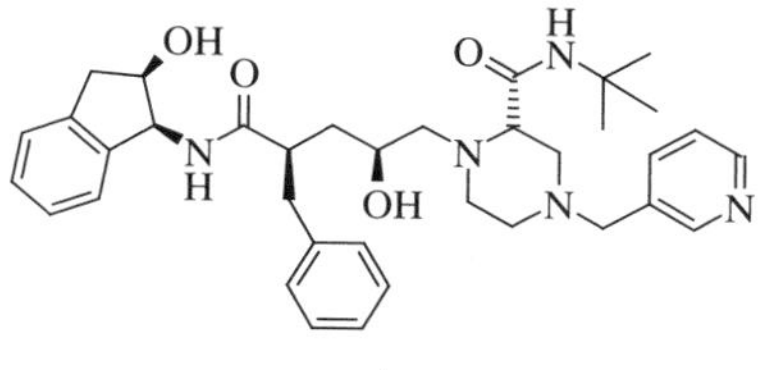

4

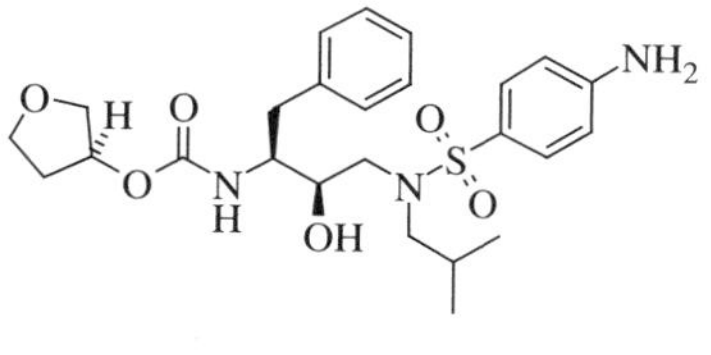

5

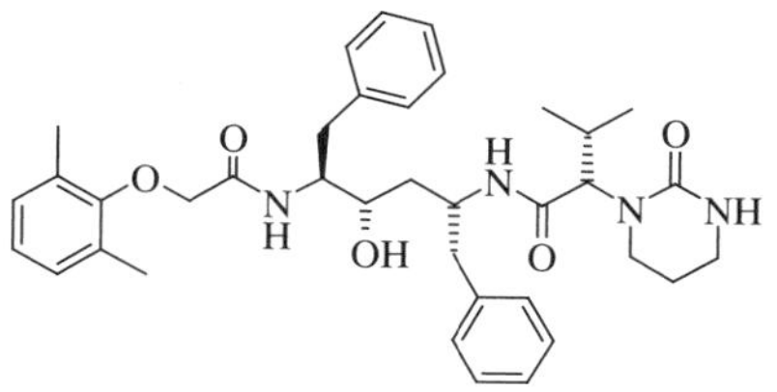

6

Structural information was important during the discovery of peptidomimetic HIV protease inhibitors, which are already in clinical use (nelfinavir (**1**), saquinavir (**2**), ritonavir (**3**), indinavir (**4**), amprenavir (**5**), and lopinavir (**6**)). Although these drugs are commercially and clinically successful they have distinct therapeutic limitations, as is indicated by the continuing search for more effective inhibitors. Poor bioavailability has been reported for saquinavir, and variable bioavailability for a number of members of this class of agents. They also suffer from moderate-to-high clearance, non-linear phamacokinetics and very significant drug–drug interactions, and are substrates for P-glycoprotein- (pgp-) efflux proteins. The HIV protease market may well accept improved inhibitors with more favorable absorption, distribution, metabolism and elimination (ADME) properties.

The HIV protease inhibitor DMP323 (**7**), with an IC50 of 0.031 nM, was discovered by structure-guided design.

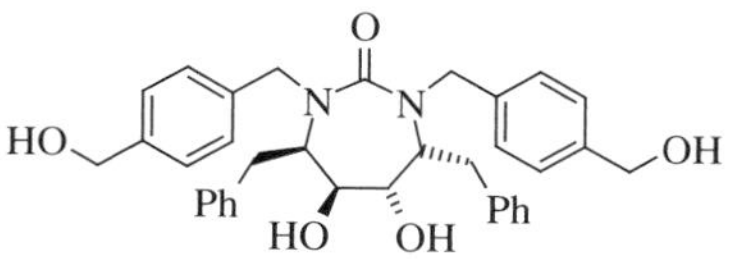

7

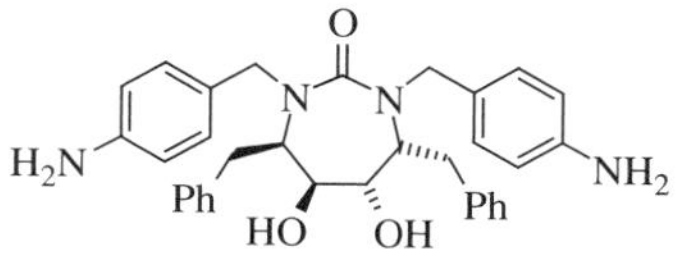

8

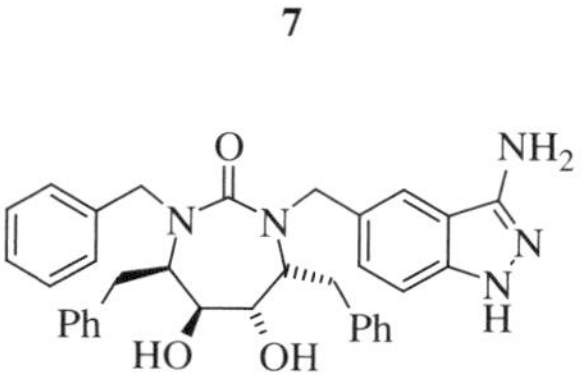

9

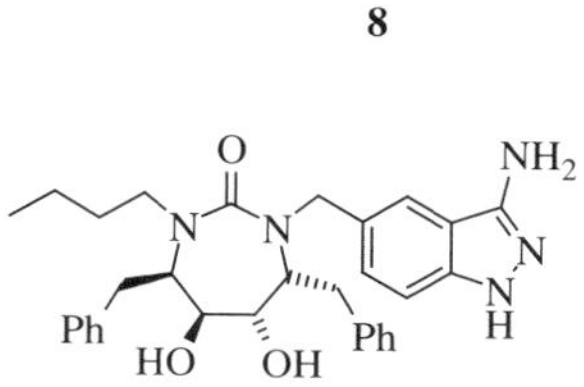

10

This compound progressed into clinical development. Its progress illustrates a number of important and recurring themes[5,6] concerning progress from a ligand to an effective drug. The clinical trials of **7** were terminated due to poor bioavailability caused by low solubility and metabolic instability associated with the benzyl alcohol groups. An excellent ligand proved to be a sub-optimal drug. A second clinical candidate entered development, DMP450 (**8**), which combined improved affinity with better solubility and good bioavailability in humans.

Upon reaching phase II, DMP450 was found to have only modest potency in patients. Dupont–Merck re-entered the discovery phase of the project, again utilizing a structure-guided approach, but this time including a potency assay in whole plasma. This aimed to address the deficiency of DMP450, which was perceived to be its high plasma-protein binding. Plasma-protein binding affects all drugs *in vivo,* and largely depends upon lipophilicity and charge. It modulates the free plasma concentration of the drug, which drives efficacy. Dupont–Merck's latest clinical candidates are DMP850 (**9**) and DMP851 (**10**), both of which have improved whole-blood potency, as well as increased solubility and bioavailability. The second and third phases of the program focussed upon incorporating drug-like properties while maintaining ligand potency. The current development progress of DMP850 and DMP851 is unknown, as, unfortunately, no development for DMP850 or DMP851 has been reported since 2002.

Similar problems were also encountered with Pharmacia–Upjohn's pyrone sulfonamide inhibitors of HIV protease.[7] Broad screening of a "diverse" subset identified warfarin as an interesting but weak inhibitor. Similarity searching identified a related compound, phenprocoumon, as a potential lead. The use of X-ray structural information led to PNU-103017 (**11**), which (although a potent HIV-protease inhibitor with excellent phamacokinetics) failed to demonstrate sufficient cellular activity due to its high plasma-protein binding. Again, optimization aimed at reducing plasma–protein binding and increasing potency resulted in an improved clinical candidate, tipranavir (**12**). Tipranavir was approved in 2005 and is now manufactured and marketed by Boehringer Ingelheim as APTIVUS for the treatment of HIV infection.

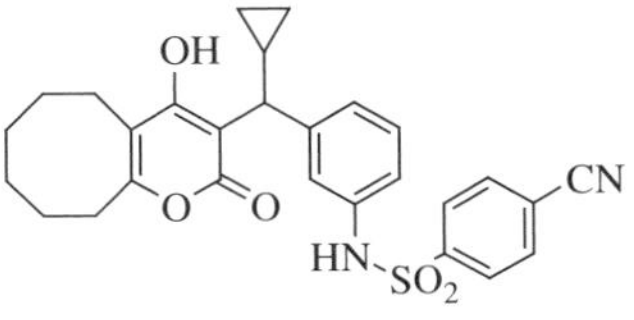

11

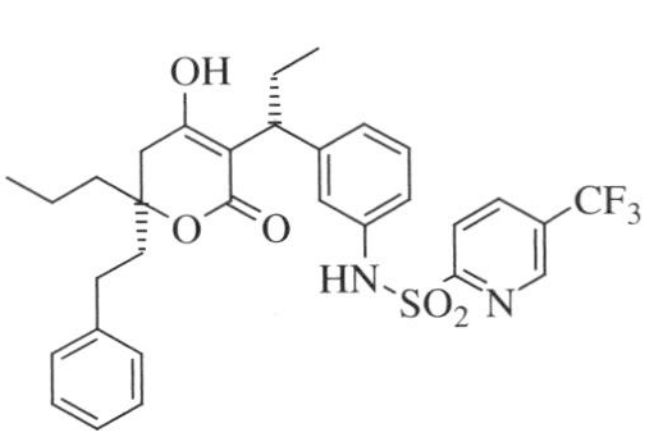

12

A second therapeutic class of drugs originating from structure-guided studies is provided by the search for neuraminidase inhibitors. Neuraminidase has been an important target for anti-influenza therapy for many years. Mark Von Itzstein and his group at Monash University used the program GRID in an attempt to identify binding hotspots in the neuraminidase active site to guide compound design.[8] GRID suggested replacement of the 4-hydroxyl in (**13**) by a basic moiety. Replacement of the hydroxyl by the basic guanidinyl group resulted in a 5000-fold increase in affinity. This compound, zanamivir (**14**), was subsequently developed by GlaxoSmithKline and marketed as the first neuraminidase-based anti-influenza treatment, Relenza. Zanamivir is a very polar molecule and is dosed topically to the lung using dry-powder inhaler technology.[9] However, Gilead Pharmaceuticals, working on a carbocyclic scaffold, was able to obtain sufficient potency without incorporation of either the strongly basic guanidine group or the polar glycerol side chain by replacement of the guanidine group by a less basic primary amine and of the glycerol chain with a 1-ethylpropoxy group.[10] This group participates in favorable hydrophobic contacts and induces movements in protein side chains, which result in the formation of an additional salt bridge between Glu276 and Arg244. This process resulted in oseltamivir (**15**), a compound with more moderate polarity and charge. The zwitterionic parent is unsuitable as an oral drug, but the ethyl ester pro-drug allows the compound to be orally absorbed. Oseltamivir is marketed by Hoffman–La Roche as Tamiflu. This was the first oral anti-influenza drug targeting neuraminidase, and in the first six weeks of sales in the USA Tamiflu took 40% of the neuraminidase inhibitor market from Relenza.[11] More balanced polarity, charge and lipophilicity in Tamiflu resulted in a more acceptable physical property profile and considerable commercial success.

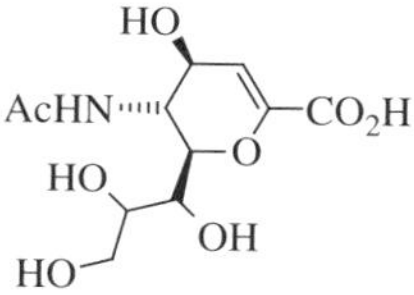

13

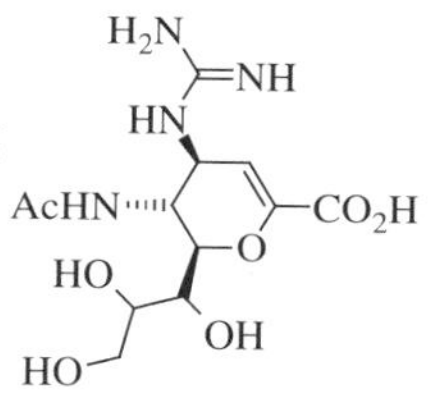

14

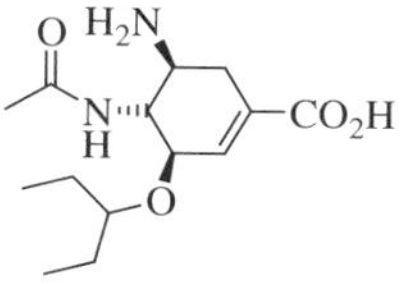

15

The enzymes thrombin and renin have been targets for protein-structure guided drug-discovery projects for many years. As targets, they have suffered similar problems, with potent inhibitors being produced, but their large peptidic-like nature limits their potential to produce oral drugs. Recently thrombin has yielded its first marketed thrombin inhibitor, melagatran.[4] Melagatran (**16**), a potent sub 500 molecular weight thrombin inhibitor which, although showing good oral absorption in dogs, showed only 3–7% bioavailability in humans. The dibasic, monoacidic nature of melagatran was identified as limiting its absorption. After a two-year search, the problem was finally overcome with the development of ximelagatran (**17**), a di-prodrug of melagatran, with a four-fold improvement in bioavailability.

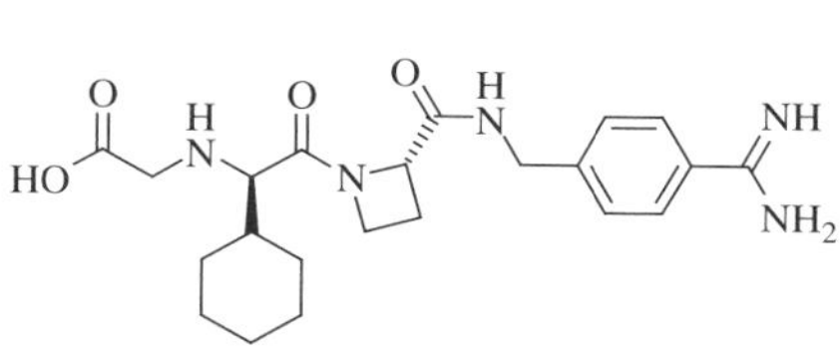

16

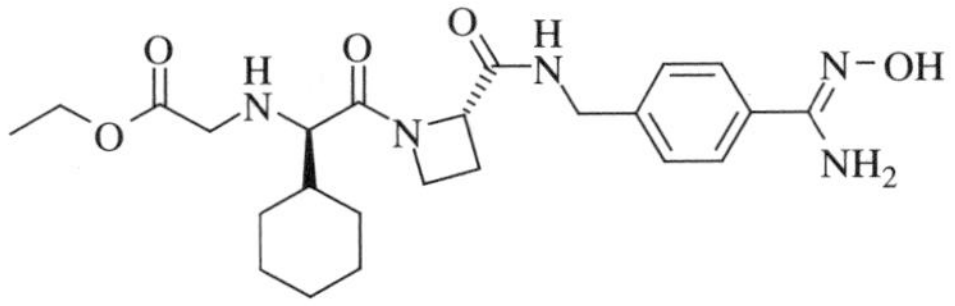

17

The first renin inhibitor has also finally reached the marketplace. Aliskiren, manufactured by Novartis, shows a plasma half-life of 23.7 hrs, making it suitable for once-a-day oral dosing.[12] The bioavailability of aliskiren (**18**) of only 2.7% means the search for improved renin inhibitors continues.

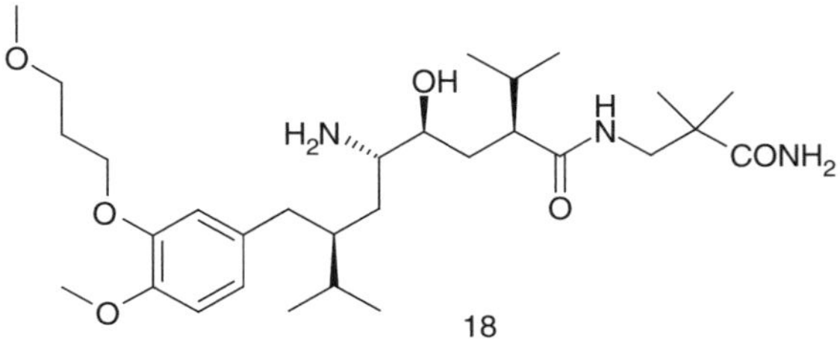

18

These and other examples[13] illustrate the distinction between ligands and drugs. The relative maturity of these areas provides opportunities to view successes and difficulties in the context of the whole journey from concept to a drug in the clinic. The HIV protease, neuraminidase, thrombin and renin case studies are instructive when viewed in the light of current understanding of drug-like properties. The use of protein structural information in ligand optimization often leads to the maintenance and incorporation of polar interactions while increasing lipophilic contacts in order to increase potency. However, the combination of these two strategies may not result in good oral drug-like properties. The use of protein structural information in conjunction with *in vitro* potency determination may tempt medicinal chemists into designing ligands that are not drugs, or make the journey to a drug longer and more complicated. High potency is not necessarily the most important requirement for a drug. The importance of ADME properties as well as potency is now recognized in most drug-discovery programs. In order to be effective, the concentration of free drug must be maintained at a level at which the binding site on the target protein is significantly occupied throughout the dosing interval. This is a function of dose, clearance, plasma-protein binding, intrinsic potency, volume of distribution and dosing interval. Furthermore, an acceptable margin is required between the maximum concentration achieved at a therapeutic dose and the concentration that produces toxic side effects. The drug's pharmacodynamics and pharmacokinetics must be such that it meets these requirements at the desired dosing frequency.

Limiting the size, charge and lipophilicity of a ligand in order to fulfil ADME requirements can limit the number of interactions made between the ligand and the residues composing the binding pocket. This limits the affinity that can be derived from interactions at the ligand–protein interface. This problem can be particularly acute if the ligand occupies a large binding site mimicking a large natural substrate, as is the case with many peptidases. The problem can sometimes be solved using small ligands, which display induced fit of the protein to the ligand. Greater affinity is obtained from a small ligand, by it intercepting or inducing a conformation of the protein, which produces a complex of lower total free energy. This is often the result of the ligand making hydrophobic interactions favorable, with residues made available as a consequence of the inherent conformational mobility of the protein. This new receptor then offers new and different opportunities for the design of ligands with improved ADME properties, and also, significantly, new opportunities for intellectual property. The application of the lead-like concept[14] to structure-guided drug design is currently a popular approach to maximizing the affinity that can be gained at the ligand–receptor interface. High-throughput crystallography or NMR screening is being used to identify small 'ultra-lead-like' fragments[15] with high ligand efficiencies[16,17] which, when subject to lead optimization, can produce smaller but potent and selective ligands with the potential for good ADME properties.

While the application of current ADME thinking to structure-guided ligand design offers the potential for smarter drug design, there are limitations to our use of protein structures that are often overlooked or even forgotten, which may limit our potential to be really smart.

4.3 Some Limitations in the Use of X-ray Data

A number of common and implicit assumptions are made by chemists who use protein structural data during structure-guided design. First, we briefly define basic crystallography terms, which aid in interpretation of crystallographically derived structures. Then we discuss possible pitfalls and caveats in the structure-determination process, which users of such structures ought to be aware of.

4.3.1 Basic Crystallography Terms

In a crystallographic (X-ray diffraction) experiment, the raw data consist of the positions and intensities of the reflections as measured in the diffraction pattern of the crystal. From these intensities, the so-called structure-factor amplitudes can be calculated (roughly as the square root of the intensities). Once the phases of the structure factors are also known (*i.e.* once the "phase problem" has been solved), Fourier transformation of the structure factors provides a map, which is a three-dimensional matrix of numbers that represents the local electron density.[18] Where there are many electrons (and, hence, heavier atoms) the density is higher than in places where (on average) there are few electrons. It is now the task of the crystallographer to interpret the electron density in terms of a discrete atomic model.[19] This is typically an iterative process, in which the crystallographer (or, nowadays often a program) builds a part of the model, and then uses a refinement program to improve this. The refinement program will make small changes to the model (by adjusting the parameters of the model, such as the atomic coordinates) that improve the ability of the model to explain the experimental data. Simultaneously, geometric and other restraints and constraints are enforced onto the model to ensure that it is chemically reasonable. With an improved model, new maps can be calculated that may reveal further details, such as previously missing or uninterpretable density for loops, ligand, solvent molecules, *etc.* The crystallographer can then add these. Simultaneously, the crystallographer should be on the lookout for possible errors in the current model, and correct them if possible.[20]

Besides coordinates, atoms in the model also tend to have a "temperature factor" (also known as B-factors or atomic displacement parameters) associated with them to model the effects of static and dynamic disorder in the crystal. Except at high resolution (typically, better than ~1.5 Å) where there are sufficient observations to warrant refinement of anisotropic temperature factors (requiring six parameters per atom), temperature factors are usually constrained to be isotropic (requiring only one parameter per atom). The isotropic temperature factor of an atom is related to the atom's mean-square displacement. In most cases, temperature factors provide a useful relative indication of the reliability of different parts of the model. If they are high, *e.g.* for a lysine side chain, this usually means that little or no electron density was observed for the atoms in that side chain, and that the coordinates are therefore less reliable. Figure 4.1 provides an example of the atomic coordinate records encountered in Protein Data Bank (PDB)[21] files of crystallographically determined structures.

An important parameter in crystallographic studies is the resolution of the data, which is expressed in Å, where *lower* numbers signify *higher* resolution. The higher the resolution, the more experimental data, and the more reliable (in terms of accuracy and precision) one may expect the resulting model to be. At high resolution (<1.5 Å), the model is probably more than 95% a consequence of the observed data.[22] However, at lower resolution (>2.5 Å), the modeling of details in protein structures is much more subjective than is widely

a)

```
HEADER    PLANT PROTEIN                               02-MAR-00   1EJG
TITLE     CRAMBIN AT ULTRA-HIGH RESOLUTION: VALENCE ELECTRON DENSITY.
...
REMARK 999 PRO/SER22:LEU/ILE25 ISOFORMS ARE MODELLED
REMARK 999 AS ALTERNATE CONFORMERS IN COORDINATE RECORDS.
REMARK 999 BECAUSE OF FORMAT RESTRICTIONS, ONLY PRO22/LEU25
REMARK 999 ISOFORM IS REPRESENTED IN THE SEQUENCE RECORDS.
```

Figure 4.1a An example of information presented on REMARK records in PDB entries. In this case, the structure of crambin has been determined (PDB entry 1EJG). Crambin exists in two isoforms that differ in two residues (either Pro 22/Leu 25 or Ser 22/Ile 25), and both forms were present in the crystal. The two sequence heterogeneities have been modelled as alternative conformations for residues 22 and 25, but due to format restrictions, only one sequence is recorded in the sequence records.

b)

```
ATOM     136  N   AVAL A   8       6.382   2.222  13.070  0.55  1.92           N
ANISOU   136  N   AVAL A   8     421    149    160     33    -23    -35         N
ATOM     137  N   BVAL A   8       6.695   2.072  13.037  0.45  1.88           N
ATOM     138  CA  AVAL A   8       5.099   2.259  12.380  0.55  2.32           C
ANISOU   138  CA  AVAL A   8     397    258    224     16    -22   -120         C
ATOM     139  CA  BVAL A   8       5.471   2.048  12.164  0.45  1.94           C
ATOM     140  C    VAL A   8       5.208   3.386  11.373  1.00  2.63           C
ANISOU   140  C    VAL A   8     380    402    213     71    -42   -209         C
ATOM     141  O    VAL A   8       4.712   3.302  10.238  1.00  2.67           O
ANISOU   141  O    VAL A   8     378    434    200     39    -31   -178         O
ATOM     142  CB  AVAL A   8       3.944   2.394  13.375  0.55  3.36           C
ANISOU   142  CB  AVAL A   8     376    635    262    186      4   -244         C
ATOM     143  CB  BVAL A   8       4.263   1.630  13.035  0.45  3.05           C
ATOM     144  CG1AVAL A   8       2.629   2.981  12.701  0.55  4.22           C
ANISOU   144  CG1AVAL A   8     326    856    420    159     34   -159         C
ATOM     145  CG1BVAL A   8       3.580   2.930  13.664  0.45  3.89           C
ATOM     146  CG2AVAL A   8       3.635   1.036  13.955  0.55  4.82           C
ANISOU   146  CG2AVAL A   8     771    684    376    230    -61   -385         C
ATOM     147  CG2BVAL A   8       3.136   1.077  12.028  0.45  4.17           C
ATOM     148  H   AVAL A   8       6.374   2.231  14.084  0.55  2.27           H
ATOM     149  H   BVAL A   8       6.648   2.059  14.045  0.45  5.91           H
...
END
```

Figure 4.1b Fragment of a PDB file from the same entry. The basic information about the atoms in the model is listed on "cards" (records, lines). These begin with ATOM for protein or nucleic acid components or HETATM for entities that are ligands, ions, metals, and solvent molecules. The second item on each line is simply a sequential index number of that atom. In the first line, atom 136 is the amide nitrogen (N) of residue valine (VAL) A8. "A" is the chain name, "8" the residue number. The "A" before the residue symbol (VAL) signifies that this atom is statically disordered. This means that this atom is observed in more than one location in the electron density, and the various instances will be labelled "A", "B", "C", etc. Indeed, the third line in the figure contains the alternative location, "B", of this atom. The three real numbers that follow the residue number, 6.382, 2.222, 13.070", are the Cartesian x, y, and z-coordinates of the atom in orthogonal Å. The fourth number is the occupancy of the atom. This is a number between zero and one, which indicates the fraction of the amide nitrogen atom of valine A8 that occurs in this location. Here, the first conformation has been given an occupancy is 0.55, and line 3 shows that the alternative conformation B accounts for the remaining 0.45. Note that quite a few programs that read and process PDB files ignore alternative conformations completely. When the occupancy of ligands and solvent molecules is refined or set to a number less than one, this implies that they occupy the position in only a fraction of the molecules in the crystal, or for only a fraction of the time, or a combination of both. The fifth number, 1.92 in line 1, is the value of the isotropic temperature factor (B factor). Line 2 reveals that this atom has been modelled anisotropically, (this involves six parameters per atom which are listed on the ANISOU card), but the isotropic equivalent value is always listed as the fifth real number of the ATOM (or HETATM) card. At the end of each card the atomic symbol of the chemical element of the atom is listed, since this cannot always be deduced unambiguously from the atom's name.

appreciated.[23] This can be understood by calculating typical data-to-parameter ratios, *i.e.* the ratio of the number of experimental observations and the number of adjustable parameters (atomic coordinates, parameters associated with the temperature factors and occupancies, among others) in the model. For an average protein structure at a resolution of 2 Å, this ratio is slightly greater than two, but at ~2.7 Å it becomes less than unity. Whereas gross errors are unlikely to persist to the publication stage if the resolution is high, once the resolution becomes >2 Å, the balance shifts and some published protein models appear to have been determined more by the crystallographer's imagination than by any experimental data.[22] In fact, in the 1980s the first reports of some of the "hottest" protein crystal structures, some of which were also prime drug targets, contained extremely serious errors.[24] Examples included HIV-1 protease, photoactive yellow protein, the small subunit of ribulose-1,5-bisphosphate (RuBP) carboxylase/oxygenase (RuBisCO), D-Ala-D-Ala peptidase, ferredoxin, metallothionein, gene V binding protein and the GTP-binding domain of Ha-ras p21.

The structure of a complex between botulinum neurotoxin type B protease and the inhibitor bis(5-amidino-2-benzimidazolyl)methane (BABIM) was published,[25] and the structure and experimental data deposited in the PDB (entry 1FQH). However, subsequent critical analysis of the electron-density maps revealed that these did not support the placement of the inhibitor as stated in the earlier paper, and the structural conclusions based on it were withdrawn by the authors.[26]

Another trap to be aware of (and one that many crystallographers have fallen into) is that of deriving "high-resolution information" from low-resolution models. For instance, in a typical 3 Å structure, the uncertainty in the position of the individual atoms can easily be 0.5 Å or more. Nevertheless, many such models have been described where hydrogen-bonding distances are listed with a precision (not accuracy) of 0.01 Å (probably because the program that was used to calculate these distances used that particular precision), and solvent-accessible surface areas with a precision of 1 Å^2.

The ability of the model to explain the experimental data is usually assessed by means of the (conventional) *R*-value, which is defined as in equation (4.1):

$$R = \{\Sigma|\,|F_{\text{obs}}| - \text{scale}^*|F_{\text{calc}}|\,|\}/\{\Sigma|F_{\text{obs}}|\} \qquad (4.1)$$

Here F_{obs} are the experimental structure-factor amplitudes, F_{calc} are the structure-factor amplitudes calculated from the model and the sums extend over all observed reflections. However, by introducing more and more parameters into the model, the *R*-value can be made almost arbitrarily small (this is called "over-fitting the data"). In 1992, Brünger[27] introduced the concept of cross-validation in crystallographic refinement, and with it the free *R*-value (R_{free}), whose definition is identical to that of the conventional *R*-value, except that the free *R*-value is calculated for a small subset of reflections that is *never used* in the refinement of the model. The free *R*-value therefore measures how well the model predicts experimental observations that are not used to fit the model.

Until a few years ago, a conventional *R*-value below 0.25 was generally considered a sign that a model was essentially correct. While this is probably true at high resolution, it was subsequently shown for several intentionally mistraced models that these could be refined to deceptively low conventional *R*-values.[24,28] Brünger suggests a threshold value of 0.40 for the free *R*-value, *i.e.* models with free *R*-values greater than 0.40 should be treated with caution.[29,30] Since the difference between the conventional and free *R*-values is partly a measure of the extent to which the model over-fits the data (*i.e.* some aspects of the model improve the conventional but not the free *R*-value and are therefore likely to fit noise rather than signal in the data), this difference ($R_{free} - R$) should be small for the final model, ideally <0.05.

4.3.2 Uncertainty in the Identity or Location of Protein or Ligand Atoms

It is often forgotten that an X-ray crystal structure is one crystallographer's subjective interpretation of an observed electron-density map expressed in terms of an atomic model. This structure is treated by chemists undertaking structure-guided design as if it were at perfect resolution, independent of the resolution at which the structure was actually determined and ignoring the interpolations, assumptions, biases and sometimes mistakes incorporated by the crystallographer.

Uncertainties can involve the identity of important atoms, like those in the binding pocket. For instance at typical macromolecular resolution (~2 Å), the relative positions of the δN and δO of asparagine and εN and εO of glutamine side chains cannot usually be determined directly from the electron density since they are isoelectronic. The decision as to which density features should be assigned to N and to O should therefore involve inspection of the local hydrogen-bonding networks. However, these decisions may have to be made before solvent molecules have been added to the model, and hence be based on incomplete hydrogen-bonding networks. Moreover, in low-resolution studies many of the solvent entities are not resolved in the electron density and can therefore not be modeled, thereby further complicating the analysis. A careful crystallographer will verify the assignment in the final model, but in general the users of the model should treat the final assignment with caution. This is also borne out by large-scale analysis of the hydrogen-bonding patterns involving histidine, glutamine and asparagine residues with the program WHAT IF,[31] as listed on the PDBREPORT site.[32] This analysis suggests that as many as one in six of all histidine, asparagine and glutamine residues in the PDB may have been modeled in a "flipped" orientation.

Uncertainties can also occur at the level of whole residues. This is the case for flexible residues, which often diffract so weakly that no clear electron density is observed for them. This is quite common for the side chains of surface residues, but may also be found in some active sites, particularly with the flexible side chains of lysine and glutamate. Analysis of real-space density fits[33] shows that the most poorly defined residues are, in order of improving average fit to the density, Lys< Glu< Arg,Gln< Asp,Asn. The crystallographer knows they are present from the amino-acid sequence, and so they are incorporated into the structure in a conformation commonly observed for that residue in databases of high-resolution structures. The final conformation of the side chain, as viewed by the chemist, can be the product of intelligent guesswork and the van der Waals term in the refinement program's force field, rather than of experimentally observed electron density. It is also quite common for whole sections of the protein to give little or no observable electron density. Sometimes these parts are mobile loops, which can have great functional significance also, by virtue of this greater mobility.[34] In other cases, entire domains may be invisble in the electron-density maps.

Similarly, ambiguities apply when considering the bound ligand. For instance, the position of pyridine nitrogen cannot usually be determined from the electron density alone. This fact will introduce uncertainty into many crystal structures containing a molecule with an asymmetrically substituted pyridine. For instance, during the study of benzo[*b*]thiophene inhibitors of thrombin, compound (**19**) was complexed and a structure built from the electron density. The C-3 pyridyl ring was arbitrarily oriented so that the nitrogen resides in the more hydrophilic of the two possible environments. This is a reasonable assumption, but not the result of direct experimental observation and so still uncertain.

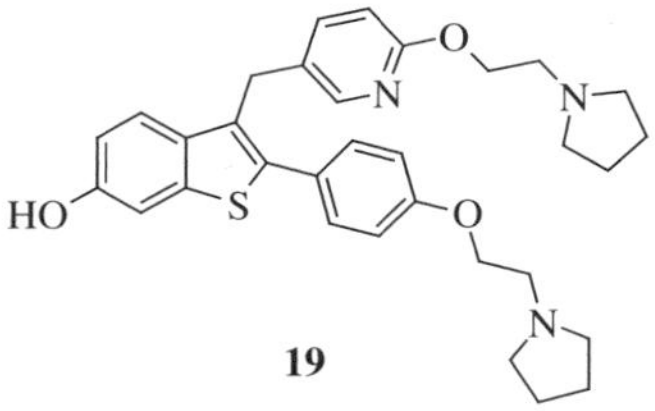

19

An example of how ambiguous X-ray crystallographic data can be without prior knowledge of the exact chemical composition of a ligand or residue was encountered recently. The exact identity of the 22nd genetically encoded amino acid, pyrrolysine (**20**),[35] present in *Methanosarcina barkeri* monomethylamine methyltransferase (MtmB) is still unknown despite a 1.55 Å resolution structure of the protein being available. The X-substituent is a methyl, ammonium, or hydroxyl group.

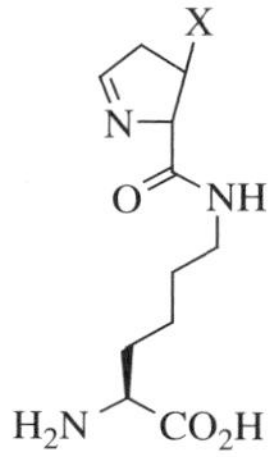

20

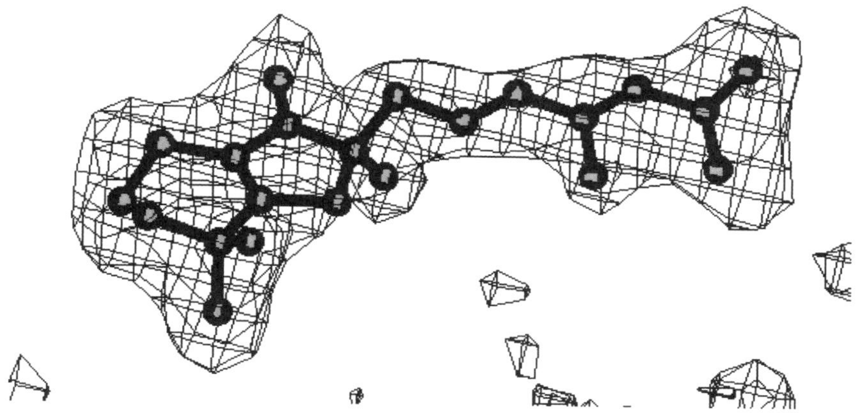

Figure 4.2 Density for, and structure of, **21** in complex with CRABP2.

On the other hand, sometimes (careful) crystallography can reveal cases of mistaken identity. For example, when the structure of cellular retinoic-acid-binding protein type 2 (CRABP2) in complex with a synthetic retinoid was solved, it was assumed that the ligand was (*E*)-4-[2-(5,6,7,8-tetrahydro-5,5,8,8-tetramethyl-2-naphthylenyl)-1-propenyl] benzoic acid TTNPB (**21**).[36] The ligand was built and fit to the density, but the maps stubbornly suggested that there was something wrong. The density failed to cover the whole ligand, and features in the map suggested that there ought to be a carbon-like atom at a distance of ~1.5 Å from C6, and that atoms C22 and C23 should be removed from the model. After double-checking the identity of the ligand with the chemists, it turned out that the ligand that was actually complexed to the protein was compound (**22**). The structure of this ligand made perfect sense in terms of the density (Figure 4.2), and the refinement of the structure could be completed successfully. However, had the resolution been 3 Å instead of 2.2 Å, the error might well have gone undetected.

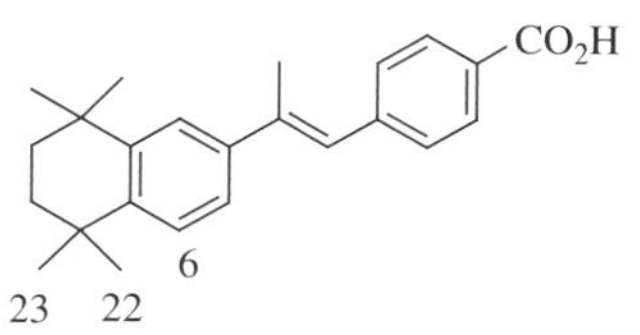

21

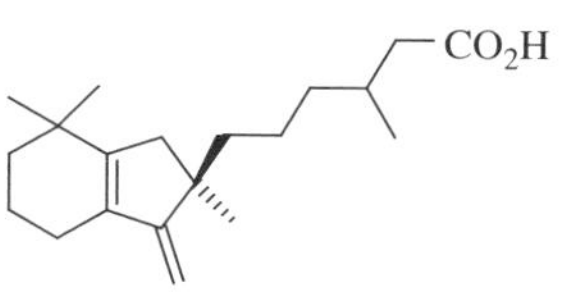

22

Since the presence of hydrogen atoms is inferred rather than observed, the tautomeric state of histidine or of bound ligands containing tautomeric groups cannot be determined directly. The latter problem is rather common when carrying out studies involving acid isosteres. Similarly, the state of ionization of ligand or protein cannot be observed. It is usually assumed that the charged state of the protein is known. However, the pK_a values of common acidic or basic side chains can be drastically different from their normal values, as measured in water, when they are located in the microenvironment of a protein active site.[37] Even when the protonation states of key active-site residues and the ligand are known, these may change upon complexation. Enthalpies of complexation measured by isothermal titration calorimetry, in aqueous buffers with different enthalpies of ionization, established that the Roche thrombin inhibitor napsagatran (**23**) binds to thrombin incorporating an additional proton.[38] A structurally similar inhibitor, CRC220 (**24**) from Behring, bound to thrombin without an additional proton. This difference in ionization, upon binding to the protein, was supported by different orientations of the ligands when the structures of the complexes with thrombin were determined by X-ray crystallography.

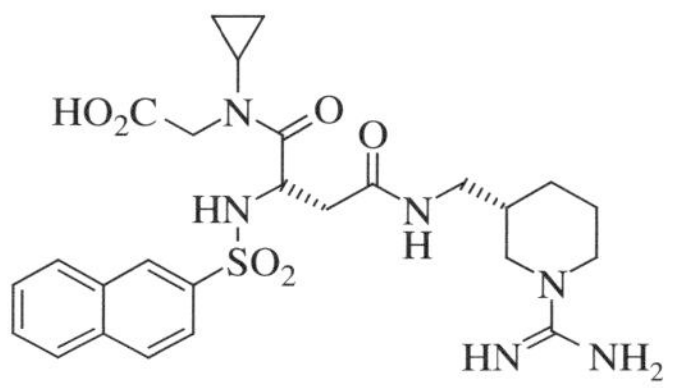

23

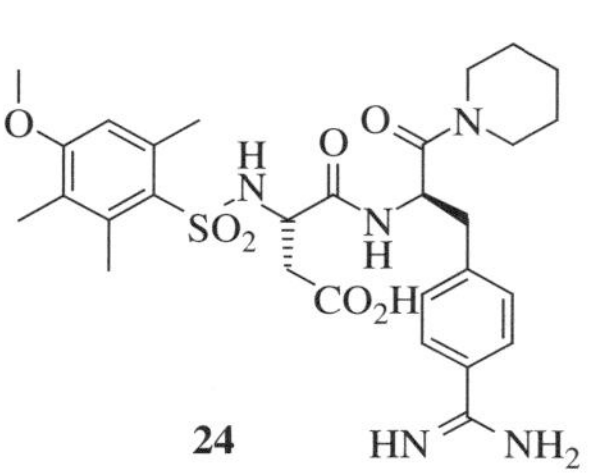

24

4.3.3 Effect of Crystallization Conditions

The conditions required to crystallize a protein or to optimize diffraction may not be the same as those employed in the biological assay. This may affect the reliability of rationalization and prediction of structure–activity relationship from sequential protein–ligand complexes. The influence of crystallization conditions is often unknown or not considered, but numerous examples highlight its importance. An unusual cubic form of trypsin was observed when it was complexed with compound (**25**) at pH 7.[39] The same ligand–protein complex crystallized at pH 8 shows a different ligand conformation, active-site conformation and crystal morphology. Normally the pH of protein crystallization has no effect upon the formation of various crystal forms, but in the case of (**24**) the pH affects the protonation state of the ligand and thereby alters its binding mode, which in turn precludes the normally observed packing of the protein.

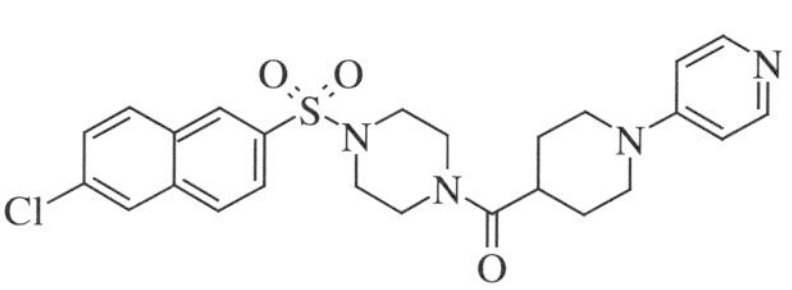

25

The terminal methylpiperazine ring of Abbott inhibitor A-70450 (**26**) was found to exist in a chair conformation in the crystal structure of secreted aspartic protease 2 crystallized at pH 4.5.[40] But in a subsequent study the methylpiperazine group was observed to assume a boat conformation when the complex was crystallized at pH 6.5.[41]

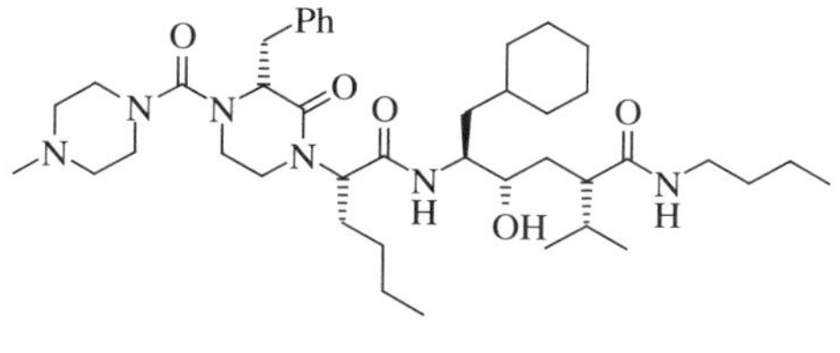

26

The recently identified genetically encoded amino-acid pyrrolysine adopts two conformations in MtmB. The occupancies of the two conformations depend upon whether the precipitating salt was sodium chloride or ammonium sulfate.[35] When ammonium sulfate is used as the precipitating agent, additional density adjacent to C-2 of the ring carbon suggests the addition of ammonia from the buffer to the imine of pyrrolysine. The change in occupancy of the two conformations appears to be controlled by new hydrogen bonds formed between this nitrogen and Glu259 and Gln333.

Two crystal forms of human pancreatic α-amylase were also studied at different pH's. The flexible loop, typical of mammalian α-amylases, was shown to exist in two conformations, suggesting that loop closure is pH-sensitive.[42] Likewise, pH-sensitive conformational changes have been observed for glycinamide ribonucleotide transformylase,[43] *Aspergillus* pectin lyase A,[44] glutathione synthetase,[45] influenza matrix protein M1[46] and ribonuclease A.[47]

4.3.4 Identification and Location of Water

Identification of water in the electron-density maps can be a problem. Water, sodium ions and ammonium ions, common buffer constituents in crystallization media, cannot always be distinguished based on their density alone, because they are isoelectronic. The local environment needs to be taken into account in order to decide how a solvent feature in the electron-density map is best interpreted. Such issues are easily missed, especially by less experienced crystallographers.

The location of water molecules can also be problematic. Unless the resolution is high, the presence or absence of water molecules cannot be determined with certainty, and it becomes a subjective matter whether a feature in the density should be ignored as noise or modeled as a water molecule. However, uncritical addition of solvent molecules (each of which introduces four adjustable parameters, *x*, *y*, and *z* coordinates and isotropic temperature factor, into the model) provides the crystallographer with an "excellent" means of absorbing problems in both the experimental data and the atomic model.[22,28] Addition of water is then simply used to reduce artificially the differences between observed and calculated structure-factor amplitudes.

Where crystallographers determine the same structure at similar resolution, their water structures are bound to reveal many discrepancies. For example, the structure of transforming growth factor-β2 was determined by two independent laboratories at similar resolutions, 1.8 Å (1TGI) and 1.95 Å (1TFG).[48] There are 58 water molecules in 1TGI with an average temperature factor of 31.8 $Å^2$ and 84 water molecules in 1TFG with an average temperature factor of 43.3 Å.[2] The 54 common water molecules in 1TFG had much lower temperature factors (average 34 $Å^2$) than the 30 extra water molecules (with an average temperature factor of 60 $Å^2$), suggesting a much lower level of reliability of the latter. The structure of human interleukin 1β was determined independently in four different laboratories at similar resolution.[49] The four models contained between 83 and 168 water molecules, but a mere 29 of these were in common between all four models. Interestingly, although all 29 belong to the first layer of solvation, not all of them are buried. In a final example, the structure of poplar leaf plastocyanin was subjected to two separate refinements by independent laboratories using the same set of synchrotron X-ray data at 1.6 Å.[50] The two groups used two different refinement protocols, and agreed not to communicate until each was convinced that their refinement calculations were complete. The structures contained 171 and 189 water molecules, respectively, but only 159 water molecules were found in common between the two structures within 1 Å. While it can be a matter of subjectivity to decide whether the electron density supports the presence of a water molecule at a particular location, a water molecule that does not form a single hydrogen bond to any other atom is almost certainly an artifact. Statistics from the protein verification tool WHAT IF,[51] found at the PDBREPORT site,[52] identify 552 179 water molecules in 16 806 structures deposited in the PDB that have no hydrogen bonds to any other atom in the structure (May 2005).

It may simply be worth remembering that at the resolution usually encountered in pharmaceutical discovery projects, the electron density for water molecules that are not well ordered is often difficult to distinguish from noise. The importance of water in binding energetics and kinetics should not be overlooked (although it sometimes is). Water is the "third party" in the ligand–receptor interaction.[51] Depending on the hydrogen-bonding environment of influential water molecules, it may be energetically favorable for a ligand to displace the water molecule, to form a hydrogen bond to it or to leave it in place and not to interact with it. With uncertainties over which water molecules are displaced and which are not displaced from the active site upon ligand binding, water molecules are often completely removed in virtual screening campaigns. This oversimplification may affect the accuracy of docking and scoring.

4.4 Macromolecular Structures to Determine Small-molecule Structures

Macromolecular structures are not a good source for small-molecule 3-D structural information. Even at high resolution the ligand may still not be well defined, as highlighted by Boström.[52] The structure in PDB entry 1PME was determined to 2.0 Å, however the planar methane-sulfoxide present in the ligand is chemically unlikely. Similarly, the 3-phenylpropylamine ligand in structure 1TNK, which was determined to 1.8 Å, contains a tetrahedral aromatic carbon bound to the propylamine chain.

One lesson is that, whereas high-quality dictionaries of acceptable bond lengths, angles and torsions are available for amino and nucleic acid model refinement, the same is not true for complexed ligands.[53] This is because of the unlimited chemical diversity of small molecules compared to amino and nucleic acids. Recently the problem has been highlighted by examination of the coordinates of one of the most frequently determined small molecules in the PDB, adenosine triphosphate. ATP is structurally very similar to adenine, a molecule present in most dictionaries. Hence crystallographers could easily have "borrowed" the restraints for adenosine and applied them to ATP. However bond angles and bond lengths were found to be considerably larger than would be expected based on standard deviations in atomic resolution structures. Surprisingly the bond angles and bond lengths of the 39 crystal structures with resolution better than 2 Å were not better than the 100 structures determined at resolution worse than 2 Å. The Hetero-Compound Information Centre, Uppsala (HIC-Up[54]) has made available ready-made dictionaries for commonly used crystallographic protein modeling software (CNS, X-PLOR, TNT and O) as an aid to the crystallographic community.[55] A similar service is provided by the PRODRG server.[56] Also accessible through the HIC-Up site is a basic validation tool, HETZE, which checks the PDB file of a ligand for acceptable ranges of bond lengths, angles and torsions. Experimental observations from small-molecule crystal structures, *ab initio*, semi-empirical or rule-based structure prediction software, such as CORINA, are also useful sources of restraints for small molecules for protein crystallographers.

4.5 Assessing the Validity of Structure Models

With the uncertainties concerning the validity of X-ray structures deposited in the PDB, that even experts disagree about, the average user does well to proceed with caution. The degree of confidence in the position of a particular atom or residue can be assessed using the temperature factors, occupancies and occasionally remarks, all of which are deposited with the atomic coordinates. If the structure factors are also deposited, electron-density maps can be calculated and superimposed on the structure. Examination of the structure together with the electron-density map is highly recommended.[57] This enables users of the refined model to assess the quality of the fit of the model to the density (data). Issues that can be addressed include the overall reliability of the model, together with the position, orientation, conformation and geometry of specific residues and ligands. This level of detailed visualization is generally only available in specialist crystallographic modeling tools, such as O, but programs such as SwissPDBViewer and ASTEXVIEWER, which are freely available on the internet, allow full visualization of PDB files together with electron-density maps.[58,59] It will not always be possible to inspect the density, since this requires that the structure factor data have been deposited with the PDB by the crystallographer. Although most journals now have strict deposition polices, a survey found that electron-density maps could be calculated for only ~30% of all crystal structures in the PDB.[60] Interestingly, it appears that more structure factors are deposited for structures with low than for those with high free R-values. This curious observation suggests that the worse the model is, the less likely it is that the crystallographer will deposit the experimental data that the structure is supposed to explain. Jones and co-workers have developed EDS, the Uppsala Electron Density Server[61] to facilitate objective assessment of the quality of the fit of the model to the electron density of any PDB entry for which structure factors are available.[33]

Before expending considerable resources on the exploitation of a protein–ligand structure, medicinal chemists and protein modelers would do well to assess the overall reliability of the model. An introductory tutorial for non-experts is available on the Internet.[22,62] Subsequently, the reliability of any crucial residues, water molecules and bound ligands needs to be assessed, either by interacting directly with the crystallographer who determined the structure, or by reading the literature. Scrutiny of the REMARK records in the PDB entry and inspection of the temperature factors and occupancies is recommended. Treating a PDB entry as a simple array of atom coordinates at perfect resolution is a gross oversimplification and can easily lead to false assumptions concerning the model.

Confidence in a model can be gained when multiple, independently determined protein–ligand complexes are available, at very good resolution, and by closely inspecting the electron-density maps. Important factors can then be assessed, such as the position of influential water molecules,[63] the degree of flexibility in residues neighboring the active site and assumptions that may influence the success of structure-guided design and docking studies.

4.6 Summary and Outlook

Structure-guided drug design has contributed to the discovery of a number of drugs and late-stage clinical candidates. It is now common for iterative ligand–protein structures to be available in discovery projects. Where several ligands have been identified, more information is usually obtained by determining complexes with dissimilar ligands than by determining several in which the ligands are structurally closely related. Perversely, the persuasiveness of structural information allied to seductively high *in vitro* potency can constitute a barrier in the journey from ligand design to drug discovery. The use of ADME data alongside primary screening is now becoming routine in the pharmaceutical industry. The traditional approach of maintaining or including polar interactions while increasing *in vitro* potency using hydrophobicity is unacceptable, if it is achieved at the expense of other drug-like properties.

The availability of X-ray-derived structural information on protein–ligand complexes is increasing and this is a useful tool in lead optimization. However, the ambiguities associated with structure models derived from X-ray data may not be fully appreciated. The process of deriving an atomic model from the electron density disguises uncertainties in the identity and position of ligand, water and protein atoms. The observed ligand and protein conformation can be affected by crystallization conditions. It can be difficult for even the most conscientious medicinal chemist to avoid drawing misleading conclusions.

These ambiguities have important consequences for the application of structure-guided design methodologies. Calculation of binding affinities is currently too imprecise to guide design in the narrow range of affinities observed during the optimization of a lead to a drug. The use of docking and scoring tools to design combinatorial chemistry libraries makes some allowance for the inaccuracies of scoring functions, and cases already exist that demonstrate an important complementarity between these technologies. Virtual and property-based screening also has utility for choosing compound subsets for low-throughput screens, which are not amenable to HTS. The prevalence of induced fit in ligand–protein interactions also adds complexity to predicting affinities, but at the same time offers new opportunities in ligand design. At present the ability to predict the protein movement and its consequences upon ligand binding is limited. However it does appear that hydrophobic residues, particularly Phe, Tyr and Trp and those residues associated with function, are often implicated.

In summary, the opportunities and need for smart structure-guided design have never been greater.

References

1. J. DiMasi, R. W. Hansen and H. G. J. Grabowski, *J. Health Econ.*, 2003, **22**, 151–185.
2. H. G. J. Grabowski and J. M. Vernon, *J. Health Econ.*, 1994, **13**, 282–406.
3. A. M. Davis, J. Dixon, C. J. Logan and D. W. Payling, in *Pharmacokinetic Challenges in Drug Discovery*, ed. O. Pelkonen, A. Baumann and A. Reichel, Berlin, Springer, 2002, pp. 1-32.
4. D. Gustafsson, R. Byland, T. Antonsson, I. Nilsson, J. -E. Nystrom, E. Eriksson, U. Bredberg and A. -C. Teger-Nilsson, *Drug Discovery Today*, 2004, **3**, 649.
5. V. de Lucca and P. Y. S. Lam, *Drugs Future*, 1998, **23**, 987–994.
6. J. D. Rodgers, P. Y. S. Lam, B. L. Johnson, H. Wang, S. S. Ko, S. P. Seitz, G. L. Trainor, P. S. Anderson, R. M. Klabe, L. T. Bacheler, B. Cordova, S. Garber, C. Reid, M. R. Wright, C. -H. Chang and S. Erickson-Viitanen, *Chem. Biol.*, 1998, **5**, 597–608.
7. P. A. Aristoff, *Drugs Future*, 1998, **23**, 995–999.
8. M. von Itzstein, W. -Y. Wu, G. B. Kok, M. S. Pegg, J. C. Dyason, B. Jin, T. V. Phan, M. L. Smythe, H. F. White, S. W. Oliver, P. M. Colman,

J. N. Varghese, D. M. Ryan, J. M. Woods, R. C. Bethell, V. J. Hotham, J. M. Cameron and C. R. Penn, *Nature*, 1993, **363**, 418–423.
9. *Physicians' Desk Reference*, 55th Edition, Medical Economics Company Inc, Montvale, NJ, 2001, p. 1454.
10. C. U. Kim, W. Lew, M. A. Williams, H. Liu, L. Zhang, S. Swaminathan, N. Bischofberger, M. S. Chen, D. B. Mendel, C. Y. Tai, W. G. Laver and R. C. Stevens, *J. Am. Chem. Soc.*, 1997, **119**, 681–690.
11. *R&D Insight*, ADIS International Ltd, Chester.
12. J. A. Staessen, Y. Li and T. Richart, *Lancet*, 2006, **368**, 1449–1456.
13. A. M. Davis, S. J. Teague and G. J. Kleywegt, *Angew. Chem.*, 2003, **24**, 2693.
14. S. J. Teague, A. M. Davis, P. D. Leeson and T. I. Oprea, *Angew Chem., Int. Ed.*, 1999, **38**, 3743–3748.
15. D. C. Rees, M. Congreve, C. W. Murray and R. Carr, *Nat. Rev. Drug Discovery*, 2004, **3**, 660–672.
16. I. D. Kuntz, K. Chen, K. A. Sharp and P. A. Kollman, *Proc. Natl. Acad. Sci. U. S. A.*, 1999, **96**(18), 9997–10002.
17. A. L. Hopkins, C. R. Groom and A. Alex, *Drug Discovery Today*, 2004, **9**(10), 430–431.
18. J. Drenth, *Principles of Protein X-ray Crystallography*, Springer-Verlag, New York, 1994.
19. T. A. Jones and M. Kjeldgaard, *Methods Enzymol.*, 1997, **277**, 173–208.
20. G. J. Kleywegt and T. A. Jones, *Methods Enzymol.*, 1997, **277**, 208–230.
21. (a) F. C. Bernstein, T. F. Koetzle, G. J. B. Williams, E. F. Meyer Jr., M. D. Brice, J. R. Rodgers, O. Kennard, T. Shimanouchi and M. Tasumi, *J. Mol. Biol.*, 1977, **112**, 535–542; (b) http://www.rcsb.org/pdb.
22. G. J. Kleywegt, *Acta Crystallogr. Sect. D: Biol. Crystallogr.*, 2000, **D56**, 249–265.
23. G. J. Kleywegt and T. A. Jones, in *Making the Most of Your Model*, ed. W. N. Hunter, J. M. Thornton and S. Bailey, SERC Daresbury Laboratory, Warrington, 1995, pp. 11–24.
24. C. I. Brändén and T. A. Jones, *Nature*, 1990, **343**, 687–689.
25. M. A. Hanson, T. K. Oost, C. Sukonpan, D. H. Rich and R. C. Stevens, *J. Am. Chem. Soc.*, 2000, **122**, 11268–11269.
26. M. A. Hanson, T. K. Oost, C. Sukonpan, D. H. Rich and R. C. Stevens, *J. Am. Chem. Soc.*, 2002, **124**, 10248.
27. A. T. Brünger, *Nature*, 1992, **355**, 472–475.
28. G. J. Kleywegt and T. A. Jones, *Structure*, 1995, **3**, 535–540.
29. A. T. Brünger, *Methods Enzymol.*, 1997, **277**, 366–396.
30. G. J. Kleywegt and A. T. Brünger, *Structure*, 1996, **4**, 897–904.
31. R. W. W. Hooft, G. Vriend, C. Sander and E. E. Abola, *Nature*, 1996, **381**, 272.
32. http://www.cmbi.kun.nl/gv/pdbreport.
33. G. J. Kleywegt, M. R. Harris, J. Zou, T. C. Taylor, A. Wahlby and T. A. Jones, *Acta Crystallogr. Sect. D: Biol. Crystallogr.*, 2004, **D60**, 2240–2249.
34. J. F. Leszczynski, G. D. Rose and S. Milton, *Science*, 1986, **234**, 849–855.

35. B. Hao, W. Gong, T. K. Ferguson, M. Carey, J. A. Krzycki and M. K. Chan, *Science*, 2002, **296**, 1462–1466.
36. G. J. Kleywegt, T. Bergfors, H. Senn, P. Le Motte, B. Gsell, K. Shudo and T. A. Jones, *Structure*, 1994, **2**, 1241–1258.
37. A. Fersht, *Enzyme Structure and Mechanism*, WH Freeman and Company, New York, 1985, pp. 155–175.
38. G. Klebe, M. Bohm, F. Dullweber, U. Gradler, H. Gohlke and M. Hendlich, in *Molecular Modelling and Prediction of Bioactivity*, ed. K. Gundertofte and F. S. Jorgensen, Kluwer Academic/Plenum Publishers, New York, 2000, pp. 103–110.
39. M. T. Stubbs, S. Reyda, F. Dullweber, M. Moller, G. Klebe, D. Dorsch, W. W. K. R. Mederski and H. Wurziger, *ChemBioChem*, 2002, **2**, 246–249.
40. S. M. Cutfield, E. J. Dodson, B. F. Anderson, P. C. E. Moody, C. J. Marshall, P. A. Sullivan and J. F. Cutfield, *Structure*, 1995, **3**, 1261–1271.
41. C. Abad-Zapatero, R. Goldman, S. W. Muchmore, C. Hutchins, K. Stewart, J. Navaza, C. D. Payne and T. L. Ray, *Protein Sci.*, 1996, **5**, 640–652.
42. V. Nahoum, G. Roux, V. Anton, P. Rouge, A. Puigserver, H. Bischoff, B. Henrissat and F. Payan, *Biochem. J.*, 2000, **346**, 201–208.
43. Y. So, M. M. Yamashita, S. E. Greasley, C. A. Mullen, J. H. Shim, P. A. Jennings, S. J. Benkovic and L. A. Wilson, *J. Mol. Biol.*, 1998, **281**, 485–499.
44. O. Mayans, M. Scott, I. Connerton, T. Gravesen, J. Benen, J. Visser, R. Rickersgill and J. Jenkins, *Structure*, 1997, **5**, 677–689.
45. K. Matsuda, K. Mizuguchi, T. Nishioka, H. Kato, N. Go and J. Oda, *Protein Eng.*, 1996, **9**, 1083–1092.
46. A. Harris, F. Forouhar, S. Qiu, S. Shihong and L. M. Bingdong, Options for the control of influenza IV, *Int. Congr. Ser.*, 2001, **1219**, 405–410.
47. R. Berisio, F. Sica, V. S. Lamzin, K. S. Wilson, A. Zagari and L. Mazzarella, *Acta Crystallogr.*, 2002, **D58**, 441–450.
48. S. Daopin and D. R. Davies, *Acta Crystallogr. Sect., D: Biol. Crystallogr.*, 1994, **D50**, 85–92.
49. D. H. Ohlendorf, *Acta Crystallogr. Sect., D: Biol. Crystallogr.*, 1994, **D50**, 808–812.
50. B. A. Fields, H. H. Bartsch, H. D. Bartunik, F. Cordes, J. M. Guss and H. C. Freeman, *Acta Crystallogr. Sect., D: Biol. Crystallogr.*, 1994, **D50**, 709–730.
51. J. E. Ladbury, *Chem. Biol.*, 1996, **3**, 973–980.
52. J. Boström, *J. Comput. Aided Mol. Des.*, 2001, **15**, 137–152.
53. G. J. Kleywegt, K. Henrick, E. J. Dodson and D. M. F. van Aalten, *Structure*, 2003, **11**, 1051–1059.
54. G. J. Kleywegt and T. A. Jones, *Acta Crystallogr. Sect., D: Biol. Crystallogr.*, 1998, **D54**, 1119–1131.
55. http://xray.bmc.uu.se/hicup.
56. D. M. F. van Aalten, R. Bywater, J. B. C. Findlay, M. Hendlich, R. W. W. Hooft and G. Vriend, *J. Comput. Aided Mol. Des.*, 1996, **10**, 255–262.

57. E. E. Abola, A. Bairoch, W. C. Barker, S. Beck, D. A. Benson, H. Berman, G. Cameron, C. Cantor, S. Doubet, T. J. P. Hubbard, T. A. Jones, G. J. Kleywegt, A. S. Kolastar, A. Van Kuik, A. M. Lest, H.-W. Mewes, D. Neuhaus, F. Pfeiffer, L. F. TenEyck, R. J. Simpson, G. Stoesser, J. L. Sussman, Y. Tateno, A. Tsugita, E. L. Ulrick and J. F. G. Vliegenthart, *Bioassays*, 2000, **22**, 1024–1034.
58. http://www.expasy.ch/spdbv.
59. M. J. Hartshorn, AstexViewerTM: a visualization aid for structure-guided drug design, *J. Comput. Aided Mol. Des.*, 2003, **16**(12), 871–881.
60. G. J. Kleywegt and T. A. Jones, *Structure*, 2002, **10**, 465–472.
61. http://eds.bmc.uu.se/.
62. http://xray.bmc.uu.se/embo2001/modval.
63. P. C. Sanschagrin and L. A. Kuhn, *Protein Sci.*, 1998, **7**, 2054–2064.

CHAPTER 5

Dealing with Bound Waters in a Site: Do they Leave or Stay?

DONALD HAMELBERG AND
J. ANDREW McCAMMON*

Howard Hughes Medical Institute, Department of Chemistry and Biochemistry, and Department of Pharmacology, University of California at San Diego, La Jolla, California, 92093-0365, USA

5.1 Introduction

Water molecules are ubiquitous to biomolecules. They play a very important role in the function and structure of proteins. They also make up an integral part of protein structures and contribute to their stability. Localized water molecules can be found in the crevices on protein surfaces, in deeper channels or ligand binding sites, and within buried hydrophilic cavities of protein molecules. There is also evidence that water molecules could sometimes bind to cavities that are mostly hydrophobic.[1–4] Also, water molecules are characteristically present in the interfaces found in protein–ligand, protein–protein, protein–carbohydrate, protein–nucleic acid, and nucleic acid–ligand complexes.[5–16] These water molecules can now be routinely detected by X-ray crystallographic and nuclear magnetic resonance (NMR) experiments.[17–19] Interfacial water molecules typically act as bridges between the two solute molecules, playing a role in recognition and specificity, and at the same time stabilizing the complex by accepting and donating hydrogen bonds. However, in some cases water molecules have been shown to direct the function of an enzyme protein, for example the catalytic abilities of protein kinases.[20,21] The highly conserved protein kinase CK2,[20] which utilizes ATP, could also efficiently make use of GTP by localizing a water molecule at the interface between GTP and the binding pocket.

The water molecules in polar ligand binding sites of proteins are often highly structured with interconnecting hydrogen bond networks and, if connected with bonds, almost resemble the shape of a bound ligand, as shown in Figure 5.1. Here, the binding site of the ligand-free streptavidin[22] is filled with ordered water

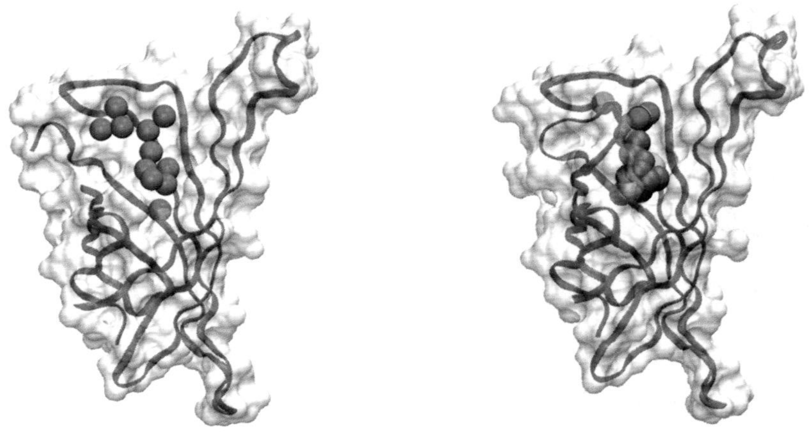

Figure 5.1 Streptavidin and streptavidin–biotin complex. An example of binding site with (right) and without (left) ligand (pdb id: 1n4j and 1n43). The binding site of the ligand-free protein is filled with water molecules that are mostly displaced once the biotin enters the binding site.

molecules that are neatly hydrogen bonded to each other. Ligand binding disrupts such networks of water molecules by breaking the hydrogen bonds and releasing the water molecules into the bulk solvent, as can be seen in the case of the streptavidin–biotin complex wherein most of the water molecules in the binding site have been displaced. This process normally contributes favorably to the free energy of binding, because in most cases the gain in entropy is more than enough to compensate for the water–water and water–protein hydrogen bonds that are lost. Also, some of the enthalpy loss is offset by new hydrogen bonds when the water molecules move into the bulk solvent. In short, since water molecules in polar binding sites are usually localized, upon ligand binding they are transferred from an ordered state to a disordered state, and in so doing contribute positively to the entropy of the system.

5.2 Localized Water Molecules in Binding Sites of Proteins

In many cases, non-covalent association of protein and ligand results in the displacement of all of the interfacial water molecules between the protein and the ligand, followed by rearrangement of the binding pocket in order to accommodate the shape of the ligand. The driving forces of the protein–ligand complex formation include protein–ligand van der Waals and hydrogen bond interactions, and solvent entropy. However, in other cases ligand binding is accompanied by partial release of water molecules, and one or more water molecules are trapped in the binding site between the ligand and protein. These highly localized water molecules effectively act as moieties of the ligand and protein molecules by optimizing the interaction of the ligand with the walls of

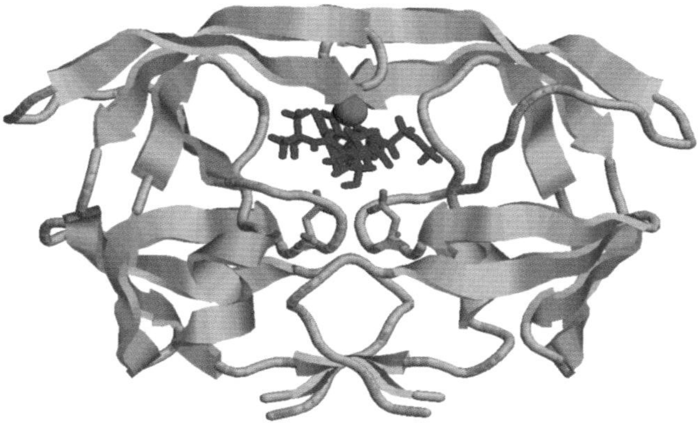

Figure 5.2 The X-ray crystal structure of the HIV-1–KNI-272 complex with a bound water molecule (red sphere) in the binding pocket interacting with the inhibitor and the flaps of the protein.

the binding site. The water molecules normally fill hydrophilic voids in the interface between the ligand and the binding site, thus taking advantage of the fact that a water molecule can act as a hydrogen bond donor and acceptor, and can easily reorient to optimize such interactions. Thus, in addition to the driving forces for binding mentioned above, enthalpy of hydration also contributes. The solute interaction enthalpy of such localized water molecules overcompensates for the loss in solvent entropy due to their restricted motion.

Two classic examples in which a water molecule bridges a bound ligand and its binding site are the human immunodeficiency type 1 (HIV-1) protease–KNI-272 complex[8,19] and the trypsin–benzamidine complex.[9] In the case of the HIV-1 protease KNI complex (Figure 5.2), a water molecule stabilizes the complex by accepting two hydrogen bonds from residues ILE50 and ILE150 that are at the tips of the flaps of the protein, while donating two hydrogen bonds to the ligand. Similarly, the stabilizing water molecule in the interface of trypsin and benzamidine (Figure 5.3) accepts one hydrogen bond from the amidine hydrogen of the ligand and donates two hydrogen bonds to residues TRP190 and VAL201 of the protein.

An interesting drug design question that is apparent from the HIV-1 and trypsin examples is how could one take advantage of the entropy gain in releasing a localized water molecule between the ligand and the protein interface? In trying to achieve this goal, the existing ligand could be modified such that a functional group or moiety of the new ligand occupies the space of the water molecule and interacts directly with the residues in the binding site. Hopefully, the free energy of binding of the new ligand would be more favorable than that of the existing ligand due to the entropy gain of releasing the water molecule. Thermodynamically, this could be realized if the free energy contribution of the entropy of releasing the water molecule surpasses any possible net enthalpy losses due to, for example, less effective hydrogen bond formation and any possible loss in conformational freedom of the new

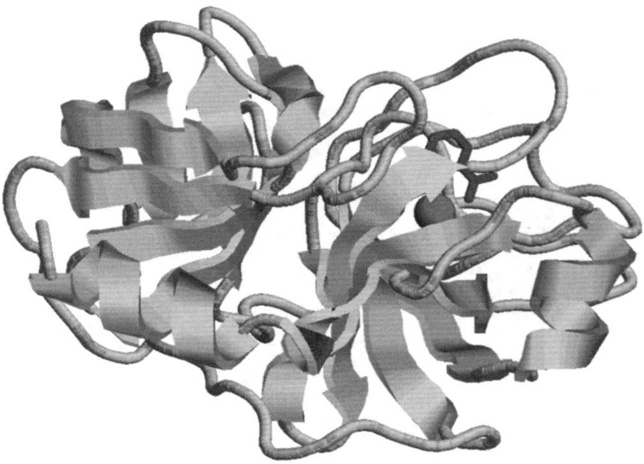

Figure 5.3 The X-ray crystal structure of the trypsin–benzamidine complex with a bound water molecule (red sphere) in the binding pocket.

functional group. Using the above thermodynamic principles and computational modeling, Lam *et al.*[8] designed a new class of cyclic HIV-1 protease inhibitors that interacted directly with the flaps and hence displaced the localized water molecule. This set of cyclic inhibitors binds more strongly than the linear analogues that accommodate the bridging water molecules. Lam *et al.*[8] designed the cyclic inhibitors such that the functional groups interacting directly with the flaps are already rigid and do not have to lose any additional degrees of freedom upon binding, thus maximizing the gain in entropy due to the release of the localized water molecule.

Water molecules also play a remarkably adaptive compensatory role in certain promiscuous binding proteins. An example of this is the OppA binding protein, which binds peptides with widely varying shapes and sizes.[14,15,23] Among other oligopeptides, this protein binds Lys–Xxx–Lys, where Xxx represents almost any of the 20 amino acids and their analogues. When a tripeptide with a large central side chain binds into the binding site of OppA, it displaces a majority of the water molecules in the binding site (Figure 5.4). However, as the size of the side chain of the central residue is decreased, the number of water molecules that remain in the binding site increases, filling up the space created by the decrease in the size of the side chain. Therefore, instead of the binding site of OppA adjusting to the sizes and shapes of the side chain of the central residue, the shape of the binding site stays relatively unchanged and water molecules fill any voids in the binding site.

In the same way, OppA can bind a dipeptide (Lys–Lys), tripeptide (Lys–Lys–Lys) and tetrapeptide (Lys–Lys–Lys–Ala).[23] In the binding site, the number of water molecules decreases around the peptide ligand as the size of the peptide increases. The protein uses extra water molecules as a way to fill up the empty space in the binding site as the peptide gets smaller. The dipeptide

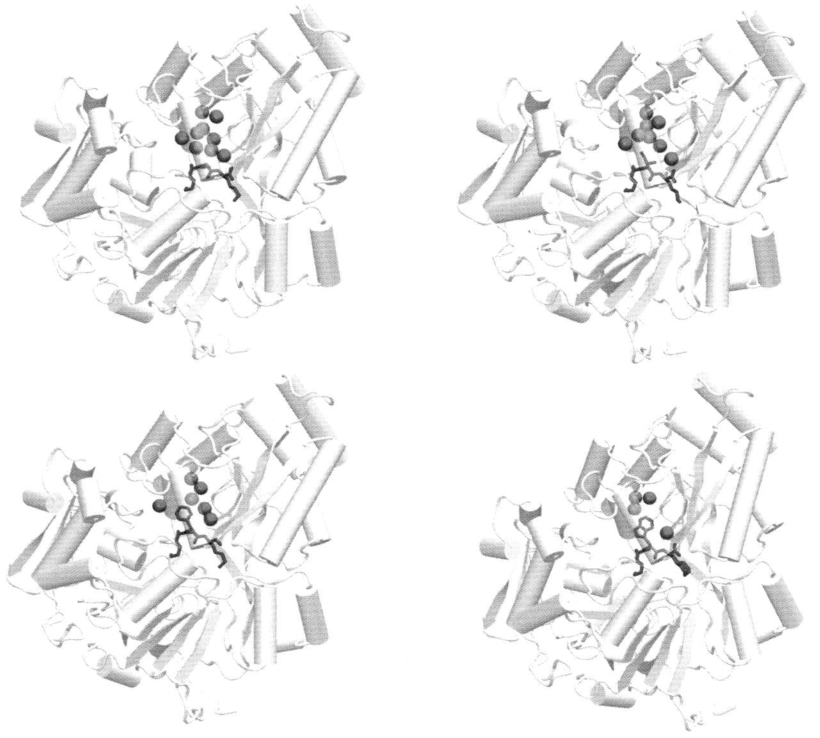

Figure 5.4 Four X-ray crystal structures (pdb id: 1b3l, 1b3g, 1jev, and 1b40) of Lys–Xxx–Lys bound to OppA. Xxx is (clockwise) Gly, Ile, Trp, and Phe. The water molecules in the binding site are shown as red spheres.

was found to have a lower binding affinity for OppA compared to that for the tripeptide and the tetrapeptide. The main reason for this is the entropy of water release. Since the Lys–Lys–Lys and Lys–Lys–Lys–Ala displace more localized water molecules from the binding site of OppA than the dipeptide, they respectively bind with a 6.6 and 11 kcal mol^{-1} extra entropic gains, compared to the dipeptide.[23] These extra entropic gains by the tripeptide and tetrapeptide are more than enough to dominate the modest increase in enthalpy of 3.5 and 8.3 kcal mol^{-1} for Lys–Lys–Lys and Lys–Lys–Lys–Ala, respectively. The dipeptide (Lys–Lys) has a lower enthalpy than Lys–Lys–Lys and Lys–Lys–Lys–Ala because of the extra hydrogen-bonding network of the highly structured water molecules around it.

5.3 Identifying Localized Water Molecules from Computer Simulations

With the determination of more protein crystal structures with higher resolution, the positions of localized water molecules are being found more frequently.

NMR is also contributing to our knowledge of bound water molecules. These water molecules are normally retained when carrying out molecular dynamics (MD) simulations with experimentally determined atomic coordinates. Explicit water MD simulations have allowed us to map out water density around protein structures.[24,25] They have also reproduced some of the positions of localized water molecules that are observed in crystal structures. An additional advantage of simulation is that it allows for dynamic properties, like residence times, to be calculated directly. The structure and interactions of the site-bound water molecules with the ligands and the binding sites can also be studied fully. However, the present sub-microsecond timescale limitation of MD simulations makes it hard to study certain dynamic properties of buried water molecules. This is particularly critical when studying water molecules that are trapped in the binding sites of proteins. In order to observe the exchange of the bound waters with bulk solvent, which can be probed by NMR spectroscopy, one has to be able to simulate binding and unbinding or partial unbinding of the ligand, which is almost impossible with present-day simulation techniques.

Experimentally solved protein structures do not always identify the locations of localized water molecules. This is more frequent with NMR structures and low-resolution X-rays crystal structures. Alternatively, with the development of new methods, computational calculations and simulations have been used to probe the presence or absence of buried water molecules in these protein cavities.[26–33] Therefore, in principle, before carrying out an MD simulation on a protein structure with poorly resolved water positions, the presence or absence of these buried water molecules could be determined by computational means. One way that this could be done computationally is by identifying the cavities in the protein and calculating the free energy of tying up water molecules in those cavities. This will give an impression of how tolerant the protein's cavities are to hydration. This technique has been applied to several proteins, including the study of hydrophobic cavities. A grand canonical ensemble-based simulation approach that allows for the variation of the number of particles in a simulation has also been used in studying the locations and properties of buried water molecules in proteins. Resat and Mezei[31,33] developed the grand canonical Monte Carlo (GCMC) simulation technique that has been used successfully to predict localized water molecules in protein-binding sites and other cavities in several proteins, including the HIV-1 protease inhibitor and trypsin–benzamidine complexes.[29,30] Moreover, a data-driven approach has also been introduced in identifying possible binding sites for water molecules. Neural networks were trained with relevant descriptors derived from structures in the protein database to recognize the binding sites of water molecules.[34]

In lieu of explicit solvent simulation, continuum (implicit) models, like the Generalized Born or Poisson Boltzmann solvation models, can be used in accounting for the solvent environment.[35] Therefore, the environments outside of the protein and cavities that are large enough to allow one or more water molecules can be represented by a high dielectric constant. One advantage of the implicit models is that they are computationally faster than explicit representations of the water molecules. However, several limitations are inescapable with such an implementation. Because the water molecules are not physically present, their properties cannot be studied, and the detailed interactions of bridging and localized water molecules with different parts of the protein molecules would be lacking. Furthermore, the interactions within protein–ligand complexes that would normally have one or more interfacial water molecules would probably not be represented properly because the implicit representation of the interfacial water molecules could allow those parts of the protein and ligand to form direct, and possibly unfavorable, contacts.

5.4 Calculation of Free-energy Cost of Displacing a Site-bound Water Molecule

A trapped water molecule in the binding site of a protein stabilizes the protein–ligand complex. This favorable contribution to the standard free energy of binding is largely enthalpic in origin and comes mainly from the hydrogen bond donating and accepting strength of each water molecule. Intuitively, the localization of a site-bound water molecule results in loss of both translational and rotational degrees of freedom, and hence a loss in entropy that represents an unfavorable contribution to the overall free energy of binding. Therefore, the enthalpy contribution due to hydrogen bonding must be large enough to offset the entropic loss. However, a balance has to be struck between enthalpy and entropy. If the bound water molecule is too restricted, then the free-energy cost due to entropy loss could be greater than that of the gain in enthalpy, thus rendering the localization process thermodynamically unfavorable, even though the shape of the cavity might be ideal for the water molecule to complement the ligand structurally. The reverse also applies. This trade-off between enthalpy and entropy is inherent in the free-energy relationship:

$$\Delta G^\circ = \Delta H^\circ - T\Delta S^\circ \tag{5.1}$$

where ΔG° is the change in free energy, ΔH° is the change in enthalpy, ΔS° is the change in entropy, and T is the temperature. The stronger the hydrogen bonds (more favorable enthalpic contribution) formed between the water molecule and the binding site, the less disordered and more highly restricted (less favorable entropic contribution) the water molecule becomes. Conversely, the weaker the hydrogen bonds formed (less favorable enthalpic contribution), the less restricted (more favorable entropic contribution) is the water molecule.

Using the standard entropies of anhydrous and hydrated inorganic salts, Dunitz[36] estimated an upper limit for the entropic cost of bound water in biomolecules. He observed that the entropic cost of transferring a water molecule from bulk water to the binding site in a salt crystal is around 7 cal mol^{-1} K^{-1}, corresponding to a free-energy loss of approximately 2 kcal mol^{-1} at 300 K. Therefore, he concluded that since most bound water molecules in proteins do not bind more strongly than water molecules in crystalline hydrated salts, then the free-energy cost of transferring a water molecule from bulk to any site in the protein should be in the range 0–2 kcal mol^{-1} at 300 K. However, this value does not take into account the enthalpic contribution, which could be favorable or unfavorable and depends on the environment of the water molecule in the binding site of the protein.

Over the years, several computational methods have been developed to study and calculate the free-energy cost of transferring water molecules from the bulk to a protein cavity. The application of statistical thermodynamics perturbation theory by Wade *et al.*[26] was one of the first to be applied to such a system. Wade *et al.*[26] found that the free energy of transferring a water molecule from bulk water to a site which had been shown to contain a water molecule in the crystal structure of the sulfate-binding protein was favorable. On the other hand, they found that the free energy was unfavorable when they tried to introduce a water molecule into a site in the same protein that had been experimentally shown to lack any water molecule.

Subsequently, similar free-energy simulation methods have been used to study the properties of protein cavities. Zhang and Hermans[28] studied the hydrophilic nature of protein cavities by calculating the free energies of introducing a water molecule into these cavities using MD free-energy simulations. They were able to distinguish empty cavities from hydrated ones based on the calculated free energy of introducing water molecules into them. Similarly, Roux *et al.*[37] used MD free-energy perturbation methods to calculate the stability of water molecules in the hydrophobic bacteriorhodopsin proton channel. Their results suggested that the transfer of four water molecules from bulk solvent to the channel is thermodynamically feasible and thereby shed some light on the mechanism of proton transfer in bacteriorhodopsin.

Alternatively, a non-free-energy perturbation approach based on the inhomogeneous fluid solvation theory has been used by Zheng and Lazaradis[38,39] to estimate the free-energy cost of bound water molecules in the binding sites of the HIV-1–KNI-272 protease inhibitor complex. They calculated the entropy loss in tying up this water molecule to be large (9.8 cal mol^{-1} K^{-1}), and the total contribution of this water molecule to the free energy of solvation to be around –15.2 kcal mol^{-1}. This approach is not as exact as using a free-energy perturbation method, but it is able to extract qualitatively the contribution of the water molecule to the energy, entropy and heat capacity of protein solvation.

Hamelberg and McCammon[40] have used the double-decoupling method to calculate the standard free energy of tying up a water molecule in the binding site of the HIV-1–KNI-272 protease and the trypsin–benzamidine complexes. The double-decoupling protocol[41] was rigorously derived from the underlying theory of statistical mechanics and made the connection to the MD free-energy perturbation. The double-decoupling method removed some of the approximations of the initial applications[42–46] of the free-energy perturbation technique to the calculation of absolute free energies. The thermodynamic analysis that underlies the double-decoupling method is shown in Figure 5.5, assuming that A is the protein–ligand complex, and B is the site-bound water molecule.

$$\mathrm{AB(sol)} \longrightarrow \mathrm{A(sol)} + \mathrm{B(gas)} \qquad \Delta G_1^\circ$$

$$\mathrm{B(sol)} \longrightarrow \mathrm{B(gas)} \qquad \Delta G_2^\circ$$

$$\mathrm{A(sol)} + \mathrm{B(sol)} \longrightarrow \mathrm{AB(sol)} \qquad \Delta G_{AB}^\circ = \Delta G_2^\circ - \Delta G_1^\circ$$

Figure 5.5 Thermodynamic analysis of the double-decoupling method. A represents the protein–ligand complex, and B is the site-bound water molecule.

Calculation of the standard free energy, $\Delta G^\circ{}_2$, of removing the water molecule B from the bulk solvent to the gas phase is straightforward, since it does not depend on the choice of the standard concentration. $\Delta G^\circ{}_2$ can be determined by simply decoupling the interactions of B from solution using free-energy perturbation simulation.[47] Therefore, $\Delta G^\circ{}_2$ can be written as:

$$\Delta G_2^\circ = \int_0^1 \left\langle \frac{\partial V(\lambda, \mathbf{r})}{\partial \lambda} \right\rangle_\lambda d\lambda \tag{5.2}$$

where $V(\lambda, \mathbf{r})$ is the coordinate-dependent potential energy function that interpolates between the energy function of the initial state ($\lambda = 0$; B in the solution phase) and the energy function of the final state ($\lambda = 1$; B in the gas phase). Calculation of $\Delta G^\circ{}_1$ is, however, not as straightforward. $\Delta G^\circ{}_1$ is defined as the free energy of decoupling B from the binding site of A during the simulation. Therefore, during the later part of the simulation when B is weakly coupled to A, B would have to explore the entire simulation box in order for $\Delta G^\circ{}_1$ to converge. This is circumvented by restraining the coordinates of B to occupy the binding site of A while decoupling the interactions of B. At the end of decoupling B from A, B is simply an ideal gas restrained in the binding site with a definite chemical potential. Therefore, after several steps that have previously been described,[40,41] $\Delta G^\circ{}_1$ can be written as:

$$\begin{aligned} \Delta G_1^\circ = & \int_0^1 \left\langle \frac{\partial U(\lambda, \mathbf{r})}{\partial \lambda} \right\rangle_\lambda d\lambda \\ & - RT \ln\left(\frac{\sigma_{AB}}{\sigma_A \sigma_B}\right) \\ & + RT \ln(C^o V_I) + RT \ln\left(\xi_I / 8\pi^2\right) + P^o(\bar{V}_A - \bar{V}_{AB}) \end{aligned} \tag{5.3}$$

where the definition of $U(\lambda, \mathbf{r})$ for AB is similar to that of $V(\lambda, \mathbf{r})$ for B, V_I is the integral over the volume element in which B is restrained to occupy, and ξ_I is the integral over the rotational space that B is allowed to sample. σ_i is the symmetry number of molecule *i*. The third and fourth terms can be viewed as correction terms due to the restraining of B during free-energy simulation. They are the change in free energy when the restrained ligand is allowed to expand and occupy a volume of $1/C^{\circ}$ and to rotate freely, respectively.

Using a harmonic restraint to confine the water in the binding sites of the HIV-1 protease–KNI-272 and trypsin–benzamidine complexes during the double-decoupling simulations while allowing them to rotate freely, Hamelberg and McCammon[40] calculated the standard free energies of tying up the water molecule in the binding site of the two complexes to be around –3.1 and $-1.9\,\mathrm{kcal\,mol^{-1}}$, respectively. These results signify that the localized water molecules help in stabilizing the protein–ligand interactions of both complexes. Also, the results confirm the assignment of these water molecules in the binding site of the X-ray crystal structures. These results also clearly show the importance of localized water molecules in the binding pockets of proteins and, because of their energetic contribution, could provide some indication of how ligands could be designed to increase their binding affinity, as was demonstrated by Lam *et al.*[8] Therefore, it can be seen that the contribution by site-bound water molecules to the binding free energy is a very important component that cannot be ignored in drug design and in understanding non-covalent protein interactions.

5.5 Inclusion of Explicit Water Molecules in Drug Discovery

The hydrogen bonding abilities of ligands, the polar and non-polar regions of the binding sites of proteins, the electrostatic potential of ligands and proteins, and the extent of flexibility of ligands and their potential receptors are some of the properties routinely considered in structure-based drug discovery.[48] Despite the fact that formation of protein–ligand complexes takes place in aqueous solution, the inclusion of explicit water molecules is commonly ignored. It is obvious from experimental and theoretical results that interfacial waters can provide significant contributions to the free energy of binding. Therefore, exclusion of some of these water molecules in drug design could lead to ligands with highly charged or very large hydrophilic or hydrophobic regions, respectively.[49] In general, as a first approximation, the effect of water molecules has been modeled implicitly into scoring functions, thus representing hydration as high dielectric continuum.

However, an implicit approximation might not be enough, because the actual presence or absence of a water molecule in the binding site of a protein could control the plasticity of the binding site. These water molecules could be considered to be an integral part of the protein structure. Therefore, most of the studies in this area have focused on three key questions. How to identify water molecules that are conserved in the binding sites of proteins after protein–ligand complex formation? How could interfacial localized water molecules be used to improve ligands in structure-based drug design? How could explicit water molecules be introduced into *de novo* drug-design processes, so as to recognize their hydrogen bonding ability?

There are water molecules in some proteins that occupy the same position after forming complexes with ligands.[50,51] Therefore, a number of machine-learning and statistical tools have been developed to predict conserved water molecules by using properties like the temperature B-factors, number of protein–water hydrogen bonds, sequence information, and polarity of close protein atoms. Ehrlich *et al.*[52] used sequence information of proteins to predict ordered water molecules by modular neural networks. Raymer *et al.*[53] used four environmental features of the water molecules to predict bound water molecules conserved between free and ligand-bound protein. The four features, similar to those listed above, were used to train a hybrid k-nearest neighbor classifier and genetic algorithm. Also, by considering some simple structural properties of crystallographic water molecules in the binding sites of proteins, Garcia-Sosa *et al.*[54] developed a method called WaterScore, based on a multivariate logistic regression analysis, to discriminate between bound and displaceable water molecules. The water's temperature B-factor, solvent-contact surface area, hydrogen bond energy, and the number of protein–water contacts were used in the analysis.

Designing ligands that are capable of targeting experimentally observed localized water molecules in the binding pockets of proteins is becoming common in the field of drug design. The resulting free energy of ligand binding could benefit from the potential free-energy gain accompanying the release of these water molecules. Hamelberg and McCammon[40] quantified the energetic importance of localized water molecules in the binding site of HIV-1 protease–KNI-272 and trypsin–benzamidine complexes for drug design purposes. They calculated the standard free energies of tying up the water molecule in the binding pockets of these complexes to be favorable. According to the rationale by Dunitz,[36] the release of a site-bound water molecule can lead to an entropic gain of as high as $7\,\mathrm{cal\,mol^{-1}\,K^{-1}}$. The enthalpic contribution will depend on the types of interactions experienced by the water molecules in the binding site. These results provided an indication of how ligands could be designed to increase their binding affinity. The model example of how this could work in drug design is that of the HIV-1 protease inhibitor complex by Lam *et al.*[8] The displacement of a site-bound water molecule localized between the flaps and the inhibitor by a modified ligand contributed to an increase in the binding affinity of the new ligand. The underlying principle lies in the fact that if a substituent is added to the ligand that displaces a bound water molecule, and if the net contribution of this substituent is greater than the modest free-energy cost of displacing the solvent, the result will be an increase in ligand affinity. In practice, this may not be too difficult to do, because the ligand will already have paid the price of translational and rotational entropy loss. Its substituent need not pay this price again, but can harvest the entropy of releasing the water molecule.

Other than trying to identify localized water molecules that are conserved in the binding sites of proteins, research studies have also turned to incorporating explicit water molecules in the design of new ligands. For example, Rarey *et al.*[55] have extended their docking method FlexX to place single water molecules explicitly during protein–ligand docking. The aim is to find favorable positions for the water molecules in the binding site which may guide the placement of the ligand by forming hydrogen bonds at the interface. Moreover, the inclusion of explicit water molecules helps in improving the plasticity of the binding site and the pool of possible drug targets. Also, water molecules have been incorporated in protein–ligand docking studies by Rao *et al.*[56] Alternatively, Pastor *et al.*[57] presented a strategy for including water molecules into a three-dimensional quantitative structure–activity relationship (QSAR) analysis used for drug design. The predictability of their QSAR model was improved due to the inclusion of water molecules. Furthermore, simulation methods like grand canonical techniques allow for water molecules to be inserted and deleted during a simulation. Therefore, the probability of a water molecule occupying a particular site in the protein-binding site could be calculated. A GCMC technique has successfully predicted water-binding sites in some protein systems.[29,30] However, one limitation of this method is that it is computationally intensive and could not at present be easily used in rapid drug screening procedures.

Acknowledgements

This work was supported in part by grants from NSF, NIH, the Center for Theoretical Biological Physics, the National Biomedical Computation Resource, San Diego Supercomputing Center, and Accelrys, Inc.

References

1. A. M. Buckle, P. Cramer and A. R. Fersht, Structural and energetic responses to cavity-creating mutations in hydrophobic cores: observation of a buried water molecule and the hydrophilic nature of such hydrophobic cavities, *Biochemistry*, 1996, **35**(14), 4298–4305.
2. J. A. Ernst, R. T. Clubb, H. X. Zhou, A. M. Gronenborn and G. M. Clore, Demonstration of positionally disordered water within a protein hydrophobic cavity by NMR, *Science*, 1995, **267**(5205), 1813–1817.
3. G. Otting, E. Liepinsh, B. Halle and U. Frey, NMR identification of hydrophobic cavities with low water occupancies in protein structures using small gas molecules, *Nat. Struct. Biol.*, 1997, **4**(5), 396–404.
4. K. Takano, J. Funahashi, Y. Yamagata, S. Fujii and K. Yutani, Contribution of water molecules in the interior of a protein to the conformational stability, *J. Mol. Biol.*, 1997, **274**(1), 132–42.
5. T. N. Bhat, G. A. Bentley, G. Boulot, M. I. Greene, D. Tello, W. Dall'Acqua, H. Souchon, F. P. Schwarz, R. A. Mariuzza and R. J. Poljak, Bound water molecules and conformational stabilization help mediate an antigen-antibody association, *Proc. Natl. Acad. Sci. U. S. A.*, 1994, **91**(3), 1089–93.

6. T. N. Bhat, G. A. Bentley, T. O. Fischmann, G. Boulot and R. J. Poljak, Small rearrangements in structures of Fv and Fab fragments of antibody D1.3 on antigen binding, *Nature*, 1990, **347**(6292), 483–485.
7. J. E. Ladbury, J. G. Wright, J. M. Sturtevant and P. B. Sigler PB, A thermodynamic study of the trp repressor-operator interaction, *J. Mol. Biol.*, 1994, **238**(5), 669–681.
8. P. Y. Lam, P. K. Jadhav, C. J. Eyermann, C. N. Hodge, Y. Ru, L. T. Bacheler, J. L. Meek, M. J. Otto, M. M. Rayner and Y. N. Wong, Rational design of potent, bioavailable, nonpeptide cyclic ureas as HIV protease inhibitors, *Science*, 1994, **263**(5145), 380–384.
9. McGrath, J. R. Vasquez, C. S. Craik, A. S. Yang, B. Honig and R. J. Fletterick, Perturbing the polar environment of Asp102 in trypsin: consequences of replacing conserved Ser214, *Biochemistry*, 1992, **31**(12), 3059–3064.
10. C. J. Morton and J. E. Ladbury, Water-mediated protein–DNA interactions: the relationship of thermodynamics to structural detail, *Protein Sci.*, 1996, **5**(10), 2115–2118.
11. B. Nguyen, D. Hamelberg, C. Bailly, P. Colson, J. Stanek, R. Brun, S. Neidle and W. D. Wilson, Characterization of a novel DNA minor-groove complex, *Biophys J.*, 2004, **86**(2), 1028–1041.
12. B. Nguyen, M. P. Lee, D. Hamelberg, A. Joubert, C. Bailly, R. Brun, S. Neidle and W. D. Wilson, Strong binding in the DNA minor groove by an aromatic diamidine with a shape that does not match the curvature of the groove, *J. Am. Chem. Soc.*, 2002, **124**(46), 13680–13681.
13. F. A. Quiocho, D. K. Wilson and N. K. Vyas, Substrate specificity and affinity of a protein modulated by bound water molecules, *Nature*, 1989, **340**(6232), 404–407.
14. J. R. Tame, G. N. Murshudov, E. J. Dodson, T. K. Neil, G. G. Dodson, C. F. Higgins and A. J. Wilkinson, The structural basis of sequence-independent peptide binding by OppA protein, *Science*, 1994, **264**(5165), 1578–1581.
15. J. R. Tame, S. H. Sleigh, A. J. Wilkinson and J. E. Ladbury, The role of water in sequence-independent ligand binding by an oligopeptide transporter protein, *Nat. Struct. Biol.*, 1996, **3**(12), 998–1001.
16. J. W. Schwabe, The role of water in protein–DNA interactions, *Curr. Opin. Struct. Biol.*, 1997, **7**(1), 126–134.
17. M. Levitt and B. H. Park, Water: now you see it, now you don't, *Structure*, 1993, **1**(4), 223–226.
18. B. P. Schoenborn, A. Garcia and R. Knott, Hydration in protein crystallography, *Prog. Biophys. Mol. Biol.*, 1995, **64**(2–3), 105–119.
19. Y. X. Wang, D. I. Freedberg, P. T. Wingfield, S. J. Stahl, J. D. Kaufman, Y. Kiso, T. N. Bhat, J. W. Erickson and D. A. Torchia, Bound water molecules at the interface between the HIV-1 protease and a potent inhibitor, KNI-272, determined by NMR, *J. Am. Chem. Soc.*, 1996, **118**(49), 12287–12290.

20. K. Niefind, M. Putter, B. Guerra, O. G. Issinger and D. Schomburg, GTP plus water mimic ATP in the active site of protein kinase CK2, *Nat. Struct. Biol.*, 1999, **6**(12), 1100–1103.
21. J. Zheng, E. A. Trafny, D. R. Knighton, N. -H. Xuong, S. S. Taylor, S. S. Ten, L. F. Eyck and J. M. Sowadski, A refined crystal structure of the catalytic subunit of cAMP-dependent protein kinase complexed with MnATP and a peptide inhibitor, *Acta Crystallogr. D: Biol. Crystallogr.*, 1993, **D49**(3), 362–365.
22. I. Le Trong, S. Freitag S, L. A. Klumb, V. Chu, P. S. Stayton and R. E. Stenkamp, Structural studies of hydrogen bonds in the high-affinity streptavidin–biotin complex: mutations of amino acids interacting with the ureido oxygen of biotin, *Acta Crystallogr. D: Biol. Crystallogr.*, 2003, **D59**(9), 1567–1573.
23. S. H. Sleigh, J. R. H. Tame, E. J. Dodson and A. J. Wilkinson, Peptide binding in OppA the crystal structures of the periplasmic oligopeptide binding protein in the unliganded form and in complex with lysyllysine, *Biochemistry*, 1997, **36**(32), 9747–9758.
24. R. H. Henchman and J. A. McCammon, Structural and dynamic properties of water around acetylcholinesterase, *Protein Sci.*, 2002, **11**(9), 2080–2090.
25. R. H. Henchman and J. A. McCammon, Extracting hydration sites around proteins from explicit water simulations, *J. Comput. Chem.*, 2002, **23**(9), 861–869.
26. R. C. Wade, M. H. Mazor, J. A. McCammon and F. A. Quiocho, Hydration of cavities in proteins: molecular dynamics approach, *J. Am. Chem. Soc.*, 1990, **112**, 7057–7059.
27. R. C. Wade, M. H. Mazor and J. A. McCammon, A molecular dynamics study of thermodynamic and structural aspects of the hydration of cavities in proteins, *Biopolymers*, 1991, **31**(8), 919–931.
28. L. Zhang and J. Hermans, Hydrophilicity of cavities in proteins, *Proteins*, 1996, **24**(4), 433–438.
29. T. J. Marrone, H. Resat, C. N. Hodge, C. H. Chang and J. A. McCammon, Solvation studies of DMP323 and A76928 bound to HIV protease: analysis of water sites using grand canonical Monte Carlo simulations, *Protein Sci.*, 1998, **7**(3), 573–579.
30. H. Resat, T. J. Marrone and J. A. McCammon, Enzyme-inhibitor association thermodynamics: explicit and continuum solvent studies, *Biophys. J.*, 1997, **72**(2 Pt 1), 522–532.
31. H. Resat and M. Mezei, Grand canonical ensemble Monte Carlo simulation of the dCpG/proflavine crystal hydrate, *Biophys. J.*, 1996, **71**(3), 1179–1190.
32. H. Resat and M. Mezei, Calculating the local solvent chemical potential in crystal hydrates, *Phys. Rev. E: Stat. Phys., Plasmas, Fluids, Relat. Interdiscip. Top.*, 2000, **62**(5 Pt B), 7077–7081.
33. H. Resatand and M. Mezei, Grand canonical Monte Carlo simulation of water positions in crystal hydrates, *J. Am. Chem. Soc.*, 1994, **116**, 7451–7452.

34. R. C. Wade, H. Bohr and P. G. Wolynes, Prediction of water binding sites on proteins by neural networks, *J. Am. Chem. Soc.*, 1992, **114**, 8284–8285.
35. P. E. Smith and B. M. Pettitt, Modeling solvent in biomolecular systems, *J. Phys. Chem.*, 1994, **98**(39), 9700–9711.
36. J. D. Dunitz, The entropic cost of bound water in crystals and biomolecules, *Science*, 1994, **264**, 670.
37. B. Roux, M. Nina, R. Pomès and J. C. Smith, Thermodynamic stability of water molecules in the bacteriorhodopsin proton channel: a molecular dynamics free energy perturbation study, *Biophys. J.*, 1996, **71**(2), 670–681.
38. L. Zheng and T. Lazaridis, Thermodynamic contributions of the ordered water molecule in HIV-1 protease, *J. Am. Chem. Soc.*, 2003, **125**, 6636–6637.
39. L. Zheng and T. Lazaridis, The effect of water displacement on binding thermodynamics: Concanavalin A, *J. Phys. Chem. B*, 2005, **109**, 662–670.
40. D. Hamelberg and J. A. McCammon, Standard free energy of releasing a localized water molecule from the binding pockets of proteins: double-decoupling method, *J. Am. Chem. Soc.*, 2004, **126**(24), 7683–7689.
41. M. K. Gilson, J. A. Given, B. L. Bush and J. A. McCammon, The statistical-thermodynamic basis for computation of binding affinities: a critical review, *Biophys. J.*, 1997, **72**(3), 1047–1069.
42. W. L. Jorgensen, J. K. Buckner, S. Boudon and J. Tirado-Rives, Efficient computation of absolute free energies of binding by computer simulations. Application to methane dimer in water, *J. Chem. Phys.*, 1988, **89**, 3742–3846.
43. S. Miyamoto and P. A. Kollman, What determines the strength of noncovalent association of ligands to proteins in aqueous solution?, *Proc. Natl. Acad. Sci. U. S. A.*, 1993, **90**(18), 8402–8406.
44. S. Miyamoto and P. A. Kollman, Absolute and relative binding free energy calculations of the interaction of biotin and its analogs with streptavidin using molecular dynamics/free energy perturbation approaches, *Proteins*, 1993, **16**(3), 226–245.
45. J. Pranata and W. L. Jorgensen, Monte-Carlo simulations yield absolute free energies of binding for guanine-cytosine and adenine-uracil base pairs in chloroform, *Tetrahedron*, 1991, **47**, 2491–2501.
46. S. F. Sneddon, D. J. Tobias and C. L. Brooks, 3rd, Thermodynamics of amide hydrogen bond formation in polar and apolar solvents, *J. Mol. Biol.*, 1989, **209**(4), 817–820.
47. T. P. Straatsma and J. A. McCammon, Computational alchemy, *Annu. Rev. Phys. Chem.*, 1992, **43**, 407–435.
48. C. F. Wong and J. A. McCammon, Protein flexibility and computer-aided drug design, *Annu. Rev. Pharmacol. Toxicol.*, 2003, **43**, 31–45.
49. B. K. Shoichet, A. R. Leach and I. D. Kuntz, Ligand solvation in molecular docking, *Proteins*, 1999, **34**(1), 4–16.
50. H. L. Carrell, J. P. Glusker, V. Burger, F. Manfre, D. Tritsch and J.-F. Biellmann, X-ray analysis of D-xylose isomerase at 1.9 A: native

enzyme in complex with substrate and with a mechanism-designed inactivator, *Proc. Natl. Acad. Sci. U. S. A.*, 1989, **86**(12), 4440–4444.
51. C. H. Faerman and P. A. Karplus, Consensus preferred hydration sites in six FKBP12-drug complexes, *Proteins*, 1995, **23**(1), 1–11.
52. L. Ehrlich, M. Reczko, H. Bohr and R. C. Wade, Prediction of protein hydration sites from sequence by modular neural networks, *Protein Eng.*, 1998, **11**(1), 11–19.
53. M. L. Raymer, P. C. Sanschagrin, W. F. Punch, S. Venkataraman, E. D. Goodman and L. Kuhn, Predicting conserved water-mediated and polar ligand interactions in proteins using a K-nearest-neighbors genetic algorithm, *J. Mol. Biol.*, 1997, **265**(4), 445–464.
54. A. T. Garcia-Sosa, R. L. Mancera and P. M. Dean, WaterScore: a novel method for distinguishing between bound and displaceable water molecules in the crystal structure of the binding site of protein–ligand complexes, *J. Mol. Model (Online)*, 2003, **9**(3), 172–182.
55. M. Rarey, B. Kramer and T. Lengauer, The particle concept: placing discrete water molecules during protein–ligand docking predictions, *Proteins*, 1999, **34**(1), 17–28.
56. M. S. Rao and A. J. Olson, Modelling of factor Xa-inhibitor complexes: a computational flexible docking approach, *Proteins*, 1999, **34**(2), 173–183.
57. M. Pastor, G. Cruciani and K. A. Watson, A strategy for the incorporation of water molecules present in a ligand binding site into a three-dimensional quantitative structure–activity relationship analysis, *J. Med. Chem.*, 1997, **40**(25), 4089–4102.

CHAPTER 6

Knowledge-Based Methods in Structure-Based Design

MARCEL L. VERDONK* AND WIJNAND T. M. MOOIJ

Astex Therapeutics Ltd., 436 Cambridge Science Park, Milton Road, Cambridge CB4 0QA, UK

6.1 Introduction

Knowledge-based approaches work on the (reasonable) assumption that the relative frequencies of different interaction geometries observed in a range of experimental structures are indicative of the relative stability of these geometries. This idea even pre-dates the start of the computerized structural databases, and was used early on to determine ideal hydrogen-bond geometries for various functional groups.[1,2] Initially, such studies focused on individual types of interactions, which meant that the results were split over various sources, making it difficult to utilize them in an automated fashion. More recently, systematic analyses of experimental structures have led to interaction databases and to software applications that use these databases to predict intermolecular interactions.

In this chapter we introduce computational methods that use Cambridge Structural Database (CSD) or Protein Data Bank (PDB) data in structure-based design. We describe the various knowledge-based potentials that have been presented in the literature. We distinguish between atom-based potentials and group-based potentials, and we discuss the choice of the 'reference state' in these potentials. Finally, applications of knowledge-based methods are described and the key difficulties and issues around these methods are discussed.

6.2 Atom-based Potentials

In atom-based approaches, statistical atom–atom distance preferences are determined from a database of known structures. Atom–atom distributions in the database are analyzed in terms of radial distribution functions (RDFs), which are often converted into pseudo-potentials. An example of two statistical

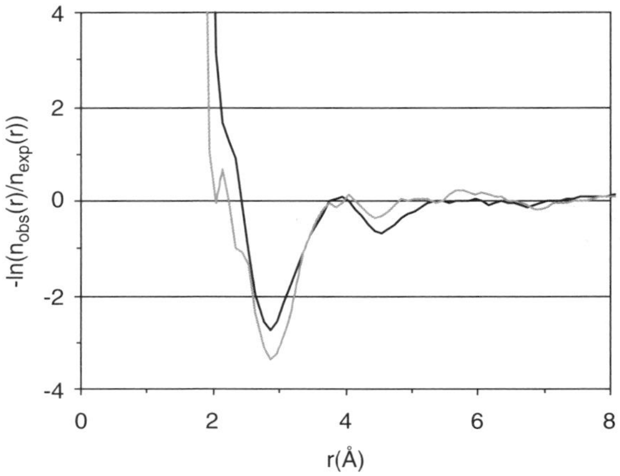

Figure 6.1 Statistical potentials (ASP[12]) for N.am (protein)–O.2 (ligand) (black line), and O.2 (protein)–N.am (ligand) (gray line) atom pairs. (N.am = amide nitrogen, O.2 = carbonyl oxygen).

atom–atom potentials is given in Figure 6.1. The individual potentials can then be combined to calculate the score of a ligand placed in a binding site. This procedure was initially described for protein folding applications,[3] and was first applied to protein–ligand complexes by Verkhivker *et al.*[4] In another early knowledge-based approach, Wallqvist *et al.*[5,6] analyzed the statistical preferences of surface burial for different atom types in 38 high-resolution crystal structures.

Over the past few years a number of general statistical potentials have been proposed that are all based on a selection of protein–ligand complexes from the PDB.[7] In addition to the differences in reference state, which are detailed in the next section, there are additional differences in terms of atom typing, grid spacing and the selection and size of the database of structures the potentials are derived from. The various statistical potentials are named PMF,[8] DrugScore,[9] BLEEP,[10] SMoG,[11] ASP,[12] and PLASS.[13] All are distance-dependent statistical pair potentials, apart from SMoG, which is a coarse-grained contact potential, although a newer version[14] introduced two distance ranges. In addition to the distance-dependent atom pair potentials, DrugScore contains a knowledge-based solvent-accessible surface (SAS) dependent term. This potential captures the tendency of protein and ligand atoms to become buried upon formation of a protein–ligand complex, rewarding burial of certain atom types, and penalizing the burial of others.

6.3 Group-based Potentials

Group-based approaches seek to derive statistical preferences for interactions between functional groups rather than atoms. As a result, these methods conserve information about the directionality of interactions. Instead of RDFs or potentials, the interaction database here consists of 3D distributions (or mathematical descriptions thereof) of the occurrence of a probe around a central functional group. Again these can be combined into probability maps or scoring functions.

A common technique to visualize these distributions is to use 3D 'scatter plots'. A scatter plot shows the distribution of one functional group 'A' (the 'probe', or 'contact group') around another functional group 'B'.[15] Klebe generated such scatter plots from the CSD[16] for groups 'A' that occur in proteins.[17] The scatter plots were then superimposed on the relevant functional groups on a protein structure, hence highlighting potential interaction sites.

It is often more convenient to convert scatter plots into 3D probability maps.[18] Singh *et al.*[19] developed a technique that was implemented in the X-SITE program.[20] In this approach, scatter plots based entirely on protein side-chain–side-chain contacts from 83 high-resolution protein structures were translated into propensity maps. For a given probe, X-SITE then superimposes these propensity maps onto a protein structure to produce a composite propensity map.

The Cambridge Crystallographic Data Centre (CCDC) produce a database, IsoStar,[21] in which non-bonded interactions are presented in the form of scatter plots that can be transformed into propensity maps. IsoStar contains scatter plots derived from the CSD and scatter plots originating from protein–ligand contacts observed in the PDB. An example of an IsoStar scatter plot and map is given in Figure 6.2. The computer program SuperStar[22–24] combines the

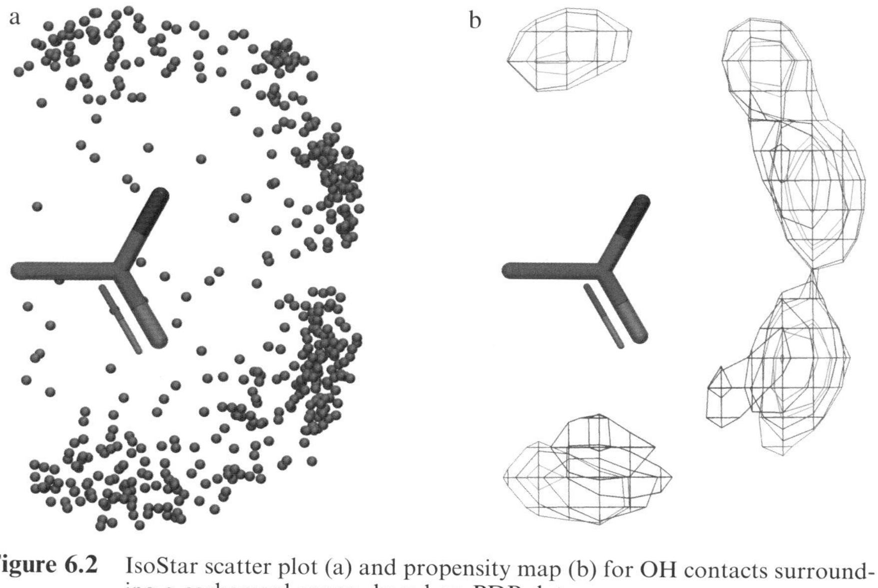

Figure 6.2 IsoStar scatter plot (a) and propensity map (b) for OH contacts surrounding a carbamoyl group, based on PDB data.

scatter plots from IsoStar into composite propensity maps (either CSD-based or PDB-based).

Instead of working with the 'raw' scatter plots (or the propensity distributions), methods have been proposed to derive mathematical descriptions of these distributions. Nissink and co-workers[25] used non-spherical Gaussian functions to describe all scatter plots in IsoStar. These functions were also implemented in SuperStar and were shown to speed up calculations with only a marginal loss of accuracy. Rantanen *et al.*[26,27] compiled a separate interaction library based entirely on protein–ligand contacts observed in the PDB. The interactions were first converted to scatter plots showing ligand probe atoms around protein fragments, and then to propensity maps, which in turn are described with mixtures of Gaussian distributions using a Bayesian approach.

Others have derived preferences for a number of orientational parameters from the 3D distributions. These parameters usually consist of one distance, and one or two angles describing, *e.g.* lone-pair directionality, and they are assumed to be independent. Such an approach is used in an early example of a knowledge-based application (HSITE[28,29]), which predicts favorable hydrogen-bonding regions on proteins, based on hydrogen-bond geometries observed in the CSD. The program AQUARIUS[30] uses a similar approach to this end, using structural data from the PDB. Chemical Computing Group (CCG) have developed a method for calculating 'probabilistic receptor potentials'[31] derived from the non-bonded contacts observed in all PDB structures with resolution ≤ 2.0 Å. Like the HSITE program, the CCG approach characterizes the geometry of each non-bonded contact by up to three spherical parameters that are assumed to be independent. All distributions of the three parameters are then fitted using various mathematical functions, and combined to produce composite probability maps for the receptor.

6.4 Methodologies

The derivation of atom-based statistical potentials starts off by converting atom–atom distributions in a database of structures into RDFs:

$$g(i,j,r) = \frac{n_{obs}(i,j,r)}{n_{exp}(i,j,r)} \tag{6.1}$$

where $n_{obs}(i,j,r)$ is the number of contacts between protein atoms of type i and ligand atoms of type j observed at separation r, and $n_{exp}(i,j,r)$ is the expected number of contacts between atom types i and j at distance r. In group-based potentials, a similar expression is used, but g is generally referred to as the 'contact propensity', and is dependent on additional (3D) geometric parameters. Particularly in atom-based potentials, the RDFs are usually converted into pseudo-potentials via:

$$StatScore(i,j,r) = -\ln g(i,j,r) \tag{6.2}$$

6.4.1 The Reference State

In eqn (6.1), the expected number of contacts, $n_{exp}(i,j,r)$, refers to a hypothetical state of no interaction, which is often referred to as 'the reference state'. The atom-based potentials that have been published in the literature differ in their definition of the reference state, and we discuss the different approaches below. One approach is to use the average contact density for each atom pair as its reference state. This is exactly what is used in PMF,[8] where the expected number of contacts between atom types i and j is calculated as:

$$n_{exp}^{PMF}(i,j,r) = \frac{\sum_{r'=0}^{R_{max}} n_{obs}(i,j,r')}{(4/3)\pi R_{max}^3 \cdot F(j, R_{max})} \cdot 4\pi r^2 \Delta r \cdot f(j,r) \tag{6.3}$$

where Δr is the bin width used and R_{max} is the maximum tabulated distance between protein and ligand atoms. The terms $f(j,r)$ and $F(j,R_{max})$ are ligand-dependent volume corrections, explained in more detail below.

DrugScore[9] adds an extra normalization to this. An RDF $g(i,j,r)$ is not used straightaway, but instead it is divided by the mean RDF $g(r)$. The score is then defined as $-\ln(g(i,j,r)/g(r))$. For comparison to the other methods we rewrite this to:

$$n_{exp}^{DrugScore}(i,j,r) = \sum_{r'=0}^{R_{max}} \left(\frac{n_{obs}(i,j,r')}{4\pi r'^2 \Delta r} \right) \cdot 4\pi r^2 \Delta r \cdot C(r) \tag{6.4}$$

for the expected number of contacts in DrugScore, with:

$$C(r) = \frac{1}{(I * J)} \cdot \sum_{i'} \sum_{j'} \frac{n_{obs}(i',j',r)/4\pi r^2 \Delta r}{\sum_{r'=0}^{R_{max}} (n_{obs}(i',j',r')/4\pi r'^2 \Delta r)} \tag{6.5}$$

The division by the mean RDF results in the term $C(r)$, which is the mean over all atom-type combinations of the ratio of the contact density at distance r and its average. It describes for the mean atom pair how much more or fewer contacts are expected at that distance than based on the overall average contact density.

BLEEP[10] uses the average over all observations for all atom pairs, instead of the mean RDF. It also includes a sparse-data correction term, which merely flattens the potential for atom pairs with few observations, resulting in a potential that is 0 everywhere for atom pairs without any observations. Ignoring this correction, the expected number of contacts in BLEEP is defined as:

$$n_{exp}^{BLEEP}(i,j,r) = \left(\sum_{r'=0}^{r=R_{max}} n_{obs}(i,j,r') \Big/ \sum_{i'} \sum_{j'} \sum_{r'=0}^{r'=R_{max}} n_{obs}(i',j',r') \right) \times \sum_{i'} \sum_{j'} n_{obs}(i',j',r) \tag{6.6}$$

In words, the expected number of contacts for a pair *i,j* at distance r is now based on the total number of contacts at that distance, multiplied by the fraction of contacts in the database that is of the combination *i,j*. A recent study[32] proposed a random reference state as the state of no interaction. This reference state was derived by random permutation of atoms in the database of protein–ligand complexes.

In group-based methods, generally only groups in contact with (or at least close to) the central groups are stored in the distributions. As a result, the average contact density cannot be used as the reference state. Instead, $n_{exp}(i,j)$ can be estimated from the number of contacts the two atoms and/or groups form with all other types in the database:

$$n_{exp}(i,j) = \frac{\sum_{i'} n_{obs}(i',j) \sum_{j'} n_{obs}(i,j')}{\sum_{i'} \sum_{j'} n_{obs}(i',j')} \tag{6.7}$$

This method properly normalizes for the fact that more contacts will be observed for an atom and/or group type with a higher abundance in the database. This normalization method was used by X-SITE,[20] for the PDB-based propensity maps in SuperStar,[22] and in SMoG2001.[14] It is also part of the reference state of PLASS.[13]

6.4.2 Volume Corrections

As stated above, PMF's reference state corrects for the volume occupied by the ligand.[33] The reasoning behind this is that not all the volume around a ligand atom *j* is actually available to a protein atom *i*, as protein atoms cannot occupy space already taken by ligand atoms. As a result, fewer contacts can be expected at those distances. This effect had previously been recognized in a molecular dynamics study.[34] To correct for it, one should use the available volume, instead of the full volume, of a spherical shell in the calculation of $n_{exp}(i,j,r)$ Hence the multiplication in eqn (6.3) by $f(j,r)$, which is the fraction of the spherical shell at distance *r* around a ligand atom of type *j* that is available to protein atoms. $f(j,r)$ equals 0 for very short distances, rising to 1 for long distances. A similar correction term, $F(j,R_{max})$, is included in the calculation of the average contact density. $F(j,R_{max})$ is the fraction of the complete sphere of radius R_{max} that is available to protein atoms.

Note that DrugScore and BLEEP correct for the less-than-average number of observations at short range, even though they do not contain an explicit ligand atom-type dependent volume correction term: their reference states are based on the number of observations for all atom pairs at a particular distance. At short range all pairs will exhibit a lower-than-average contact density, as a result of volume already occupied by bonded and neighboring atoms. Hence, an overall correction for the reduced available volume at shorter atomic separation is incorporated.

The argument to correct for the volume available to protein atoms around ligand atoms can also be applied to the volume available to ligand atoms around protein atoms. After all, *e.g.* around a side-chain lysine NH_3 there tends to be more space available for ligand atoms than around a backbone NH. The expected number of contacts should be corrected for such differences.

ASP[12] incorporates available volume corrections for protein as well as for ligand atoms. Here, the expected number of contacts for a given atom pair *i* and *j* is defined as:

$$n_{exp}^{ASP}(i,j,r) = \left\langle \frac{n_{obs}(i,j,r')}{f_p(i,r')f_l(j,r')4\pi r'^2 \Delta r} \right\rangle_{r'=6.0}^{r'=8.0} \cdot f_p(i,r) \cdot f_l(j,r) \cdot 4\pi r^2 \Delta r \quad (6.8)$$

The expected number of contacts is the product of an average contact density with the doubly corrected volume of a sphere shell at distance *r*. The average contact density is taken to be the long-range average instead of the overall average for the sphere with radius R_{max}. At this long range, atoms are thought to be making no specific interactions, and are therefore an appropriate choice to provide the reference state. This also ensures that the scores will be close to 0 within this distance range. It was shown[12] that the protein volume correction is especially important for backbone atom types.

Other reference states were recently proposed[32] that incorporate corrections dependent either on the protein atom types or on the ligand atom types. These corrections are based on the observed propensities for any ligand atom around each protein atom type, or on those for any protein atom around each ligand atom type.

6.5 Applications

6.5.1 Visualization and Interaction 'Hot Spots'

The interaction fields produced from knowledge-based potentials can be used to identify interaction 'hot spots' to guide the optimization of lead compounds. In Figure 6.3 a propensity map for HIV-protease is displayed. The donor and acceptor 'hot spots' correspond well to the experimental binding mode for the ligand. It has been shown that techniques like SuperStar and DrugScore are able to identify ligand interaction sites quite successfully.[23,35] CCG's probabilistic receptor potentials[31] were used in the analysis of a number of cyclooxygenase-2 (COX-2) inhibitors.[36] Interaction hot spots derived from DrugScore, SuperStar and GRID were recently combined to define a pharmacophore pattern that was used in the identification of novel inhibitors of human carbonic anhydrase.[37] Using a similar approach, the same research group identified sub micro-molar leads against tRNA-guanine transglycosylase[38] and low micro-molar leads against aldose reductase.[39]

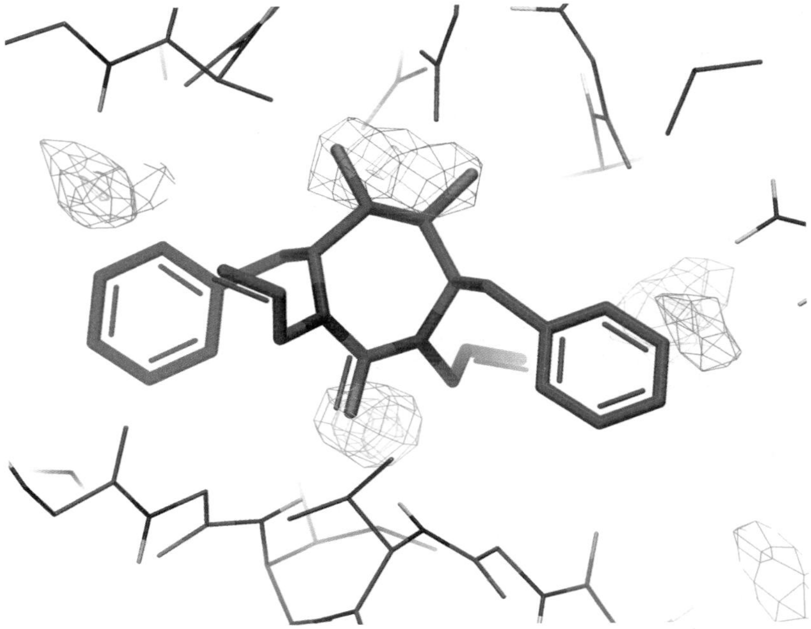

Figure 6.3 SuperStar propensity maps for donor (blue) and acceptor (red) contacts for HIV-protease (PDB-code 1hwr). Maps generated using an implementation of the SuperStar methodology in AstexViewer™.[73]

6.5.2 Docking and Scoring

DrugScore, PMF and ASP have all been implemented as energy functions in protein–ligand docking programs. PMF was implemented as a scoring function in DOCK,[40] and improved success rates for docking were reported for matrix metalloproteinase-3 (MMP3) inhibitors.[41] When DrugScore was tested in AutoDock,[42] a success rate of 44% was achieved for flexible docking of a test set of 41 complexes,[43] similar to the success rate achieved with the original AutoDock function; the success rate is defined as the percentage of complexes for which the top-ranked solution was within 2.0 Å root mean square deviation (RMSD) of the experimental binding mode. On the CCDC/Astex validation set[44] DrugScore-like pair potentials were seen to result in significantly lower success rates than Chemscore or Goldscore.[45] ASP, on the other hand, gave success rates similar to those two scoring functions on the same test set.[12] An example of a docking result using ASP is displayed in Figure 6.4, where also the various 'hot spots' in the ASP maps are depicted.

Knowledge-based potentials have also been used to re-score and re-rank the dockings produced by a docking program. Re-scoring approaches are a form of consensus docking and have been shown to improve docking success rates.[46–48]

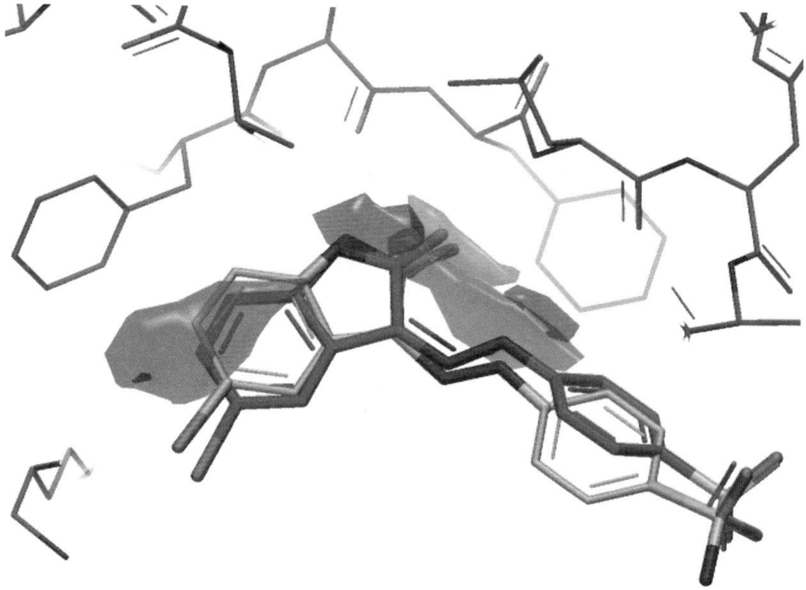

Figure 6.4 Predicted binding mode for the cdk2 ligand from PDB-entry 1fvt (docked against the protein from PDB-entry 1di8). This is the top-ranked solution produced by GOLD, using the ASP potential.[12] In orange is the overlaid experimental ligand structure (root mean squared deviation: 0.85 Å). In green is displayed the ASP field for the C.ar atom type, contoured at −9.5, in red the O.2 field (−14.0) and in blue the N.c2 field (−15.0); (picture created with AstexViewer™).[73]

DrugScore was shown to improve docking success rates when it was used to re-score docking poses generated by FlexX[49] and DOCK;[9,35] BLEEP was used to re-score DOCK solutions for 15 protein–ligand complexes and provided a small improvement in the docking accuracy.[50] Recently, SuperStar fields were added as an extra term (another form of consensus docking) to the Goldscore function in GOLD,[51,52] and improved binding-mode predictions were observed.[53]

All statistical potentials have been tested for their ability to estimate binding affinities for known ligands. For all potentials, reasonable correlation between experimental binding affinity and the statistical score has been reported.[8,10,13,14,35] However, correlation can vary wildly over different protein classes. For example, for DrugScore the R^2 varies from 0.22 for arabinose-binding proteins to 0.86 for serine proteases.[35] A similar spread of R^2 over various classes was reported for PMF.[8] As a result, comparing the various reported R^2 values is fairly meaningless unless they refer to exactly the same dataset, and unfortunately many different test sets are used. In an evaluation of a large selection of scoring functions,[54] the regression-based scoring function Chemscore[55] was reported to give the best correlation with binding affinities for a large test set of protein–ligand complexes, clearly outperforming knowledge-based as well as empirical scoring functions.

Many studies have investigated the performance of knowledge-based potentials in virtual screening.[12,45,54,56,57] Not surprisingly, results vary strongly with the protein target. PMF was reported to work well for neuraminidase[57] and thymidine kinase,[58] DrugScore for the estrogen receptor, especially for its agonists,[45,57] but for many other targets these functions behaved poorly. In a study comparing three implementations of the reference state, ASP was seen to give the most consistent results across a range of targets;[12] nevertheless, for each target studied, either Goldscore or Chemscore performed better than the three knowledge-based potentials tested.

A limited number of true drug-discovery applications using knowledge-based scoring functions in virtual screening have appeared in the literature. In three separate applications, DrugScore was used by Klebe and colleagues to rank and select compounds from virtual screening, which led to lead compounds against carbonic anhydrase,[37] aldose reductase[39] and tRNA-guanine transglycosylase.[38] Recently, Evers and Klabunde[59] used the PMF scoring function to rank dockings produced by GOLD in a virtual screening application against a homology model of the alpha 1A receptor, which lead to the identification of nM inhibitors against this receptor.

6.5.3 *De Novo* Design

The scatter plots described by Klebe[17] were, in a modified form, used in the *de novo* design program LUDI to guide the placement of functional groups.[60] LUDI is now a widely used structure-based design tool, and various applications have been described in the literature.[61] SMoG uses a metropolis Monte Carlo molecular growth algorithm in combination with a knowledge-based scoring function.[11] This software was used to predict novel binders for the Src SH3 domain and for CD4,[62] but the predicted compounds were not tested for their potencies against these targets. CombiSMoG, a reincarnation of the SMoG program, was later used to identify highly potent inhibitors for human carbonic anhydrase II.[63]

6.5.4 Targeted Scoring Functions

Since statistical scoring functions are based on known structures, they seem ideally suited to serve as a basis for targeted scoring functions. Indeed, the first application of statistical potentials to protein–ligand complexes was a targeted potential derived only from HIV complexes.[4] More recently, Gohlke and Klebe developed a quantitative structure–activity relationship (QSAR) procedure to tailor DrugScore potential maps based on affinities for protein–ligand complexes.[64] The procedure starts by masking the protein's DrugScore fields with the positions of the atom types in the known binders, followed by a QSAR adaptation of these fields to the known affinities. Starting from potential fields centered on the protein instead of the ligand they brand the method Adaptation of Fields for Molecular Comparison (AFMoC), *i.e.* 'reverse' Comparative Molecular Field Analysis (CoMFA). An increased predictive power for affinity prediction was demonstrated for thermolysin and glycogen phosphorylase b, as well as for Factor Xa,[65] and DOXP-reductoisomerase.[66]

We recently described how experimental structural data alone, without any affinity data, can be used to develop targeted scoring functions.[12] Statistical potentials are based on a database consisting of only protein–ligand complexes for the targeted scoring function under consideration, combined with the general database of all protein–ligand complexes in the PDB. These two sources of data are mixed, dependent on the amount of data present in the target specific database. Four increasingly large target-specific databases were used for the target cdk2. Using these targeted scoring functions, docking success rates as well as enrichments are significantly better than for the general ASP scoring function. Results improve with the amount of structures used in the construction of the target scoring functions, thus illustrating that these targeted ASP potentials can be continuously improved as new structural data become available.

6.6 Discussion

Ramachandran and coworkers[1] were the first to suggest that the geometric distributions derived from crystal structures should follow the Boltzmann distribution. However, this theory was later criticized by Burgi and Dunitz,[67] who argued that a set of crystal structures does not represent a canonical ensemble of geometric states as there is no interaction (*i.e.* equilibrium) between the structures. More recently, Grzybowski *et al.*[68] claimed to have shown that geometry distributions derived from a set of independent crystal structures do follow the Boltzmann distribution and that true energies can be derived from such distributions. In our opinion, it is not hugely important to answer the question whether the Boltzmann distribution applies or not. What all authors *do* appear to agree on is that the geometry distributions derived from crystal structures roughly correlate with the relative energies of the different geometries.

Various methods have been suggested in the literature to combine the individual atom-based or group-based potentials to produce a composite potential map for a complete binding site on a protein. In all atom-based potentials, the individual pseudo-potentials are simply summed. This seems a reasonable approach, particularly if the pseudo-potentials are regarded as (pseudo-)energies. In X-SITE, for each position in the binding site, the normalized contact densities (which are not converted into pseudo-potentials) are averaged over all the contributing (*i.e.* nearby) fragments to generate the composite propensity maps. The CCG receptor potentials take the maximum contact propensity from the groups surrounding each grid point. In SuperStar the contact propensities are multiplied to generate the overall map; this is equivalent to summing (pseudo-)energies, which intuitively seems the most obvious approach. However, a reason for not multiplying contact propensities is that, unless great care is taken with overlapping fragments, 'double counting' can occur.

Atom-based potentials and group-based methods in which the central functional groups are small are sensitive to 'double counting'. Imagine a ligand carbon atom in contact with a protein phenyl ring. When the complete phenyl ring is considered as the central group, only the score for a single contact to a phenyl group will be added. When an atom-based method is used, six scores, for contacts to all phenyl carbon atoms, will be added. This is related to the problem of 'secondary contacts' that can occur in the atom-based RDFs or group-based propensity maps. For example, the RDF for contacts between two amide oxygen atoms shows a strong peak around 4.9 Å. This peak is not caused by a direct interaction between the oxygen atoms, but by the adjacent amide NH forming a hydrogen bond to an amide oxygen atom. This leads to double counting, as the secondary contact is scored in addition to the primary contact that is causing it. To address such problems, apart from a distance cut-off, CCG used several additional rules to define non-bonded contacts; among these rules is a 'line-of-sight' check to ensure that there are no atoms 'between' the two atoms forming the contact. Nissink and Taylor applied a distance-based filter on the IsoStar scatter plots to remove secondary contacts.[69] Another possible solution to this problem would be to generate the Voronoi polyhedra for the protein–ligand complexes and only include atom pairs that have contact 'face' between them.

Compared to atom-based potentials, some of the above issues are less problematic for group-based potentials, particularly if the central groups used are relatively large. Another advantage of group-based methods is that they contain information on the directionality of interactions, which is lost in atom–atom methods, although Gohlke *et al.* argue that the directionalities are implicitly contained in the RDFs.[9] However, group-based potentials are much more likely to suffer from a lack of data, and hence a low statistical significance. Even for atom-based potentials, lack of data can be an issue, particularly for less common atom types. It is therefore important to ensure that sparsely populated atom-based RDFs or group-based scatter plots do not translate into noisy maps; this can be achieved straightforwardly by applying a minimum-number-of-observations cut-off, or by using a sparse-data correction.[3] Also, it is important to achieve the correct balance between the number of defined atom types and the amount of data: when two atom types are merged, the resulting potential will be the average of the two separate potentials. Therefore, all atoms combined in one type should be similar in order not to loose valuable information in averaging.

For distributions derived from protein–ligand complexes, automatic assignment of atom types is challenging. Various approaches have been described,[70,71] but certain atom types remain problematic, *e.g.* it is not always clear whether an imidazole nitrogen is a donor or an acceptor. When donor and acceptor atoms are combined in one atom type, this will result in weaker donor or acceptor features in the derived potential. When CSD data are used this problem is overcome because the bond types are stored in this database. There are several other advantages to the use of CSD data: (i) the hydrogen atoms are present in most structures and could, in theory, be used in the potentials; (ii) the data precision is usually high; (iii) it covers a more diverse range of chemical types than the PDB. On the other hand, the use of CSD data may appear less biologically relevant to protein–ligand complexes. Boer *et al.*[22] showed that interaction geometries in the CSD are similar to those observed in protein–ligand interfaces, but that the relative propensities of different types of interactions are quite different; lipophilic interactions are more common in the PDB and polar interactions occur more often in the CSD. In SuperStar, this was initially addressed in a rather crude fashion by up-weighting CSD-based propensity maps for interactions deemed lipophilic.[23] Recently, Nissink and Taylor[69] used a much more elegant approach to correct the CSD-based maps, based on octanol–water partition values.

6.7 Conclusion

Knowledge-based potentials are useful tools for structure-based drug design, in similar ways to equivalent theoretical approaches like GRID.[72] However, in terms of docking and scoring, knowledge-based potentials are yet to provide significant improvements over existing empirical and force-field-based scoring functions. It needs to be pointed out, though, that other (non-knowledge-based) scoring functions have not provided any major breakthroughs in recent years either. Knowledge-based scoring functions do provide a natural mechanism to implement targeted scoring functions and this could well prove to be the key strength of these potentials in future applications.

References

1. R. Balasubramanian, R. Chidambaram and G. N. Ramachandran, *Biochim. Biophys. Acta*, 1970, **221**, 196.
2. R. Chidambaram, R. Balasubramanian and G. N. Ramachandran, *Biochim. Biophys. Acta*, 1970, **221**, 182.
3. M. J. Sippl, *J. Mol. Biol.*, 1990, **213**, 859.
4. G. M. Verkhivker, K. Appelt, S. T. Freer and J. E. Villafranca, *Protein Eng.*, 1995, **8**, 677.
5. A. Wallqvist, R. L. Jernigan and D. G. Covell, *Protein Sci.*, 1995, **4**, 1881.
6. A. Wallqvist and D. G. Covell, *Proteins*, 1995, **25**, 403.
7. H. M. Berman, J. Westbrook, Z. Feng, G. Gilliland, T. N. Bhat, H. Weissig, I. N. Shindyalov and P. E. Bourne, *Nucleic Acids Res.*, 2000, **28**, 235.
8. I. Muegge and Y. C. Martin, *J. Med. Chem.*, 1999, **42**, 791.
9. H. Gohlke, M. Hendlich and G. Klebe, *J. Mol. Biol.*, 2000, **295**, 337.
10. J. B. O. Mitchell, R. A. Laskowski, A. Alex, M. J. Forster and J. M. Thornton, *J. Comput. Chem.*, 1999, **20**, 1177.

11. R. S. Dewitte and E. I. Shakhnovich, *J. Am. Chem. Soc.*, 1996, **118**, 11733.
12. W. T. M. Mooij and M. L. Verdonk, *Proteins*, 2005, **61**, 272.
13. V. D. Ozrin, M. V. Subbotin and S. M. Nikitin, *J. Comput. Aided Mol. Des.*, 2004, **18**, 261.
14. A. V. Ishchenko and E. I. Shakhnovich, *J. Med. Chem.*, 2002, **45**, 2770.
15. R. E. Rosenfield and P. Murray-Rust, *J. Am. Chem. Soc.*, 1982, **104**, 5427.
16. F. H. Allen, *Acta Crystallogr. Sect. B: Struct. Sci.*, 2002, **58**, 380.
17. G. Klebe, *J. Mol. Biol.*, 1994, **237**, 212.
18. R. E. Rosenfield, S. M. Swanson, J. Meyer, H. L. Carrell and P. Murray-Rust, *J. Mol. Graphics.*, 1984, **2**, 43.
19. J. Singh, J. Saldanha and J. M. Thornton, *Protein Eng.*, 1991, **4**, 251.
20. R. A. Laskowski, J. M. Thornton, C. Humblet and J. Singh, *J. Mol. Biol.*, 1996, **259**, 175.
21. I. J. Bruno, J. C. Cole, J. P. Lommerse, R. S. Rowland, R. Taylor and M. L. Verdonk, *J. Comput. Aided Mol. Des.*, 1997, **11**, 525.
22. D. R. Boer, J. Kroon, J. C. Cole, B. Smith and M. L. Verdonk, *J. Mol. Biol.*, 2001, **312**, 275.
23. M. L. Verdonk, J. C. Cole and R. Taylor, *J. Mol. Biol.*, 1999, **289**, 1093.
24. M. L. Verdonk, J. C. Cole, P. Watson, V. Gillet and P. Willett, *J. Mol. Biol.*, 2001, **307**, 841.
25. J. W. M. Nissink, M. L. Verdonk and G. Klebe, *J. Comput. Aided Mol. Des.*, 2000, **14**, 787.
26. V. V. Rantanen, K. A. Denessiouk, M. Gyllenberg, T. Koski and M. S. Johnson, *J. Mol. Biol.*, 2001, **313**, 197.
27. V. V. Rantanen, M. Gyllenberg, T. Koski and M. S. Johnson, *J. Comput. Aided Mol. Des.*, 2003, **17**, 435.
28. D. J. Danziger and P. M. Dean, *Proc. R. Soc. Lond., Ser. B: Biol. Sci.*, 1989, **236**, 115.
29. D. J. Danziger and P. M. Dean, *Proc. R. Soc. Lond., Ser. B: Biol. Sci.*, 1989, **236**, 101.
30. W. R. Pitt and J. M. Goodfellow, *Protein Eng.*, 1991, **4**, 531.
31. P. Labute, 2005, *Probabilistic Receptor Potentials*, Chemical Computing Group, http://www.chemcomp.com/Journal_of_CCG/Features/cstat.htm.
32. A. M. Ruvinsky and A. V. Kozintsev, *Biophys. Chem.*, 2005, **115**, 255.
33. I. Muegge, *J. Comput. Chem.*, 2001, **22**, 418.
34. T. Astley, G. G. Birch, M. G. B. Drew, P. M. Rodger and R. H. Wilden, *J. Comput. Chem.*, 1998, **19**, 363.
35. H. Gohlke, M. Hendlich and G. Klebe, *Perspect. Drug Discovery Des.*, 2000, **20**, 115.
36. G. Ermondi, G. Caron, R. Lawrence and D. Longo, *J. Comput. Aided Mol. Des.*, 2004, **18**, 683.
37. S. Gruneberg, M. T. Stubbs and G. Klebe, *J. Med. Chem.*, 2002, **45**, 3588.
38. R. Brenk, L. Naerum, U. Gradler, H. D. Gerber, G. A. Garcia, K. Reuter, M. T. Stubbs and G. Klebe, *J. Med. Chem.*, 2003, **46**, 1133.
39. O. Kraemer, I. Hazemann, A. D. Podjarny and G. Klebe, *Proteins*, 2004, **55**, 814.

40. T. J. A. Ewing, S. Makino, A. G. Skillman and I. D. Kuntz, *J. Comput. Aided Mol. Des.*, 2001, **15**, 411.
41. S. Ha, R. Andreani, A. Robbins and I. Muegge, *J. Comput. Aided Mol. Des.*, 2000, **14**, 435.
42. D. S. Goodsell and A. J. Olson, *Proteins*, 1990, **8**, 195.
43. C. A. Sotriffer, H. Gohlke and G. Klebe, *J. Med. Chem.*, 2002, **45**, 1967.
44. J. W. M. Nissink, C. W. Murray, M. J. Hartshorn, M. L. Verdonk, J. C. Cole and R. Taylor, *Proteins*, 2002, **49**, 457.
45. M. L. Verdonk, V. Berdini, M. J. Hartshorn, W. T. M. Mooij, C. W. Murray, R. D. Taylor and P. Watson, *J. Chem. Inf. Comput. Sci.*, 2004, **44**, 793.
46. D. Hoffmann, B. Kramer, T. Washio, T. Steinmetzer, M. Rarey and T. Lengauer, *J. Med. Chem.*, 1999, **42**, 4422.
47. N. Paul and D. Rognan, *Proteins*, 2002, **47**, 521.
48. M. L. Verdonk, J. C. Cole, M. Hartshorn, C. W. Murray and R. D. Taylor, *Proteins*, 2003, **52**, 609.
49. M. Rarey, B. Kramer, T. Lengauer and G. Klebe, *J. Mol. Biol.*, 1996, **261**, 470.
50. I. Nobeli, J. B. O. Mitchell, A. Alex and J. M. Thornton, *J. Comput. Chem.*, 2001, **22**, 673.
51. G. Jones, P. Willett and R. C. Glen, *J. Mol. Biol.*, 1995, **245**, 43.
52. G. Jones, P. Willett, R. C. Glen, A. R. Leach and R. Taylor, *J. Mol. Biol.*, 1997, **267**, 727.
53. J. W. M. Nissink, 2005, personal communication.
54. P. Ferrara, H. Gohlke, D. J. Price, G. Klebe and C. L. Brooks III, *J. Med. Chem.*, 2004, **47**, 3032.
55. M. D. Eldridge, C. W. Murray, T. R. Auton, G. V. Paolini and R. P. Mee, *J. Comput. Aided Mol. Des.*, 1997, **11**, 425.
56. I. Muegge, Y. C. Martin, P. J. Hajduk and S. W. Fesik, *J. Med. Chem.*, 1999, **42**, 2498.
57. M. Stahl and M. Rarey, *J. Med. Chem.*, 2001, **44**, 1035.
58. C. Bissantz, G. Folkers and D. Rognan, *J. Med. Chem.*, 2000, **43**, 4759.
59. A. Evers and T. Klabunde, *J. Med. Chem.*, 2005, **48**, 1088.
60. H. J. Bohm, *J. Comput. Aided Mol. Des.*, 1992, **6**, 61.
61. H. J. Boehm, M. Boehringer, D. Bur, H. Gmuender, W. Huber, W. Klaus, D. Kostrewa, H. Kuehne, T. Luebbers, N. Meunier-Keller and F. Mueller, *J. Med. Chem.*, 2000, **43**, 2664.
62. R. S. Dewitte, A. V. Ishchenko and E. I. Shakhnovich, *J. Am. Chem. Soc.*, 1997, **119**, 4608.
63. B. A. Grzybowski, A. V. Ishchenko, C. Y. Kim, G. Topalov, R. Chapman, D. W. Christianson, G. M. Whitesides and E. I. Shakhnovich, *Proc. Natl. Acad. Sci. U. S. A.*, 2002, **99**, 1270.
64. H. Gohlke and G. Klebe, *J. Med. Chem.*, 2002, **45**, 4153.
65. H. Matter, D. W. Will, M. Nazare, H. Schreuder, V. Laux and V. Wehner, *J. Med. Chem.*, 2005, **48**, 3290.
66. K. Silber, P. Heidler, T. Kurz and G. Klebe, *J. Med. Chem.*, 2005, **48**, 3547.

67. H. B. Burgi and J. D. Dunitz, *Acta Crystallogr. Sect. B: Struct. Sci.*, 1988, **44**, 445.
68. B. A. Grzybowski, A. V. Ishchenko, R. S. Dewitte, G. M. Whitesides and E. I. Shakhnovich, *J. Phys. Chem. B*, 2000, **104**, 7293.
69. J. W. M. Nissink and R. Taylor, *Org. Biomol. Chem.*, 2004, **2**, 3238.
70. M. Hendlich, F. Rippmann and G. Barnickel, *J. Chem. Inf. Comput. Sci.*, 1997, **37**, 774.
71. R. Sayle, 2001, *PDB: Cruft to content (Perception of molecular connectivity from 3D coordinates)*. Daylight user meeting MUG01 (http://www.daylight.com/meetings/mug01/Sayle/m4xbondage.html).
72. P. J. Goodford, *J. Med. Chem.*, 1985, **28**, 849.
73. M. J. Hartshorn, *J. Comput. Aided Mol. Des.*, 2002, **16**, 871.

CHAPTER 7

Combating Drug Resistance – Identifying Resilient Molecular Targets and Robust Drugs

CELIA A. SCHIFFER

Department of Biochemistry and Molecular Pharmacology, University of Massachusetts Medical School, Worcester, Massachusetts, USA

7.1 Introduction

Drug resistance is a major obstacle in modern medicine. The occurrence of drug resistance negatively impacts the lives of millions of patients by limiting the longevity of many of our most promising new drugs. Resistance occurs any time rapid growth and evolution exists under the selective pressure of a drug that, while restricting growth, is not entirely inhibiting it. Therefore most antibiotic,[1] anti-viral,[2,3] anti-malarial,[4,5] anti-tuberculosis[6] and chemotherapeutic[7] agents become obsolete before their time.

Amazingly, modern drug design rarely considers and may inadvertently be facilitating the occurrence of resistance. Most novel drugs are discovered and developed by utilizing a high-throughput screen with an inhibition assay in combination with medicinal chemistry and possibly structure-based design. These techniques, however, do not focus on the biologically relevant interactions of the macromolecular target, but rather focus only on disrupting the target's activity. Disrupting the target's activity is necessary but not sufficient for developing a drug that is *robust* against resistance.

The rationale for why modern drug design may facilitate drug resistance is that, by not focusing on the detailed atomic basis for the target's macromolecular function, many of the inhibitors found by traditional drug design are likely to contact regions within the target that can mutate and confer resistance without significantly impairing function. Rather than developing better strategies for inhibitor design, current efforts to combat resistance are primarily limited to developing a larger repertoire of inhibitors, each of which in turn can lose its effectiveness because of drug resistance. To reduce susceptibility to drug resistance and thus design new *robust* inhibitors, a new strategy should be adopted. This strategy would involve the identification of key biological macromolecular targets that are chosen because of their unique position in a functional pathway that would make them less susceptible to drug-resistant mutations and therefore more resilient. These *resilient macromolecular targets* must be identified and a detailed atomic understanding of how a particular target interacts with its functionally important partners is required in order to discover successfully those inhibitors that will be effectively *robust* against resistance.

7.2 Resilient Targets and Robust Drugs

To specifically discover inhibitors for pathogenic microbes or for invasive cancers that have the tendency to evolve quickly, both the macromolecular target and the specific inhibitor must be identified carefully. A biological macromolecular target that is likely to be *resilient* against resistance is one that cannot easily tolerate change and maintain function. Potentially *resilient targets* can likely be identified by meeting one of two criteria:

(i) Having diverse protein or nucleic acid substrates or binding partners. The rationale is that a macromolecule target that is functionally required to interact in many diverse yet overlapping ways could make it potentially more difficult to evolve resistance without disrupting function.
(ii) Having a relatively unique small molecule ligand whose recognition cannot easily be altered. Once again such a macromolecular target could not tolerate mutation and maintain the binding of its unique ligand, especially if such a ligand was a small organic molecule, which itself could not mutate.

Thus a resilient macromolecular target must have the potential of being unable to tolerate change.

However, identification of these potentially resilient targets alone is not enough. Ideally many high-resolution crystal structures of the macromolecular target would be available in a variety of functionally important biological conformations and complexes. Such structures would elucidate the structural basis for the target's biological function and would therefore allow the specific identification of critical regions within a binding site or at an interface that an inhibitor could bind. Specifically, a robust inhibitor or drug, defined as one that does not quickly lose effectiveness due to resistant mutations, would bind *only* to critical regions within the target's active site or binding interface that would be essential for function and thus intolerant to change. By identifying *resilient molecular targets* and exploring the atomic basis for their function a general paradigm can be developed for identifying *robust inhibitors* and thereby circumvent drug resistance.

7.3 Example of HIV-1 Protease: Substrate Recognition *vs.* Drug Resistance

This novel theory of resilient targets and robust drugs was developed from research in our laboratory on human immunodeficiency virus (HIV) protease. HIV protease is potentially a resilient target as it permits viral maturation by recognizing and cleaving nine diverse substrate sequences. Nevertheless, drug resistance in HIV protease has arisen widely.[2,3] This appears to be primarily an inadvertent consequence of not considering the atomic details of substrate recognition when the presently used inhibitors were developed.

HIV protease was the first macromolecular drug target for which structure-based drug design was extensively used to successfully complement more traditional methods of drug design.[8] Both academic and pharmaceutical groups have determined hundreds of crystal structures of inhibitor complexes with HIV-1 protease,[9] resulting in the current nine Food and Drug Administration (FDA) approved competitive, active-site protease inhibitors. However none of this research focused on how HIV-1 protease specifically recognized its diverse substrate sequences. We determined for the first time the crystal structures of a variety of substrate complexes with an inactive variant of HIV-1 protease.[10–13] Through a comparison of these substrate structures we discovered that substrate specificity appears to be due to a conserved shape (Figure 7.1a and 7.1b), which we defined as the "substrate envelope" that is conserved regardless of the diversity of the substrate sequence.[11] The molecular recognition dilemma is how can drug resistance occur against the competitive active-site protease inhibitors while retaining the ability for the protease to continue to recognize and cleave its nine substrates and therefore allow viral maturation.

The solution for this recognition dilemma became evident in investigating the structural mechanism for drug resistance of a common active site mutation, V82A, that confers multi-drug resistance affecting many of the protease inhibitors.[14] In determining the structures of three substrate and two inhibitor complexes with this variant, we found that the three substrates did not make extensive contact with Val82, while the inhibitors did. Thus when the valine mutates to an alanine the extensive van der Waals contact is selectively lost to the inhibitors, thereby directly impacting their binding affinity, while substrate recognition is retained.

In fact, although nine protease inhibitors are currently being utilized therapeutically, multi-drug resistance is common as eight of the nine inhibitors, although chemically different, occupy a very similar volume within the active site of HIV-1 protease (Figure 7.1c and 7.1d). Thus, for the most part, these inhibitors contact the active site in analogous manners and at similar residues. In fact, as was the case for Val82, most drug-resistant mutations in the active site of HIV protease occur where the inhibitors protrude beyond the region necessary for substrate recognition (Figure 7.1e).[11,12,15] These include other common sites of resistance (D30, G48, I50 and I84), as well as some newer sites that are becoming more prevalent with more recent inhibitors, such as L23[16] and V32.[17] Although mutations at residues outside the active site also contribute to drug resistance, these sites within the active site are considered the primary sites for drug resistance. Therefore, those residues that contact the inhibitors where they protrude from the substrates are more important for inhibitor binding than for substrate recognition and thus prime sites for resistant mutations to occur. Drug resistance occurs in a manner that solves

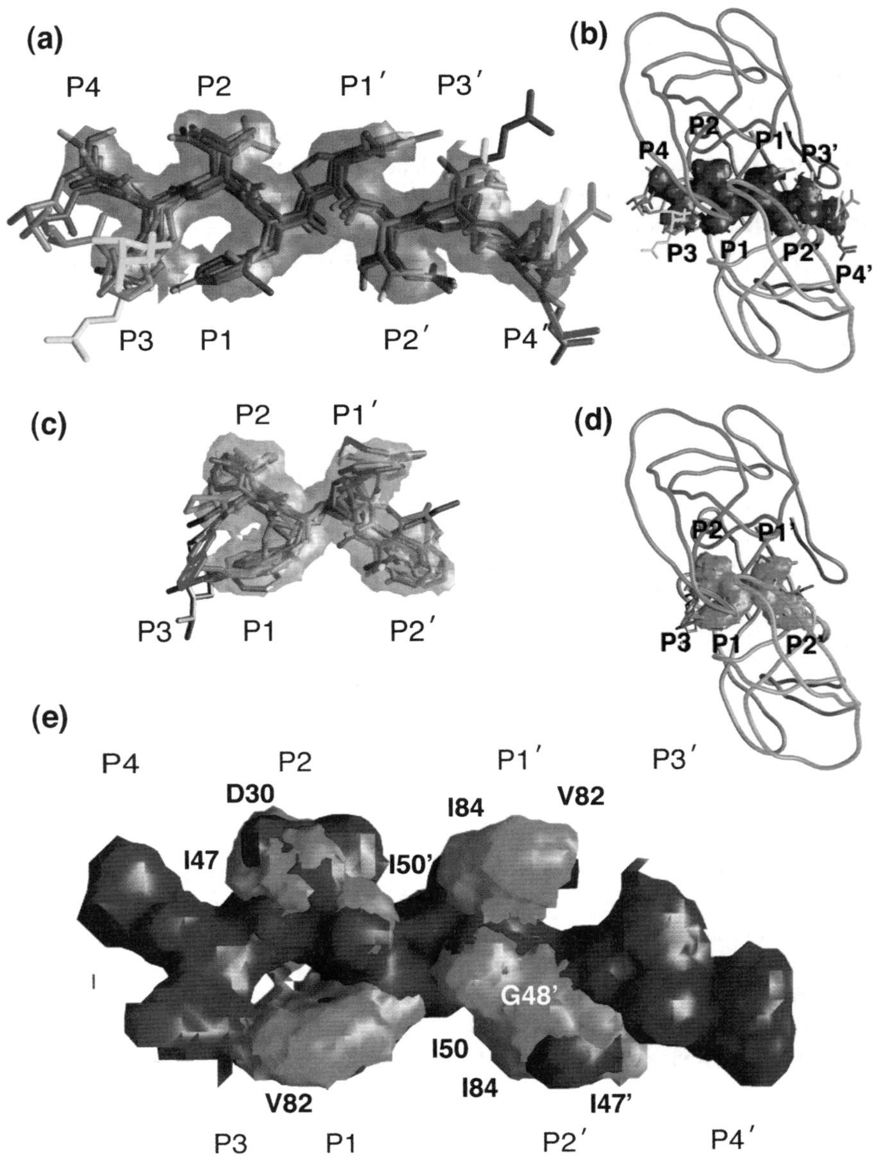

the molecular recognition dilemma by retaining substrate recognition and protease activity while weakening inhibitor binding.

Nevertheless, HIV-1 protease is still potentially a resilient drug target, as it recognizes and cleaves nine diverse substrate sequences. However, in order for a drug to be robust against resistance it needs to be positioned in a manner such that no mutation can occur that would impact the inhibitor recognition without simultaneously strongly impacting substrate recognition. Such an inhibitor of HIV-1 protease would be one that fits fully within the substrate envelope. contacting only those residues that are also necessary for substrate recognition. For such an inhibitor the only way a residue could directly confer resistance is if it simultaneously affected the recognition of more than half the substrates. While it may be possible for the virus to simultaneously co-evolve one or two of the substrate sites with the viral protease, the likelihood that four or more substrates would be able to co-evolve would be statistically much less probable. Such an inhibitor that fits within the substrate envelope would therefore be less susceptible to drug resistance and thereby take advantage of HIV protease's potential resilience as a macromolecular target.

While this paradigm for avoiding resistance seems good in theory, it also needs to be practical for new, more robust drugs to be developed. For HIV-1 protease this should be attainable as two of the more recent FDA-approved inhibitors amprenavir and darunavir both fit fairly well within the substrate envelope and have tight binding to the viral protease, subnanomolar and picomolar, respectively.[18] While both are less susceptible to most drug-resistant variants, darunavir, that was just FDA approved in June 2006,[19,20] is proving an effective therapy for patients who are otherwise failing their anti-viral protease inhibitor regime.[21] Thus a combination of high affinity and fitting

---

Figure 7.1 Substrate and inhibitor envelopes of HIV-1 protease. (a) The substrate envelope calculated with GRASP[22] from the overlapping van der Waals volume of four or more substrate peptides. The colors of the substrate peptides are: red is matrix–capsid, green is capsid–p2, blue is p2–nucleocapsid, cyan is p1–p6, magenta is reverse-transcriptase–ribonucleaseH, and yellow is RnaseH–integrase. (b) The substrate envelope[11] as it fits within the active site of HIV-1 protease. The alpha carbon trace is of the CA–p2 substrate peptide complex.[10] (c) The inhibitor envelope calculated from overlapping van der Waals volume of five or more of eight inhibitor complexes. The colors of the inhibitors are: yellow is nelfinavir (NFV), gray is saquinavir (SQV), cyan is indinavir (IDV), light blue is ritonavir (RTV), green is amprenavir (APV), magenta is lopinavir (LPV), blue is atazanavir (ATV) and red is darunavir (DRV). (d) The inhibitor envelope as it fits within the active site of HIV-1 protease. (e) Superposition of the substrate envelope (in blue) with the inhibitor envelope (in red). Residues that contact the inhibitors where the inhibitors protrude beyond the substrate envelope and confer drug resistance when they mutate are labeled. (Reprinted from *Chemistry & Biology*, Vol. 11, King *et al.*, Combating susceptibility to drug resistance: lessons from HIV-1 protease, pp. 1333–1338, Copyright (2004), with permission from Elsevier.[15])

within the substrate envelope is a practical method for developing an effective drug that is robust against resistance.[15] Although this theory was developed from studying HIV protease, this new paradigm for drug design is general and extends to many biological macromolecular targets, having broad implications for circumventing drug resistance.

7.4 Implications for Future Structure-based Drug Design

When structure-based drug design was first described nearly 20 years ago, it was generally thought that it would revolutionize the pharmaceutical industry, allowing for the rational design of inhibitors and thereby making other techniques obsolete. In fact very few drugs were successfully developed without initial input from more traditional avenues of drug design, and issues of bioavailability prevented completely novel scaffolds from being utilized as effective therapeutics. The practical reality is that the role of structural biology for most inhibitor development is in terms of structure-assisted drug design, where crystallographic inhibitor complexes complement computational modeling efforts, working side-by-side with medicinal chemists. This collaboration is effective for developing inhibitors into drugs, but it does not actively develop inhibitors that should be more robust against drug resistance. In order to develop more effective inhibitors that are less susceptible to drug design a more detailed focus on the structural basis for biological function is necessary. These structures are needed in order to evaluate the potential of a macromolecular target to be resilient to drug resistance. With the increase in resistance to so many of our new drugs, the pharmaceutical and medical community can no longer consider determining biologically interesting complexes as strictly of "academic" interest. Rather, determination of more biologically relevant complexes of macromolecular target are a necessary step in the development of inhibitors that are robust against drug resistance.

Acknowledgements

This research was supported by the National Institutes of Health (P01-GM66524 and R01-GM64347).

References

1. M. S. Wilke, A. L. Lovering and N. C. Strynadka, *Curr. Opin. Microbiol.*, 2005, **8**, 525.
2. R. W. Shafer and J. M. Schapiro, *J. Antimicrob. Chemother.*, 2005, **55**, 817.
3. R. W. Shafer, *J. Infect. Dis.*, 2006, **194**(Suppl. 1), S51.
4. J. P. Daily, *J. Clin. Pharmacol.*, 2006, **46**, 1487.
5. A. R. Renslo and J. H. McKerrow, *Nat. Chem. Biol.*, 2006, **2**, 701.
6. M. Strong and C. W. Goulding, *Methods Biochem. Anal.*, 2006, **49**, 245.

7. H. Ji, N. E. Sharpless and K. K. Wong, *Cell Cycle*, 2006, **5**, 2072.
8. A. Wlodawer and J. W. Erickson, *Ann. Rev. Biochem.*, 1993, **62**, 543.
9. J. Vondrasek, C. P. van Buskirk and A. Wlodawer, *Nat. Struct. Biol.*, 1997, **4**, 8.
10. M. Prabu-Jeyabalan, E. Nalivaika and C. A. Schiffer, *J. Mol. Biol.*, 2000, **301**, 1207.
11. M. Prabu-Jeyabalan, E. Nalivaika and C. A. Schiffer, *Structure*, 2002, **10**, 369.
12. M. Prabu-Jeyabalan, E. A. Nalivaika, N. M. King and C. A. Schiffer, *J. Virol.*, 2004, **78**, 12446.
13. M. Prabu-Jeyabalan, E. A. Nalivaika, K. Romano and C. A. Schiffer, *J. Virol.*, 2006, **80**, 3607.
14. M. Prabu-Jeyabalan, E. A. Nalivaika, N. M. King and C. A. Schiffer, *J. Virol.*, 2003, **77**, 1306.
15. N. M. King, M. Prabu-Jeyabalan, E. A. Nalivaika and C. A. Schiffer, *Chem. Biol.*, 2004, **11**, 1333.
16. E. Johnston, M. A. Winters, S. Y. Rhee, T. C. Merigan, C. A. Schiffer and R. W. Shafer, *Antimicrob. Agents Chemother.*, 2004, **48**, 4864.
17. M. Prabu-Jeyabalan, N. M. King, E. A. Nalivaika, G. Heilek-Snyder, N. Cammack and C. A. Schiffer, *Antimicrob. Agents Chemother.*, 2006, **50**, 1518.
18. N. M. King, M. Prabu-Jeyabalan, E. A. Nalivaika, P. Wigerinck, M. P. de Bethune and C. A. Schiffer, *J. Virol.*, 2004, **78**, 12012.
19. P. E. Sax, *AIDS Clin. Care*, 2006, **18**, 71.
20. J. S. James, *AIDS Treat News*, 2006, **Jan–Jun**, 2.
21. E. Poveda, F. Blanco, P. Garcia-Gasco, A. Alcolea and V. Briz, and V. Soriano, *AIDS (London)*, 2006, **20**, 1558.
22. A. Nicholls, K. Sharp and B. Honig, *Proteins: Struct. Funct. Genet.*, 1991, **11**, 281.

Section 3
Docking

CHAPTER 8

Docking Algorithms and Scoring Functions; State-of-the-Art and Current Limitations

GREGORY L. WARREN, CATHERINE E. PEISHOFF* AND MARTHA S. HEAD

Computational, Analytical, and Structural Sciences, GlaxoSmithKline Pharmaceuticals, 1250 S. Collegeville Rd., Collegeville PA, USA

8.1 Introduction

The past three decades have seen an explosion in the amount of data required for structure-based design of pharmaceutically relevant molecules. This data expansion encompasses both an exponential increase in the number of available protein structures and an increase in the number of real and hypothetical drug-sized molecules available for virtual screening. Automated molecular docking algorithms are a cornerstone technology of structure-based design. While we have seen a substantial increase in the number of molecular docking algorithms and scoring functions (at least 30 docking programs and roughly twice as many scoring functions are available as of this writing), we have not seen a concomitant improvement in predictive performance. Although significant improvements in the speed and efficiency of the conformational search capabilities of docking algorithms have been obtained, only incremental improvements in the predictive capabilities of the scoring functions used by these algorithms have occurred. As a result, docking algorithms are fast, but not particularly predictive on their own and must be used in conjunction with knowledge of protein structure and/or mechanism and chemical intuition in order to extract useful insights from the computational results.

While it is the consensus within the computational chemistry community that scoring functions are to blame for the lack of predictiveness of docking algorithms, historically it has been difficult to run appropriate comparative studies to determine the true causative factors or the extent of the limitations of automated docking. These difficulties derived both from lack of access to multiple docking algorithms and lack of robust data sets. Within the past six years, however, an increasing number of publications have reported results from evaluations of automated molecular docking algorithms and scoring functions; these evaluation studies were aimed at quantitatively determining performance across a number of criteria. These studies are largely the result of corporate efforts where access to large proprietary datasets and larger budgets has facilitated the work. From this body of literature, statements about the state-of-the-art for docking and scoring are, for the first time, based on a substantial amount of empirical data. We do not suggest these studies have been perfect and there are certainly gaps; however, sufficient comparative data has been generated to allow an assessment of the technology as it stands today and to identify major goals yet to be achieved.

As part of a discussion about the strengths and limitations of docking algorithms and scoring functions it is worth looking at the types of tasks for which these docking algorithms are used in the context of computer-aided drug discovery. In general these tasks fall into three broad categories: binding mode prediction, virtual screening for lead identification, and potency prediction for lead optimization, with the vast majority of the published work to date addressing the first two tasks.

8.1.1 Binding Mode Prediction

Binding mode prediction is akin to virtual crystallography and involves a two-step process. In the first step, a set of ligand conformations are generated. In some docking algorithms, a database of small-molecule conformations is created and that conformer database subsequently docked into a protein binding site; in other cases, the ligand conformations are generated and modified within the environment of the binding site. In the second step, a scoring function is applied to a set of ligand poses in the protein-binding site in order to select the pose predicted to be most likely. If both the search and scoring steps of the process are working as intended, the selected pose will be similar to the pose that would be seen crystallographically. Although binding mode prediction is described here as a two-step process, in actuality the steps are coupled: scoring functions are used to drive the placement of ligand poses within the binding site and are also used to rank the set of poses returned to the user. This dependence of both steps on scoring functions can complicate analysis of the performance of docking algorithms for binding mode prediction. The scoring functions used for this purpose have typically been optimized to reproduce binding modes and ligand affinities for publicly available protein–ligand crystal structures. We therefore expect that docking algorithms should be able to identify successfully small-molecule binding modes that are sufficiently correct to be useful for structure-based design, at least for those protein classes and ligand chemotypes included in the derivation of these scoring functions.

8.1.2 Virtual Screening for Lead Identification

Molecular docking algorithms are regularly used to virtually screen large numbers of compounds to identify potential leads for a particular protein target. For the previously described task of predicting a binding mode for a single compound, the amount of time required for the computation is not of primary importance; if a more correct pose can be identified then the extra computational cost is justified. In contrast, for the task of lead identification, the practitioner would generally like to screen a database of 10^5–10^6 compounds in a time commensurate with that required for an experimental high-throughput screen. In virtual screening mode, therefore, parameters of the docking algorithm are often modified to optimize the speed of the calculation.

In the virtual screening process, a large database of small molecules is docked into the protein and a single pose for each molecule selected and scored by the same scoring functions used for calculations of binding mode prediction. At the end of the screening process, a specific number of compounds are selected for testing in experimental assays. While one could select the molecules for testing using a specific cutoff score, in practice the number of compounds selected is determined by the capacity of the experimental assay. Underlying this use of docking as a virtual screening tool is the assumption that compounds with poor shape and chemical complementarity to the protein-binding site will score poorly and will be at the bottom of the list of scored compounds. The early part of the list with the best-scoring compounds is therefore expected to be enriched with a larger number of active compounds. Accordingly, although most scoring functions have been designed to reproduce binding modes and affinities of true actives, for virtual screening the scoring function does not necessarily need to account correctly for all features conferring affinity. For a successful virtual screen that identifies lead compounds, it may be sufficient for the scoring function merely to identify compounds that are likely to be inactive.

8.1.3 Potency Prediction for Lead Optimization

The holy grail of docking has been to successfully order the rank of a set of related compounds by affinity. Even modest progress toward that goal would allow docking to be applied to lead optimization in a manner that is more quantitative and predictive rather than qualitative and descriptive. This use of docking to rank-order compounds relies on one primary assumption: if a sufficiently accurate binding mode is identified, a scoring function can count up the favorable and unfavorable interactions and thereby predict relative potency. It is generally acknowledged, however, that this assumption is not a good approximation of all of the physical events involved in the binding of a small molecule to a protein target. Current docking algorithms and typical scoring functions do not rigorously account for contributions such as loss of conformational entropy upon binding, residual conformational entropy in the bound state, solvation effects on binding, and so on. We therefore expect that single docked poses and simple counting-based scoring functions will not be adequate for reliable and universal predictions of relative binding affinities. It is often suggested that we would be more predictive if we had sufficient computing power and applied computational methods at a higher level of theory – free energy perturbation calculations, for example, or quantum mechanical scoring schemes. We cannot comment on the validity of this assertion because it has not been tested across a wide range of protein–ligand systems in the manner that we have seen in recent evaluations of simpler docking technologies. However, we do expect that the greatest potential for immediate impact on drug discovery lies in the development of docking algorithms and scoring functions that enable computational times consistent with the time required for chemical synthesis and computational predictiveness accurate enough for lead optimization.

8.2 A Brief Review of Recent Docking Evaluations

This review is neither exhaustive nor complete, but touches on seven recent publications evaluating docking algorithm performance with an emphasis on the experimental design and type of data used.

Kellenberger *et al.*[1] evaluated binding mode prediction and virtual screening accuracy in a study of eight docking algorithms. Particular emphasis was placed on virtual screening efficiency and search parameters for slow algorithms were modified so that the total calculation time was equivalent for all eight algorithms evaluated. For their evaluation of binding mode prediction, the authors tested the eight docking algorithms against 100 protein complexes from the Research Collaboratory for Structural Bioinformatics (RCSB); docking calculations were carried out using high-throughput virtual screening parameters rather than being optimized for reproduction of correct binding modes. When using a $\leq 2\,\text{Å}$ root mean square deviation (RMSD) success measure, the success rate for the top-ranked pose reproducing the experimentally determined binding mode ranged from 25 to 55%. The virtual screening evaluation was performed on a single target – thymidine kinase – using 10 active compounds salted into 990 "drug-like" and putatively inactive decoys from the Available Chemical Directory (ACD). The enrichment factors at 10% of the database screened ranged from 0 to 10, with 10 being the theoretical limit. Based on the results of their evaluation, Kellenberger *et al.*[1] concluded that algorithms that most accurately reproduced the experimentally determined binding mode were also the most successful at enrichment in the virtual screening experiment.

Kontoyianni *et al.*[2,3] published two evaluation papers covering binding mode prediction and virtual screening performance separately. The binding mode prediction evaluation examined five docking algorithms applied to 69 protein–ligand complexes selected from the RCSB and representing 14 protein families. The success rate for a top-ranked pose reproducing the experimentally determined pose within $2.0\,\text{Å}$ ranged from 0 to 25%. Based on the results for these 14 protein families, the authors conclude that algorithm performance corresponded to the protein target type: individual docking programs were better able to predict binding modes for some target types than for others. The data set for the virtual screening evaluation paper[3] consisted of six protein targets with 8–10 active compounds embedded in a database of 996 decoys from the ACD and MDL Drug Data Report (MDDR) selected using drug-likeness filters. The enrichment factors at the 10% of the database screened mark ranged from two to nine. Variability in performance by algorithms across the protein targets tested was observed. For the four docking algorithms evaluated, a correlation between virtual screening performance and accurate reproduction of the experimental binding mode was found.

Cummings *et al.*[4] concentrated on the use of docking algorithms as virtual screening tools. They evaluated five protein targets with 5–14 active compounds per target; decoys were selected from the MDDR using drug-likeness filters. Success at reproducing the experimentally determined binding mode within 2.0 Å ranged from 6 to 32% for the top-ranked pose. Though this was a virtual screening evaluation the authors were less concerned with computational speed. The authors took special care with protein target preparation, binding site definition, and parameter selection to ensure that the experiments were as similar as possible for each docking algorithm. The enrichment factors at the 10% screening mark ranged from 0 to 9 with the theoretical limit being 10. On a target-by-target basis, the algorithms evaluated were able to identify known active compounds at levels greater than expected at random for at least one target. However, all of the algorithms performed at or worse than random on at least one protein target. The authors noted a correlation between accurate reproduction of the experimental binding mode and enrichment in virtual screening. However, there was one example where an algorithm produced no binding modes within 2.0 Å RMSD and yet still identified known actives at a rate higher than random.

The docking algorithm evaluation published by Perola *et al.*[5] examined binding mode reproduction and virtual screening. The data set for binding mode prediction consisted of 200 complexes from the RCSB for which K_i/K_d was available and for which the ligands met drug-like criteria. The authors evaluated three docking algorithms and found that algorithms correctly identified poses, within 2.0 Å of the experimentally determined pose, at levels ranging from 45 to 61%. The virtual screening data set contained actives for three protein targets. The authors selected 142–247 Vertex compounds for each of the three targets; the activity of these compounds ranged from nanomolar to high micromolar. The enrichment factors for 3% of the database screened range from 1 to 18. Target-dependent variability in performance was observed. While no information about the characteristics of the Vertex compounds was provided, this active set is likely to more closely match the character of a typical corporate collection. The decoy data set was taken from unspecified commercial databases to make a total of 10 000 compounds. The conclusion from the virtual screening results was that algorithm and scoring function performance was system-dependent. No correlation between accurate reproduction of the observed binding mode and enrichment was observed.

Wang *et al.*[6] tested 11 scoring functions for their ability to correctly identify experimentally determined binding modes. The authors selected 100 complex structures from the RCSB; in many cases the ligands in these complexes were not drug-like. The authors generated hundreds of poses for each protein–ligand complex; these poses were then ranked by the 11 scoring functions. The success rate at correctly identifying the binding mode within 2.0 Å ranged from 26 to 76%. This evaluation also looked at using consensus scoring to improve the results, with only modest success. No virtual screening evaluation was performed in this study. Correlations between the docking score and ligand affinity were observed and the coefficients ranged from 0.25 to 0.70.

Chen *et al.*[7] evaluated binding mode prediction and virtual screening for four docking algorithms. The binding mode prediction data consisted of 164 public-domain complex structures from the RCSB. The success of the top-ranked pose at reproducing the experimentally observed binding mode within 2.0 Å ranged from 42 to 91%. The virtual screening data set contained 2734 active compounds for four protein targets with activity derived from public data[8] and for eight targets from AstraZeneca's high-throughput screening database; the activity of these compounds ranged from nanomolar to low micromolar. The decoy database contained 20 000 "drug-like" compounds from an unspecified commercially available database. Upon tautomer and stereoisomer expansion the database contained approximately 40 000 compounds. The enrichment values at the 10% mark of the database screened ranged from 0 to 10. Algorithm performance varied depending on the protein target being evaluated. No correlation between good binding mode prediction and enrichment was noted. This evaluation compared virtual screening performance for docking algorithms with 2- and 3D ligand-based methods. Of particular interest is that the performance of the 3D ligand-based method was comparable to or better than the performance of all but one docking algorithm.

Warren *et al.*[9] evaluated 10 docking algorithms and 37 scoring functions for three tasks: binding mode prediction, virtual screening, and rank-ordering compounds by affinity. The evaluation was carried out against eight proteins of seven target types. For each target type, the data set consisted of seven cognate sets of 140–210 compounds each. The number of chemical series in each cognate set ranged from two to five for a total of 21 chemical series; for each chemical series there was at least one protein–ligand crystal structure. The binding mode prediction subset contained 136 crystal structures, 90% of which are not in the public domain. For the top-ranked pose, performance at reproducing the observed binding mode within 2.0 Å ranged from 0 to 43%. The seven cognate data sets were merged to obtain the virtual screening data set of 1303 compounds. Enrichment factors at the 10% mark of the database screened ranged from 0 to 7.2. The virtual screening performance was found to be highly target dependent. Each individual cognate data set was used to assess the ability of the docking algorithms to rank-order compounds by affinity. The affinity range of the cognate data sets spanned at least 3.5 log units and all but one cognate set contained inactive representatives of each chemical series. The correlation coefficients between docking score and affinity ranged for 0.0 to 0.57. For all portions of the evaluation, no restrictions on time or effort were placed on the individuals doing the calculations. Surprisingly, even though the virtual screening set was derived from 21 congeneric series, the shape similarity of the molecules is low. Performance by the algorithms and their scoring functions for all three tasks were highly target dependent. In addition there was no observed correlation between good binding-mode prediction performance and virtual screening performance. Last, no scoring function was able to predict affinity or rank-order compounds by affinity.

It is important to note that, with one exception – the binding mode identification evaluation published by Wang *et al.*[6] – the experimental design and data analysis of all the evaluations is similar though not identical; direct comparison of the results is therefore impossible. In addition, there were significant differences in the data sets used. In most cases binding mode prediction evaluations were done against publicly available data, making it difficult to determine if the results are a true prediction or a prediction of the members of the training set. Many of the virtual screening data sets, by nature of their origin, do not contain chemical, activity, or shape decoys of the active molecules.

8.3 What these Evaluations Tell us about the Performance of Docking Algorithms

Given this brief review of selected recent evaluations of docking and scoring let us now look at what these papers tell us about the state of docking and scoring technology. We would like to draw specific conclusions about how well docking is performing from these reported evaluations. However, there are inherent difficulties in comparing amongst the evaluations, primarily because the authors of the reviewed studies did not carry out their evaluations for the same set of data. Furthermore, even in those instances where the same protein target was used, multiple studies in some cases produced different results. The reasons for these differing results include selection of a different protein structure for a given target, docking of a different set of ligands into that protein structure, use of different versions of a specific docking program, and differences in the program parameters used for the docking calculations. Despite the complexity caused by these differences, it is nevertheless possible to identify general conclusions that are supported across the range of evaluations.

8.3.1 Binding Mode Predictions

Given sufficient computational time, most of the frequently used docking programs can generate a correct binding pose, one that closely corresponds to the crystallographically identified pose. Indeed, many of the recent publications in this area assume or explicitly state that docking algorithms can reproduce the binding mode "in a majority of cases"[7] if 100 poses are analyzed. And, in general, docking algorithms can generate the correct binding mode

>50% of the time if ligands of low conformational complexity are docked into binding pockets of small-to-medium size. However, a close examination of the data from our evaluation[9] suggests that results degrade significantly as the size and complexity of the test system increases. Across all of the proteins and ligands included in our evaluation, docking algorithms generated a correct pose for 31–94% of the ligands for a particular protein target. When examined more closely, however, we discover that docking algorithms could not generate poses within 2.0 Å of the correct pose when applied to large binding sites, flexible ligands (greater than eight rotatable bonds), or systems with few specific non-hydrophobic interactions between ligand and protein. Overall, however, docking algorithms were able to successfully search conformational space to generate sets of protein–ligand conformations which include a correct pose. In practice, however, we do not know the correct binding pose *a priori*; in the absence of such knowledge of the binding mode, we rely on the scoring function to identify which of the generated poses is sufficiently correct to be useful for structure-based design. Scoring functions are not particularly successful at identifying which of the generated poses is most similar to the crystallographically determined bound conformation. In the case of our own evaluation, success rates ranged from 31–94% for the best RMSD over all poses; however, for the RMSD of the pose with the best score, success rates dropped to 8–74% depending on the target type and scoring function used. Based on both the results of our evaluation and on the reported results of other evaluations, we conclude that the search problem is largely solved – docking algorithms can generate good docking poses – and further conclude that scoring functions limit the predictiveness of docking algorithms.

8.3.2 Virtual Screening

For virtual screening a number of evaluations have stressed the importance of speed over accuracy, either by setting parameters of the docking algorithms to optimize execution time or by placing hard limits on the amount of computation time allowed for any one algorithm. However, in an age of distributed computing on ever-faster chips, this focus on computation speed seems unnecessary and may limit improvements in docking accuracy. Molecular docking is a trivially parallelizable task, and while 24 hours of computation would be undesirable for docking a single ligand, it is important to consider whether we need to aim for seconds per ligand if minutes would give us a significantly more accurate answer. Unfortunately, no evaluation to date has systematically varied the time allowed for a calculation to determine the effect of computational time on accuracy or enrichment.

The good news in the virtual screening literature is that enrichment above random is observed across algorithm and scoring function types and across all types of data sets. While the enrichment numbers are higher for data sets with decoy sets that are significantly different from the actives in shape and chemistry, enrichment is still observed for data sets with more similar active and decoy ligands. The bad news is that the results are inconsistent for individual docking algorithms across protein types. For example, no algorithm shows enrichment for all kinases, nor is there any one algorithm that performs well across the full range of protein types examined to date. This inconsistency is observed in every published evaluation of which we are aware. To examine this inconsistency in more detail, it would be useful to examine whether performance varies depending on the specifics of the data set used. However, to date there has been no evaluation of the performance of individual algorithms using data sets for a single protein target with either different active ligands or different decoy sets.

Two assumptions in virtual screening have been challenged by the evaluation results. First, producing an experimentally validated binding pose is neither necessary nor sufficient for virtual screening success. For example, in our evaluation Genetic Optimization for Ligand Docking (GOLD) successfully docked ligands of methionyl tRNA synthetase – 74% within 2.0 Å of the correct binding mode for the best-scoring pose – but had essentially random performance when virtually screening this protein. Conversely, GOLD was only able to dock 10% of Factor Xa ligands within 2.0 Å, but still had quite good enrichment in virtual screening. Second, there is a misconception that virtual screening is successful because the scoring functions are effectively ranking compounds on their positive attributes. It is clear from these evaluations that this is not happening; however, the scoring functions are sensitive enough to distinguish compounds that may be possible binders from those that have no chance at all. As a result of this two-state distinction, many inactive compounds will be eliminated from consideration, which by default generates a compound set with better-than-random probability of containing active compounds.

Finally, but of critical importance, these studies suggest that it is not possible *a priori* to know which molecular docking algorithm will produce better results for a new protein target. Since many of the algorithms yield reasonable enrichment, however, one could argue that it doesn't matter which algorithm is applied as long as a sufficient number of compounds are experimentally evaluated.

8.3.3 Affinity Prediction

There have been few evaluations comparing the ability of docking algorithms and their scoring functions to rank order or predict the affinity of ligands. However, there have been a number of publications on rank-order predictions within chemical series versus a protein target. These publications span a range from simple scoring methods used to rank compounds in a single congeneric series[10] to more complex scoring functions tested for general applicability. A representative example includes a publication by Muegge[11] describing development of the scoring function Potentials of Mean Force (PMF) 2004. No correlation was found for the Protein Data Bank (PDB) bind database[12] when scoring molecules using PMF2004. However, if the data were divided by protein class, 13 of the 34 protein classes had an R^2 correlation of 0.50 or greater. Of concern is that some classes of proteins that showed a correlation with the original PMF scoring function did not show a correlation with PMF2004, *e.g.* the human immunodeficiency virus-1 (HIV-1) protease data set of Holloway *et al.*[10] Another series of publications in this area are by Wang *et al.*[6] and Yang *et al.*[13]; these publications describe the authors' attempts to develop a broadly applicable scoring function using published affinity for complex structures deposited in the RCSB. From this work two scoring functions have been developed, X-Score and M-Score. X-Score was tested on a data set of 100 protein–ligand complex structures for which binding affinity data were available and yielded an R^2 correlation coefficient of 0.44 for the experimentally determined binding mode and 0.49 for the top-ranked docking pose. The more recent M-Score had an R^2 correlation coefficient of greater than 0.5 for six of the 17 protein families tested, but an overall R^2 coefficient of 0.24 for the complete data set. An evaluation of DrugScore by Gohlke *et al.*[14] for metalloprotease and serine protease data sets demonstrated good R^2 correlation coefficients when the experimentally determined poses were used (0.70 and 0.86, respectively), but a significant reduction in performance when docking poses were used (0.42 and 0.56). Ferrara *et al.*[15] assessed the ability of nine scoring functions to rank-order the affinity of 189 complex structures. The performance by all scoring functions – empirical scoring functions to more physics-based Molecular Mechanics Poisson–Boltzmann Solvation Area (MMPBSA) scoring procedures – demonstrated modest results at best, with ChemScore having an R^2 correlation coefficient of 0.43 for a data set of 119 complex structures that were not part of the data set used to calibrate the scoring function. Of particular concern was the observation by the authors that the serine protease and metalloprotease subsets demonstrated a high correlation between the log MW and affinity – R^2 of 0.81 and 0.58, respectively – which was roughly equivalent to the correlations observed for the scoring functions. The observation that correlation between log MW and affinity exists was first expressed by Kuntz *et al.*[16] when they demonstrated a linear correlation between affinity and the number of heavy atoms up to 15; for molecules larger than 15 heavy atoms the correlation plateaus. The Glaxo-Smith-Kline (GSK) evaluation of affinity prediction[9] found no statistically significant correlation between docking score and measured affinity by any of the scoring functions for most of the eight proteins examined. Chk1 kinase was the single example for which any correlation was seen, although that correlation was modest at best ($R \approx 0.7$); this is also the only congeneric series in the complete GSK data set for which there was a correlation between log MW and affinity.

It is generally assumed that a broadly applicable scoring function should be the development goal if scoring functions are to have significant impact during the lead optimization phase of drug discovery. However, it is appropriate to question whether empirical scoring functions can provide a broadly applicable scoring function that gives meaningful correlations with measured affinity; if not, then these methods may never be reliably applicable to lead optimization. Furthermore, in the development of such scoring functions, particular care needs to be taken to ensure that correlations observed are the result of correctly characterizing interactions between the protein and ligand rather than the result of serendipitous correlation with 2D properties of ligands in the training set.

8.4 How an Ideal Evaluation Data Set Might be Structured

The multiple performance needs imposed on scoring functions by the different docking tasks may necessitate multiple varieties of test and training data sets. Binding mode predictions require discrimination between good and bad interactions. For virtual screening the function must behave as a classifier and a physically inaccurate presentation is acceptable if the classification is accurate. For ligand binding and its subsequent affinity there are multiple and complex factors occurring in the binding event that must be accounted for if a scoring function is to predict affinity accurately. One limitation of using data from published docking evaluations is the lack of consistency in the data set design or data acceptance criteria in these data sets. Another approach to this problem is to predefine data set criteria and then to build a data set that will test docking algorithm and scoring function performance in the least biased and most practical way possible.

Irrespective of the task – binding mode prediction, virtual screening, affinity prediction – to test in an unbiased manner the evaluation data must be unrelated to and distinct from the data used to train and optimize algorithms or scoring functions. This distinction should occur, at least in part, both in ligand chemotype and in protein receptor space. This is most likely to occur if the data used to for evaluation are unpublished or yet-to-be published data.

8.4.1 Binding Mode Prediction

There are three components to a binding mode evaluation or development data set we consider: a difference between the evaluation data and data used for scoring function development, drug-likeness of the data, and a range of affinities. Whether the data used to evaluate are unpublished or from a public database like the RCSB, complex structures used for an evaluation can have no overlap with complex structures used during algorithm and scoring function development if the evaluation is to be unbiased. This is particularly true when evaluating empirical scoring functions. Second, if the goal for an evaluation is to assess performance as applied to drug discovery, then some attention to the "drug-likeness" of the data set is of importance. By drug-likeness we are not referring to calculated properties, but to compounds that have been shown to have either *in vitro* or *in vivo* biological activity. However, during development, as Graves *et al.*[17] eloquently point out, the use of small simple systems is of great importance because it is in simple cases that an understanding of the how and why of performance during molecular docking occurs. Ultimately docking algorithms and scoring function will be applied to drug discovery and the demonstrated performance on molecules that look like marketed drugs and are biologically active is of utmost importance. Last, attention needs to be paid to the affinity of the ligand in binding mode development and evaluation data sets and not just to the resolution of the complex crystal structure. The structure observed in a receptor–ligand complex is a time-averaged environmentally dependent state, *i.e.* low energy crystalline state and, in the case of low affinity ligands (high μM to mM), it is not a good assumption that there is a structure–activity relationship (SAR). To date there have been very few evaluations or development examples where molecules used for pose prediction are of a single chemotype with a range of affinities versus singletons extracted from the RCSB.[9]

8.4.2 Virtual Screening

Docking and scoring, during the virtual screening exercise, perform well when the scoring function is able act as a classifier – is this molecule active or inactive. To date two database architectures have been used to evaluate virtual screening performance. The first and most frequent in the literature is a database of largely unrelated compounds (chemically, by shape, by activity, or by function) into which a few active singletons are salted.[3–5,18] As a result the decoy compounds bear little or no chemical or shape similarity to the active compounds. In practice, the characteristics of these databases are what would be expected if a purchasable compounds database was screened. The second architecture is a database that is made up of a number of congeneric series such that the composition of the database mimics the composition of a typical pharmaceutical corporate collection – chemical diversity but with chemical, shape, and activity similarity.[7,9] Theoretically it is expected that the second database type would be the harder of the two for docking algorithm scoring functions to identify or classify active compounds from, since the number of similar decoy compounds is much higher. However, to date this hypothesis or assumption remains to be tested. Both database architectures are equally valid because both mimic the types of databases virtual screening will be performed on: external purchasable compound collection with few known actives and corporate databases containing a number of related biologically active molecules. There is a widely held belief that good top-ranked binding modes are important for good performance and/or enrichment. However, in practice this requirement is not necessary since, as a classifier, the scoring function may only need to discern whether a molecule fits in the active site or not.

8.4.3 Affinity Prediction

We propose that a data set to test the ability of a scoring function to rank-order or predict the affinity of compounds should have four components. First, it should contain several series of closely related compounds of varying affinity. This allows for testing to determine if a scoring function performs well for a particular target while testing for compound series dependencies within that target. Second, a wide range of affinities is important, but within a pharmaceutically relevant range (100 μM to 10 pM – pAffinity 4–11). When compound affinities are less than 100 μM it is difficult to prove specific binding. Another potential problem with a very large range of affinities is statistical in nature. When compound data sets span 10 or more log units, an accurate prediction of data at the ends of the range will result in a strong correlation coefficient. However, the same correlation function would show very little or no correlation in the middle of the range. Third, an affinity prediction data set should include experimentally determined binding modes for representatives of each chemical series as a prerequisite. Additional structures that span a range of affinities for a chemotype or across chemotypes could provide important insights (see Figure 12 in Warren *et al.*[9]). The fourth and last component of an affinity prediction evaluation data set is that the biological data be measured using a single constant assay protocol and preferably by a single group in a single location. This helps to eliminate as much experimental variability as possible. Multiple assay values for each compound allow for a direct measure of the experimental variability and allow the evaluator and developer to determine the appropriate level of accuracy. An additional factor to consider during target selection that will reduce confusion in data analysis is to choose only systems and/or targets with Michaelis–Menten-like kinetics. Systems that do not demonstrate traditional mass-action kinetics will have factors affecting ligand affinity other than the protein–ligand binding event.

8.5 Concluding Remarks

As part of a conclusion to our discussion of the current state of docking and scoring let us look back at some of the tasks and themes discussed so far – pose prediction, virtual screening, and potency prediction – with emphasis on the state-of-the-art and possible directions for improvements.

8.5.1 Binding Mode Prediction

We will define success in pose prediction as achieving a greater than 90% rate of generating and identifying a docking pose within 2.0 Å of the crystallographically determined pose; this 90% accuracy rate should further be seen across diverse targets and compound classes. Given this definition of success, we can say that pose generation may not be and pose identification certainly is not a solved problem. Identifying whether the lack of success overall is due to problems with pose generation or identification is difficult because of the tight coupling between generating and scoring poses. As discussed previously, the success rate reported in the literature ranges from 0 to 90%. This variability presents in two forms: variability by a single docking algorithm across multiple target types, and variability by different docking algorithms on the same target. When faced with a new protein type or new ligand chemotype, the docking practitioner does not know if his or her favorite algorithm will generate the correct answer, nor which answer generated by a suite of algorithms is the correct one. Thus it is the unpredictable nature of the variability that affects fundamental usability and forces the use of less-efficient analysis methods such as visual inspection.

In truth we do not know the actual state-of-the-art for pose generation and prediction. The true test of pose prediction has yet to be performed since it involves the use of evaluation data not currently in the public domain, ligand chemotypes different from those in the public domain, and protein–ligand complex structures for protein targets underrepresented in the public domain. While the evaluation performed at GSK is a step in the correct direction – 90% of the experimentally determined binding modes were not in the public domain – more effort needs to be devoted to providing the computational community with data of this sort.

Scoring functions drive the search for poses and the results produced, yet scoring functions are currently being asked to perform well on tasks that may be mutually exclusive. There is no reason to believe that the components required for optimum performance at predicting a binding mode are the same as those for classifying compounds as active or inactive. A question that needs to be asked is, should development focus on two distinct scoring functions – one for pose and affinity predictions and one for virtual screening? There are hints that this is the direction some developers are taking. There has been little evidence until recently[19] of the use of negative or inactive data in the development of scoring functions in pose prediction, virtual screening, or rank-order by affinity. Yet negative data have a strong potential to inform as to why incorrect predictions are being made.[17]

8.5.2 Virtual Screening

There have been a number of evaluations of virtual screening performance and in all cases there were examples where at least one of the docking algorithms and/or scoring functions could identify active compounds versus inactive compounds. This was the case both when the data set consists of active compounds salted into drug-like decoy data sets and when the data set includes inactive congeneric members in a background of decoys active against other targets. What is not known is what level of enrichment will occur if the decoy data set is carefully constructed to include both molecular shape and chemical decoys.

Since the emphasis in virtual screening is on active compounds, it is important to recognize that there are four possible outcomes when using automated molecular docking: (1) the binding mode is correct and the scoring function enriches, (2) the binding mode is correct and the scoring function does not enrich, (3) the binding mode is incorrect and the scoring function enriches, and (4) the binding mode is incorrect and scoring function does not enrich. The fourth case is of no interest because it is one of complete failure. In the first case the scoring function performs well for binding mode prediction and for active–inactive classification. It would be reasonable to assume that this scoring function might perform well at rank-ordering by affinity. In the second case the scoring function works well for pose prediction, but cannot classify active versus inactive. It is, however, possible that the scoring function might still work at rank-ordering closely related molecules since the three-dimensional information about binding is correct. In the third case the scoring function works for virtual screening, but it is not clear that the docking algorithm is getting the right answer for the right reason. It is possible that the scoring function is correctly removing compounds with no chance to be active, but such a scoring function is not expected to be able to rank-order compounds by affinity since three-dimensional information about binding is incorrect.

In addition, we need to ask how good the currently used performance measures are. In practice we do not care how much of the data set needs to be screened to find 90% of the actives; instead we would like to identify at least one representative of all active chemotypes within the number of compounds that can be experimentally screened. Nevertheless, the 90%-of-actives measure provides a reasonable snapshot of completeness versus good initial enrichment. Beyond this single 90%-of-actives measure, there is a need for a more robust measure that identifies early enrichment while still addressing completeness or comprehensiveness. To date no such measure has appeared in the literature, though one possibility would be to use a Receiver Operator Curve (ROC) method biased toward the early portion of the virtual screening results list.

8.5.3 Rank Order by Affinity

There have been few assessments published on the ability of scoring functions to predict or rank-order by affinity; for those few assessments the results are mixed. One concern is that the correlation for predicted score versus affinity roughly tracks that for log MW versus affinity. As such, we don't know whether the read out is simply related to compound molecular weight or reflects target–ligand interactions. While there are hints that computationally rigorous methods may provide better target- and chemotype-independent correlations,[15,20,21] there is also a potential to improve current empirical methods. Those improvements can only occur if data are used for development that is different in character and quantity from that used to date.

What should these data look like? There is a need to use data with little or no correlation between affinity and simple 2D descriptors. There is a need for a continuous stream of data sets distinct from publicly available data to be used, first for evaluation and subsequently for further development. Published data tend to be highly sanitized, *e.g.* data is published because current computational methods work and are predictive. If highly sanitized data are used for algorithm development and/or optimization then a good answer or model is the likely outcome. However, when this type of data is used the practitioners have no knowledge of the algorithm's "real-world" capabilities and resulting computational models are less likely to be broadly successful. There is a need for high-quality data with a range of similarities from very homogenous series – more likely to be successfully predicted – to series of nonhomologous molecules. While the use of a different type of data during development is no guarantee for success, it is a worthy endeavor while computationally efficient physically rigorous methods are being developed.

8.5.4 The State-of-the-art

Docking programs can generate binding mode predictions that are very similar to experimentally determined binding modes. However, the scoring functions in every study published to date except one[7] are not able to identify those poses reliably. So what is the generally accepted best practice or what practice is most likely to yield a positive result? A skilled computational chemist uses protein target and ligand SAR knowledge in conjunction with the scoring function to identify a pose or multiple poses out of a set generated by the docking algorithm that are consistent with all the accessible information. That this method produces the best result should not come as a surprise. Scoring functions as currently implemented are either empirical counting schemes or simplified physical representations. Until more sophisticated or more physically accurate representations become available that do not require inordinate amounts of computer time, using human intuition to integrate disparate pieces of data to generate a prediction will remain a cost-effective method for obtaining good predictions.

Docking algorithms can be used as virtual screening tools. A number of published evaluations demonstrate that scoring functions can be used to identify active compounds near the top of a docking-score-ordered list, with enrichment at levels greater than random. What has not been demonstrated is consistent high enrichment by a single docking algorithm or scoring function across a number of protein targets and/or ligand data sets. The implication of this observation is that, in the absence of knowledge about the performance of the algorithm or scoring function to be used for a particular protein target and ligand data set, it is impossible to know whether that particular algorithm will perform well. Best practice therefore suggests that the practitioner should run a small validation study to verify that the target–algorithm pair being considered can identify active ligands at a level greater than random.[5,9]

Current scoring functions are rarely capable of predicting compound affinity or rank-ordering compounds by affinity. A limited number of publications have shown significant ($R^2 > 0.7$) correlations between a scoring function and compound affinity. However, in most cases the data used was a congeneric series with limited chemical diversity. An additional concern stems from observations of equally significant correlations between compound affinity and log MW. One difficulty in validating the rank-order or affinity-prediction application of scoring functions is the scarcity of data, with evaluators solely reliant until recently on published data. Published data have an inherent bias because success tends to be published, failure does not.

There is a need for high-quality data. By high quality we mean data that have assay consistency and reliability along with experimentally determined binding modes for representatives of each chemical class. There is a need for data sets containing compounds with low correlations between affinity and 2D descriptors. Last there is a need for data sets that have been designed to be statistically robust. With high-quality data, advances in all areas of docking and scoring – but particularly in the area of potency prediction – have a better chance of occurring.

References

1. E. Kellenberger, J. Rodrigo, P. Muller and D. Rognan, Comparative evaluation of eight docking tools for docking and virtual screening accuracy, *Proteins: Struct., Funct., Bioinformatics*, 2004, **57**, 225–242.
2. M. Kontoyianni, L. M. McClellan and G. S. Sokol, Evaluation of docking performance: comparative data on docking algorithms, *J. Med. Chem.*, 2004, **47**, 558–565.
3. M. Kontoyianni, G. S. Sokol and L. M. McClellan, Evaluation of library ranking efficacy in virtual screening, *J. Comput. Chem.*, 2004, **26**, 11–22.
4. M. D. Cummings, R. L. DesJarlais, A. C. Gibbs, V. Mohan and E. P. Jaeger, Comparison of automated docking programs as virtual screening tools, *J. Med. Chem.*, 2005, **48**, 962–976.
5. E. Perola, W. P. Walters and P. S. Charifson, A detailed comparison of current docking and scoring methods on systems of pharmaceutical relevance, *Proteins: Struct., Funct., Bioinformatics*, 2004, **56**, 235–249.
6. R. Wang, Y. Lu and S. Wang, Comparative evaluation of 11 scoring functions for molecular docking, *J. Med. Chem.*, 2003, **46**, 2287–2303.
7. H. Chen, P. D. Lyne, F. Giordanetto, T. Lovell and J. Li, On evaluating molecular-docking methods for pose prediction and enrichment factors, *J. Chem. Inf. Modell.*, 2006, **46**, 401–415.
8. M. Stahl and M. Rarey, Detailed analysis of scoring functions for virtual screening, *J. Med. Chem.*, 2001, **44**, 1035–1042.
9. G. L. Warren, C. W. Andrews, A. M. Capelli, B. Clarke, J. LaLonde, M. H. Lambert, M. Lindvall, N. Nevins, S. F. Semus, S. Senger, G. Tedesco, I. D. Wall, J. M. Woolven, C. E. Peishoff and M. S. Head, A critical assessment of docking programs and scoring functions, *J. Med. Chem.*, 2006, **49**, 5912–5931.
10. M. K. Holloway, J. M. Wai, T. A Halgren, P. M. D. Fitzgerald, J. P. Vacca, B. D. Dorsey, R. B. Levin, W. J. Thompson and L. J. Chen, *A priori* prediction of activity for HIV-1 protease inhibitors employing energy minimization in the active site, *J. Med. Chem.*, 1995, **38**, 305–317.
11. I. Muegge, PMF scoring revisited, *J. Med. Chem.*, 2006, **49**, 5895–5902.
12. R. Wang, X. Fang, Y. Lu, C. Y. Yang and S. Wang, The PDBbind database: methodologies and updates, *J. Med. Chem.*, 2005, **48**, 4111–4119.
13. C. Y. Yang, R. Wang and S. Wang, M-Score: a knowledge-based potential scoring function accounting for protein atom mobility, *J. Med. Chem.*, 2006, **49**, 5903–5911.

14. H. Gohlke, M. Hendlich and G. Klebe, Predicting binding modes,binding affinities and "hot spots" for protein–ligand complexes using a knowledge-based scoring function, *Perspect. Drug Discovery Des.*, 2000, **20**, 115–144.
15. P. Ferrara, H. Gohlke, D. J. Price, G. Klebe and C. L. Brooks III, Assessing scoring functions for protein–ligand interactions, *J. Med. Chem.*, 2004, **47**, 3032–3047.
16. I. D. Kuntz, K. Chen, K. A. Sharp and P. A. Kollman, The maximal affinity of ligands, *P. Natl. Acad. Sci. U. S. A.*, 1999, **96**, 9997–10002.
17. A. P. Graves, R. Brenk and B. K. Shoichet, Decoys for docking, *J. Med. Chem.*, 2005, **48**, 3714–3728.
18. C. Bissantz, G. Folkers and D. Rognan, Protein-based virtual screening of chemical databases. 1. Evaluation of different docking/scoring combinations, *J. Med. Chem.*, 2000, **43**, 4759–4767.
19. T. A. Pham and A. N. Jain, Parameter estimation for scoring protein-lig and interactions using negative training data, *J. Med. Chem.*, 2006, **49**, 5856–5868.
20. S. P. Brown and S. W. Muchmore, High-throughput calculation of protein–ligand binding affinities: modification and adaptation of the MM-PBSA protocol to enterprise grid computing, *J. Chem. Inf. Modell.* 2006, **46**, 999–1005.
21. H. Fujitani, Y. Tanida, M. Ito, G. Jayachandran, C. D. Snow, M. R. Shirts, E. J. Sorin, and V. S. Pande, Direct calculation of the binding free energies of FKBP ligands, *J. Chem. Phys.*, 2005, **123**, 084108/1–084108/5 (online).

CHAPTER 9

Application of Docking Methods to Structure-Based Drug Design

DEMETRI T. MOUSTAKAS

Department of Chemistry and Chemical Biology, Harvard University, Cambridge MA, USA

9.1 Introduction

This chapter describes the use of small-molecule docking methods in drug design research. Docking is the search for spatial transformations that fit two molecules together in energetically favorable configurations. The goal of the docking problem is to predict the structure of a ligand–receptor complex accurately with respect to experiment, enabling prediction of the binding free energy of the complex.

The creation of the Protein Data Bank (PDB)[1–3] in 1971 as a repository of protein structures determined by X-ray crystallography and nuclear magnetic resonance (NMR) stimulated the development of computational methods in structural biology, including docking techniques. The groundbreaking DOCK program[4–10] developed by Kuntz *et al.* provided the first implementation of rigid body docking, in which a negative image of the receptor volume was represented as a set of spheres against which the ligand atoms were aligned. This procedure searched over the six orientational degrees of freedom to identify ligand poses that fit the receptor shape well. DOCK could recreate a protein–ligand complex structure given the complexed conformations of the receptor and ligand molecules, and it was used to predict binding modes of rigid ligands to protein targets.

Increasing computer power and large numbers of available macromolecular structures enabled the development of more sophisticated docking methods. Energy scoring functions and energy minimizers were incorporated into docking programs, enabling rigid docking methods to sample large numbers of ligand poses that could subsequently be evaluated according to binding energy. Methods were later developed that searched over ligand conformations as well as orientations in the receptor. While greatly increasing the number of degrees of freedom that must be searched, ligand flexibility enabled docking using uncomplexed ligand structures, which previously had only been possible for systems with rigid ligands. Most recently, docking methods that search over receptor conformations have been developed, enabling accurate docking of uncomplexed ligand and receptor structures, considered to be the most realistic test of docking methods. There are a number of excellent reviews of docking programs and algorithms[11–18] that describe the field of docking programs in a level of detail that exceeds the scope of this chapter.

Docking can be applied during drug design to direct experimental efforts based on receptor–ligand binding predictions. Docking has enabled small research groups without large chemistry resources to search through databases of available compounds for ligands to use as therapeutics, or in chemical biology studies. Groups with significant chemistry resources can use docking predictions to conduct expensive experiments, such as high-throughput screening (HTS) and chemical synthesis, in a more efficient fashion. A number of drugs have been developed using docking methods at some stage of development,[19–21] and with continuing advances in computational power and in structural biology, docking will likely play an increasingly important role in drug design.

9.2 Docking Methods, Capabilities and Limitations

Docking is the search for the most energetically favorable binding pose of a ligand to a macromolecular receptor. Accordingly, it should accurately predict the structure of a ligand–receptor complex with respect to experiment and calculate a binding energy that can be used to accurately rank-order different ligands relative to experimentally measured binding affinities. Docking can be thought of as a combination of spatial sampling methods with energy scoring functions, that search over many possible interactions between a ligand and a receptor molecule in order to identify a set of ligand poses that represent local minimum-energy positions of the ligand. If the ligand pose sampling is adequate, and the energy scoring function is sufficiently accurate, then the global minimum-energy position of the ligand in the receptor can be selected from the set of local energy minima. While docking is most often applied to systems of small molecule ligands and protein receptors, it has also been used to dock proteins to proteins, nucleic acids to proteins, and small molecules to nucleic acids. The remainder of this section describes the most common sampling and scoring methods, and their use in different docking applications.

9.2.1 Molecule Preparation

Prior to docking, structures for both the ligand and receptor need to be obtained and appropriately prepared. Generally, there are two steps involved in structure preparation. The first is to create a complete molecular model from a molecular structure file, and the second is to process the structure in a manner specific to the docking program being used. This second step generally involves applying techniques that spatially characterize the receptor site, such as calculating molecular surfaces or generating potential energy grids, to name just a few. This section describes the more general first step of generating molecular models of the ligand and receptor.

The PDB is the largest public source of protein and nucleic acid structures. It contains receptor structures determined by experimental methods such as X-ray crystallography or NMR spectroscopy, and those predicted by computational methods such as homology modeling or protein folding. Modeled protein structures can be used as docking receptors, though the inaccuracy of side-chain structures can significantly decrease docking accuracy. In order to build a model from a PDB structure, a modeling program such as Sybyl,[22] MOE[23] or Insight II[24] is used to remove water molecules, add missing hydrogen atoms, build any missing atoms or bonds, ensure that titratable groups are in the proper protonation states and assign partial electrostatic charges to the receptor atoms. Force-field partial charges derived from quantum mechanics calculations are typically assigned for protein residues.

A number of databases of commercially available compounds provide 1D or 2D chemical structures of hundreds of thousands of small molecules. The first step to preparing a database of compounds for docking is to generate 3D chemical structures for each molecule from the 1D structure, using a program such as OMEGA,[25] CONCORD[22] or CORINA.[24] Any missing hydrogen atoms are added, and the proper protonation states of titratable groups are determined. Finally, partial electrostatic charges are calculated using as accurate a charge model as possible. Due to the large numbers of compounds, quantum mechanics calculations on compound libraries are computationally prohibitive, necessitating the use of approximations such as the RESP[26] or AM1-BCC[27] charge models.

9.2.2 Sampling Methods

Sampling methods generate multiple positions of the ligand within the receptor, exploiting the principle that by placing the ligand using the constraints of the receptor shape, the ligand poses should be near minimum-energy positions. The three major classes of sampling methods used in docking are rigid ligand docking, flexible ligand docking and flexible receptor docking.

9.2.2.1 Rigid Ligand Docking

Rigid ligand docking, or rigid docking, is the most basic type of docking method, which treats both the receptor and ligand molecules as conformationally rigid. It searches for rigid body transformations that best fit the ligand into the receptor. Although most interesting biological systems involve both flexible ligand and receptor molecules, rigid body docking is nonetheless an important method that serves as the foundation for flexible ligand and flexible receptor docking methods.

The first class of rigid docking methods performs an exhaustive search of ligand orientations within the receptor. These methods perform an enormous amount of sampling, and therefore require that a very fast scoring function be used during a coarse search step, followed by a fine focused search using a more computationally expensive, and more accurate, scoring function. The programs EUDOCK,[28] FRED[29] GLIDE[30,31] and FTDOCK[32] fit into this category. These programs use a number of methods, including systematic grid based searches, fast affine and fast Fourier transforms to search over all potential mappings of the ligand to the receptor.

The second class of rigid docking methods uses the shapes of the ligand and of the receptor to only sample those orientations where the ligand fits into the receptor. The programs DOCK,[4–10,33] FlexX,[34] FLOG[35] and HAMMERHEAD[36] fit into this category, as they use shape descriptors of the ligand and receptor surfaces to generate orientations of the ligand in the receptor. By restricting the sampling of the ligand to regions of orientation space known to be complementary to the receptor, these methods can employ more computationally expensive scoring functions during their primary search step. A variant of this approach uses the spatial distribution of chemical functionalities on the ligand and receptor to generate ligand poses that are chemically complementary to the receptor. This pharmacophore docking method, implemented in the PhDOCK program,[37,38] reduces the number of ligand orientations sampled to those that are presumably biased towards more favorable binding energies.

9.2.2.2 *Flexible Ligand Docking*

Flexible ligand docking treats the ligand as a conformationally flexible molecule, by searching over both ligand conformations and rigid body transformations to identify the best fit of the ligand in the receptor, which is treated as a rigid body. While the induced fit of the receptor to the bound ligand is a very important contribution to both the complex structure and binding energy, the rigid receptor assumption is often used because it enables much faster docking.

The first class of flexible ligand docking methods involves explicitly generating a set of different conformations of a ligand prior to docking, and docking each ligand conformation individually to the receptor. This approach, known as the conformationally expanded database approach (or the flexibase approach), is implemented in the FLOG program,[35] and can be applied using any rigid docking method. It simply requires a pre-docking step to conformationally expand the ligand molecules, and a post-docking step to identify the best conformations of a single molecule from the docking results. This approach has the advantage of performing the conformation generation step for the ligand molecules once, prior to any dockings, thereby saving the effort of generating ligand conformations if the same ligands are docked to different receptors. If a library of ligands is to be docked to many different receptors, this approach can be quite beneficial. One disadvantage to this class of methods is that a rigid docking must be performed for each conformation of a ligand molecule, which can be quite computationally expensive for compound libraries that are highly conformationally expanded. The PhDOCK[37,38] and NWU DOCK[39,40] programs implement a variant of the flexibase method in which ligand conformations are spatially clustered and overlaid to form ensembles of multiple ligand conformations. The entire ensemble is rigidly docked to the receptor, which reduces the total number of dockings that need to be performed.

The second class of flexible ligand docking methods generates ligand conformations during the search of ligand orientations. A rigid substructure of the ligand is rigidly docked to the receptor, and conformations of the remaining portions of the molecule are generated and evaluated within the context of the receptor. This so-called incremental construction approach, implemented in the DOCK,[4–10,33] FlexX[34] and HAMMERHEAD[36] programs, has the advantage of sampling only relevant ligand conformations that fit in the receptor, thereby lowering the total number of ligand poses that must be scored.

The last class of flexible ligand docking methods employs global energy-minimization methods to search over ligand orientations and conformations for minimum-energy positions of the ligand. The programs AutoDock,[41] GOLD,[42] ICM,[43] Prodock[44] and QXP[45] fit into this category. These programs combine techniques such as genetic algorithms, simulated annealing, Monte Carlo and Brownian dynamics with very accurate scoring functions to identify energetically favorable ligand conformations. These methods perform the least amount of sampling; however, by using accurate energy scoring functions with robust energy minimizers they attempt to sample only the most relevant regions of conformational space.

9.2.2.3 Flexible Receptor Docking

Flexible receptor docking treats both the ligand and receptor as conformationally flexible molecules, providing the greatest accuracy[46,47] of any docking methods and incurring the greatest computational expense. Since the conformational space of a typical protein receptor is enormous, flexible receptor docking methods sample conformations that are close to the experimentally determined receptor structure. At a minimum, these methods implement receptor side-chain flexibility, often using a rotamer library to search through side-chain conformations. Some flexible receptor docking implementations permit a limited amount of backbone flexibility, usually by allowing loops in the receptor site to be flexible. Receptor flexibility requires the use of accurate molecular mechanics force fields, to ensure that the receptor structure does not adopt energetically unfavorable conformations.

The simplest implementation of flexible receptor docking uses a softened potential energy function to account for the induced fit of a receptor for a ligand. This is typically implemented by reducing the van der Waals radii of the receptor and ligand atoms, to alleviate ligand–receptor clashes that receptor flexibility might easily overcome. If a particularly bulky receptor residue is obstructing the target site, it can be mutated *in silico* to a less obtrusive side chain.

The next class of flexible receptor docking methods integrates the receptor conformational search with the ligand orientational and conformational searches. The programs AutoDock,[41] GOLD,[42] ICM[43] and Prodock,[44] as described previously, use global energy-minimization methods to search for low-energy ligand orientations and conformations. They also allow for conformations of receptor side chains in the target site to be included in the global energy minimization, which performs a simultaneous optimization of ligand and receptor conformations.

The last class of flexible receptor docking methods iteratively couples a flexible ligand docking method with a method that conformationally optimizes the ligand–receptor complex. One implementation of this approach is the use of docking programs with molecular dynamics programs, such as AMBER,[48,49] CHARMM[50] and NAMD,[51] to post-process docking results. The docking program GLIDE[30,31] has been used in conjunction with the protein-minimization program PRIME[52] to perform flexible receptor docking with high accuracy. While these methods represent the ideal level of sampling for reproducing experimental structures to atomic detail, they typically require on the order of hours of computation time to dock a single ligand, rendering them infeasible for docking large numbers of compounds.

9.2.3 Scoring Methods

Energy scoring functions are used to compute, to some degree of approximation, the binding free energy of a ligand to a receptor. The level of accuracy in the scoring function will inversely affect the speed of the calculation, and therefore it is important to determine what level of accuracy is required for a given docking method to best balance performance with accuracy.

9.2.3.1 Molecular Mechanics Based Scoring Functions

One class of energy scoring functions is based on molecular mechanics force fields, and implements a first-principles approach to calculate the binding energy of a ligand–receptor complex. The AMBER,[48,49] CHARMM[53] and MMFF[54,55] force fields are well parameterized for protein receptors binding to small molecule organic ligands and for protein–protein docking; force field parameters for nucleic acids[56] have also been developed. In order to use a force-field based scoring function, both the ligand and receptor molecules must have atom and bond types assigned according to the force-field definitions, and atom-centered partial charges must be assigned to all atoms in the system. The accuracy of the partial charge assignments greatly affects the quality of the energy score values; therefore it is worthwhile to use the most accurate methods possible, as described in Section 2.1.

The complete molecular mechanics energy expression includes intramolecular and intermolecular contributions. The intramolecular energy expression consists of bond energy and non-bonded interaction terms, which are evaluated independently for the ligand and the receptor molecules. The bond energy includes contributions from bond stretching, bond angle, torsion angle and improper torsion angle terms. The non-bonded interaction energy includes contributions from a Coulombic electrostatic energy term and a Lennard-Jones term for the van der Waals energy. The intermolecular energy expression consists of the same non-bonded energy terms as in the intramolecular energy expression, evaluated between all ligand–receptor atom pairs. Additionally, some force fields include explicit hydrogen bonding and metal coordination energy terms in the intermolecular energy expression.

There are two modifications to the standard molecular mechanics energy expression to account for the absence of explicit water molecules in docking systems. The first is that a surface area (SA) term is added to approximate the free energy of hydration of non-polar atoms. The second is that the Coulombic electrostatic term must be modified to account for the high dielectric field due to the solvent. One approach is to scale the Coulombic term by a distance-dependent dielectric term, $\frac{1}{\varepsilon} = \frac{1}{4*r}$, which is a very approximate correction that can be rapidly computed. A more sophisticated approach is to use an implicit solvent model, such as the Generalized Born (GB)[57,58] or Poisson–Boltzmann (PB)[59] models. The GB method corrects the Coulombic energy by adding the Born self-energy, which is the self-energy of a partial charge solvated by a high dielectric medium. The PB method entails numerically solving the non-linear Poisson–Boltzmann equation to determine the potential field resulting from a charge distribution in a non-uniform dielectric environment. Both the GB and PB approaches incur a significant increase in computational expense, while providing a much more accurate electrostatic energy than Coulomb's law with distance-dependent dielectric scaling.

In order to increase the speed of energy scoring, some docking implementations exclude terms from the complete free-energy expression. In the absence of receptor flexibility during docking, the intramolecular energy for the receptor is constant, and will not make any contribution to the binding free energy. Therefore the receptor internal energy is frequently omitted from energy scoring functions. The non-polar SA term is often omitted as well, while the distance-dependent dielectric electrostatic solvation correction can be used in place of the more accurate GB and PB methods. A very common approach is to use a precomputed potential energy grid composed of the Lennard-Jones and Coulombic potential values contributed by the receptor atoms. This enables rapid intermolecular scoring by eliminating an enormous number of pair-wise function evaluations when computing the intermolecular energy.

The major advantage of these energy scoring functions is that they are based on first principle molecular mechanics force fields, and they should be applicable to a wide range of molecular systems. The disadvantage of these scoring functions is that they are more computationally expensive than the knowledge-based and empirical energy-scoring functions described next.

9.2.3.2 Empirical Scoring Functions

Empirical scoring functions consist of molecular mechanics force-field energy terms weighted by a set of coefficients. The coefficients are optimized to fit the calculated molecular mechanics energies of a set of training data to energies from quantum mechanics calculations and from thermodynamic measurements. The principle underlying this method is that the coefficients compensate for errors in the molecular mechanics formulation of the energy function. The best-known implementations of this approach are the ChemScore energy score,[60] the Piecewise Linear Potential (PLP) energy score[61] and the OPLS force field.[62–64] The advantage of these scoring functions is that they produce more accurate energy values than "pure" molecular mechanics scoring functions, while the potential disadvantage is that accuracy for a given ligand–receptor complex will depend on how well the ligand and receptor were represented in the training data set used to fit the coefficients.

9.2.3.3 Knowledge-Based Scoring Functions

Knowledge-based scoring functions are composed of multiple weighted terms corresponding to ligand–receptor molecular descriptors. Examples of these descriptors can include the number of hydrogen bonds in a ligand–receptor complex, atom–atom contact energies and the number of rotatable bonds, just to name a few. A statistical approach, such as linear regression analysis, is used to fit the coefficients to best reproduce the binding energies in a set of training data. Neural network and Bayesian network approaches can also be used by mapping the molecular descriptors to either inputs of a neural network or to nodes of a Bayesian network, and then fitting network parameters such as edge weights to a set of training data. Many knowledge-based scoring functions are described in the literature; some of the better-known functions include the SMOG scoring function,[65] PMF,[66] DrugScore[67–69] and the SCORE1 and SCORE2 scoring functions in LUDI.[70,71] The advantage of scoring functions in this class is that they are extremely fast to evaluate. Additionally, the parameterization step is usually performed to reproduce the free energies of binding, so the scoring function should return complete binding free energies. The major disadvantage with this class of scoring functions is that the physical meaning of the scores is obscured and, as with the empirical scoring functions, the score will only be relevant for ligand–receptor complexes well represented in the training set.

9.2.4 Managing Errors in Docking

Due to the very large phase space of ligand–receptor complex configurations, approximate sampling and energy scoring methods are used when docking. The resulting errors in the sampled complex structures and binding energies can cause both complex structure prediction and docking applications based on ligand database ranking to fail. In order to use docking methods successfully, it is important to appropriately use higher accuracy sampling and scoring functions when necessary.

9.2.4.1 Complex Structure Prediction Failures

When a ligand is docked to a receptor, a list of ligand poses is generated and from this list a ligand pose with the most favorable binding energy score is selected. A standard metric of success in complex structure prediction is the root mean square deviation (RMSD) between the predicted ligand pose and the experimentally determined ligand pose. If this distance is within some threshold, usually about 2 Å, then the docking is considered successful. Structure prediction failures can be classified as sampling failures or scoring failures. Sampling failures occur when the sampling algorithm fails to sample any ligand poses near the native ligand pose; increasing the amount of sampling usually eliminates these failure modes.

Scoring failures occur when the predicted ligand pose has a more favorable energy score than the native ligand pose. These occur because of energy scoring-function errors, the most glaring of which is the frequent omission of a binding entropy term. Scoring failure ligand poses tend to be quite entropically unfavorable, however energy scoring functions that only calculate enthalpic contributions to the binding free energy cannot identify these ligand poses.

When a ligand–receptor docking results in a scoring failure, there is a good chance that a properly docked ligand pose was also sampled during the docking. It is therefore important to consider a number of high-ranking docked ligand poses in order to increase the probability of identifying the proper near-native ligand pose. Frequently, a set of ligand poses that all fall within some energy score window (*e.g.* the top 5 $kcal\,mol^{-1}$) are saved for further processing. Spatial clustering is then performed to identify clusters of nearly identical ligand poses, and discard all but the highest ranked from the set.[16,72] Rescoring the remaining ligand poses with higher accuracy scoring functions should correct the scoring failure if a sufficiently accurate energy function is used. During rescoring, the energy scoring function should at a minimum include a full molecular mechanics energy expression with either the GB/SA or PB/SA implicit solvent models. If it is computationally feasible to do so (*i.e.* only a small number of ligands are being docked), then receptor flexibility should be used while rescoring the top-ranked poses. This often alleviates small ligand–receptor bumps or clashes causing the native ligand pose to score poorly. If an extremely accurate binding free-energy value is required, then a molecular dynamics ensemble method such as MM–GBSA or MM–PBSA[73] should be used. These methods use a molecular dynamics trajectory to generate a Boltzmann-weighted ensemble of a ligand–receptor complex's conformations, which is used to compute the average energy for the entire thermodynamic ensemble. While these methods are quite computationally expensive, they produce very accurate binding free energies.

9.2.4.2 *Ligand Database Ranking Failures*

Due to errors inherent in the scoring functions, ranking a library of compounds according to their predicted binding energy will produce a correct global ordering, but the accuracy of the local ordering will depend on the magnitude of the scoring function error. Scoring function errors arise from a number of sources, including errors due to terms that are omitted from the energy scoring function. Scoring functions based on molecular mechanics without solvation correction terms contain two major sources of error. The first is that larger ligands tend to score more favorably than smaller ligands, due to the lack of a binding entropy correction or a hydration free-energy correction to counteract the greater enthalpy of binding for a larger ligand. The second is that ligands with greater formal charge tend to score more favorably than ligands with lesser formal charge, again due to the lack of solvation correction of the electrostatic energy term. Scoring functions based on molecular mechanics with GB/SA or PB/SA implicit solvation correction produce more accurate ligand–receptor binding energies, however both methods omit the configurational entropy lost by both the ligand and receptor upon binding. The molecular dynamics ensemble methods, MM-GBSA and MM-PBSA, which account for configurational entropy, are among the most accurate scoring functions that are commonly used in drug design.

In light of these scoring function errors, it is common to use a hierarchical approach when docking a library of compounds, allowing the use of the fastest energy scoring functions on a large number of compounds to identify a small number of compounds that should be re-evaluated with increasingly accurate scoring functions.

For example, when docking a library of 10^5–10^6 compounds to a protein target, in the first step all of the compounds would be flexibly docked to the receptor using a grid-based molecular mechanics energy score. For each compound, all poses that rank within the top 5 kcal mol^{-1} would be rescored using a molecular mechanics energy score with PB/SA solvation correction. The compound library would be ranked using the PB/SA energy score, and the compounds ranked within the top 5 kcal mol^{-1} would be docked and ranked, again allowing both ligand and receptor flexibility. Compounds that are ranked within the top 2 kcal mol^{-1} after this second docking would be selected for screening, or submitted for further computational studies, such as MM-PBSA calculations of more accurate binding free energies.

By estimating the magnitude of the scoring function error, it is possible to design hierarchical strategies that apply increasingly demanding binding energy calculations to shrinking numbers of compounds, enabling the accurate ranking of large compound libraries.

9.3 How is Docking Applied to Drug Design?

Drug design is the search for a chemical compound that possesses a desirable medical activity. In order for a compound to be a drug candidate, it must satisfy a large number of criteria. The compound must bind specifically and selectively to the biomolecular target, producing the desired biological activity (*e.g.* enzyme inhibition or biological signaling). The biological activity must in turn produce the desired medical effect. The compound must have favorable pharmacokinetic properties so that potent concentrations of the drug can be delivered to the relevant physiological locations, and it must not have any biological activities that cause adverse medical effects. Finally, it must be tractable to scale the synthesis of the compound up to production yields. In order for a single compound to meet all these criteria, a very diverse set of data, including chemical, biological, physiological and medical, must be collected.

Typically, a very large number of compounds are included at the start of the drug design process to maximize the probability of designing a suitable drug. Since some types of experiments are much more expensive than others (in terms of money, materials, time and equipment availability), it is necessary to apply a hierarchical approach in which lower cost experiments are applied to large numbers of compounds, and higher cost experiments are applied to smaller numbers of compounds. The key to this strategy is to correctly choose which compounds to eliminate from the drug design process at each stage, based on the data obtained in the previous stages.[74] Docking methods can be used to inform this process, and a number of excellent reviews[75–79] discuss the uses of docking in structure-based drug design approaches.

Docking can predict the both the bound structure and free energy of binding of a ligand to a macromolecular receptor. The degree of accuracy of these predictions depends upon the specific docking methods used; as a rule, more accurate results require greater computational expense. Both the accuracy of the docking predictions and the computational cost of calculating them should be considered in the context of the goals of the drug design process. This section of the chapter describes a hypothetical drug design process, excluding the preclinical and clinical development steps, and illustrates how docking can be applied.

9.3.1 Drug Target Selection and Characterization

Structure-based drug design approaches require that the molecular biology of a disease is understood and well characterized before the search for active compounds can begin. Typically this entails knowing, at a minimum, the identities of the genes and proteins involved in the disease, the structure of the biological network that they comprise and the features that differentiate the disease state from a healthy state. This research establishes the basic mechanism of a disease, and generally includes biochemistry, molecular biology, enzymology, bioinformatics, genomics and proteomics data. Structural biology studies are often initiated once the relevant macromolecules are identified; however, the structures of all the macromolecules and the complexes they form often require substantial time and effort to obtain. Both X-ray crystallography and NMR spectroscopy require large quantities and high concentrations of the target molecules, which can require extensive molecular biology efforts to achieve. NMR spectroscopy is hindered by the inability to resolve very large molecular structures, which precludes identifying the structures of large macromolecular complexes. Although X-ray crystallography can resolve very large molecular structures, identifying crystallization conditions for macromolecular complexes can be quite difficult. Homology modeling can be used to predict individual protein structures with a reasonable level of accuracy,[80] although it is much harder to predict complex structures using homology methods. Consequently, at this early stage of drug design, it is common for some apo macromolecule structures to be experimentally determined, for others to be computationally predicted and for some to be unknown. Macromolecular complex structures are rarely known at this stage.

Docking can be used in this early stage of drug design to predict molecular interactions that elude experimental determination, thereby helping to complete a model of the biological network. Small molecule docking can be used to predict enzyme–substrate interactions and complex structures,[81,82] and protein docking can be used to study protein–protein interactions.[83] Docking has also been used to predict nucleic acid interactions with proteins and small molecule ligands.[84] The purpose of docking at this stage of the drug design process is to predict answers to hypothesis-driven biological questions (*i.e.* do species A and B bind, and if so, does the complex AB bind species C?), in order to aid in the understanding of the molecular mechanism of the target disease.

Once the biological mechanism of the disease is understood, specific macromolecular receptors, usually proteins, are selected and validated as drug targets. The goal of this step is to identify molecular targets for which chemical intervention is most likely to result in a medically effective therapy. While the methods employed at this stage of the drug design process are outside the scope of this chapter, some excellent reviews discuss them.[85–90]

Once a molecular target has been selected and validated, it needs to be fully characterized. The functions of the target (*e.g.* enzymatic, signaling, structural) are identified and the functional residues are mapped, usually using a combination of site-directed mutagenesis, functional assays and bioinformatics techniques. The goal of this work is to identify the area on the target receptor where ligand binding will most likely have a biological effect. This information is very useful in the search for lead compounds, as it can be used to assist the development of ligand binding assays, and to direct virtual screening efforts.

Docking can be used to characterize the receptor in several ways. The first is a technique called computational solvent mapping, developed by Vajda *et al.*[91–93] which is used to predict the active site of a protein based on the protein structure alone. This method docks and energy-minimizes small organic solvent molecules, presenting a variety of functional groups, against the entire protein surface. The fragment poses tend to cluster during energy minimization, and protein active sites tend to contain many overlapping clusters. Solvent mapping essentially identifies protein active sites by identifying spatially clustered chemical functional groups on the protein surface. Similar methods have been described that use short peptides[94] and other small organic fragment libraries[95] rather than organic solvents.

Docking can also be used to characterize the interactions of a protein target and a known ligand, such as a substrate molecule or a natural inhibitor, if the complex structure eludes experimental determination. In this scenario, if the ligand-binding site is known or predicted, docking can be performed to predict the protein–ligand complex structure. The purpose of this prediction is to identify which receptor atoms and residues are involved in binding the known ligand, so that experimental studies can be directed to the predicted binding site. Many docking programs have been characterized for their ability to predict complex structures[28–31,33,34,41,42,45,96–98] using a wide variety of sampling and scoring techniques. In general, when they are successful the state-of-the-art methods can predict the ligand-binding pose to within 2.0 Å of the crystallographically determined ligand-binding pose, in many cases achieving predictions within 1.0 Å of the experimentally determined structure. At these levels of accuracy, the receptor residues involved in binding can be readily identified with a high degree of confidence.

The final docking application discussed in this section involves drugs that target macromolecular complexes. The assembly of a macromolecular complex can be disrupted by small molecule ligands that bind to the uncomplexed macromolecules and prevent multimerization. This requires identifying the binding interface, and ideally the residues that most contribute to the free energy of binding. Small molecule ligands can then be targeted to the region of the interface that is most likely to disrupt complex formation. As previously described, protein docking can predict protein complex structures when experimentally determined structures are not available. The goal of this complex structure prediction is to direct mutagenesis studies to the predicted interface region to validate the target. A number of protein docking programs have been evaluated in the Critical Assessment of Prediction of Interactions (CAPRI) complex structure prediction competition[99–101] and, when successful, they generally predict the interface residues to within 2.0 Å of the crystallographically determined structure. As with the protein–ligand docking described in the previous paragraph, at this level of accuracy it is easy to identify which residues make contact with each other.

In order to understand how to properly select the docking methods and parameters for the applications described in this section, it is important to consider what the motivations for performing the dockings are. The predictions made by the docking applications described in this section are used to inform and direct experiments that characterize the molecular mechanism of a disease, and that identify and characterize a target site on a macromolecular receptor. Ideally, while the docking calculations will require a certain amount of time to complete, they should reduce the number of experiments that need to be performed, thereby decreasing the cost in both time and dollars. Incorrect predictions will result in misdirected experiments, which can greatly increase the cost in both time and money. Therefore, at this stage accuracy is very important for the docking calculations, and justifies increased computational expense. This means that docking methods should consider full ligand flexibility, and receptor side-chain flexibility. The energy scoring function used should calculate an approximate free energy of binding for the ligand to the receptor, since it may be necessary to compare binding energies of one ligand to several receptors. There should be a molecular mechanics internal energy term, a van der Waals potential term and an electrostatic potential term. The electrostatic potential term should account for electrostatic solvation effects through the use of an implicit solvation model, such as the GB or PB models, and there should be a term for the free energy of solvation of non-polar atoms. Finally, docking results that conflict with experimentally determined data should be eliminated in order to increase the probability that the remaining results are correct. In summary, docking at this early stage of drug design should employ high levels of conformational sampling, using high-accuracy scoring functions and careful validation of docking results against empirical data.

9.3.2 Lead Compound Discovery

Once a macromolecular target, most often a protein, is identified, the search for lead compounds can begin. A lead compound is generally considered to be any compound that has at least moderate biological activity and is also likely to meet the remaining criteria required for a drug. A lead compound needs to be small enough so that when it is derivatized during lead optimization, the resulting molecules will not be too large to have favorable drug delivery properties. It also needs to have some likelihood of leading to compounds with favorable absorption, distribution, metabolism, elimination and toxicity (ADMET) properties. A general approach that has been widely adopted by industrial screening groups has been to create a compound screening library by eliminating from a much larger pool of potential compounds the undesirable compounds that do not satisfy their lead compound criteria. By conducting rapid functional or binding assays of the library against the target receptor, a number of diverse lead compounds can hopefully be identified. This ensures that if one lead compound is further developed as a drug candidate and fails, there will be another lead to fall back to that should have different biological properties. This section describes strategies that are commonly used to screen for lead compounds, and discusses how docking methods are used in conjunction with these.

HTS[102,103] is a widely adopted approach that uses robotics and other automation technologies to screen large numbers of compounds for biological activity. Commercial HTS systems can perform hundreds of thousands of individual assays each day, which equates to tens of thousands of IC_{50} measurements per day. HTS requires an assay for biological activity that is amenable to automation: it must be able to work in a small-volume multiwell plate, it must produce an optical readout and it must be robust to the conditions in a robotic system, such as vibrations and shocks from handling and temperature fluctuations. The primary advantage of HTS is the extremely large number of experiments that it can perform. The major drawback of HTS is that it has a high error rate, producing a large number of false positive results which increase the cost of identifying lead compounds. Additionally, HTS is expensive to operate, as it requires a large amount of chemical synthesis in order to replenish the screening library compounds as they are consumed. Despite these drawbacks, the vast majority of industrial drug design and discovery groups use HTS methods to identify lead compounds.

The most important component of HTS is the composition of the screening library, which determines the amount of information that can be gleaned from HTS results.[74,104,105] Some library design strategies emphasize chemical and structural diversity of the screening compounds to maximize the probability of finding biologically active compounds. Other library design strategies emphasize designing targeted libraries with compounds that share similar scaffolds or functional group pharmacophores, if there is information to guide scaffold or pharmacophore selection.

Combinatorial synthesis can generate thousands of compounds in parallel through the simultaneous addition, in one reaction pot, of multiple substituents to the same substitution sites on a molecular scaffold.[106] Mixtures containing many compounds are assayed for biological activity, and any mixture that yields a positive assay result is further analyzed to determine which specific compounds confer the biological activity. The use of combinatorial synthesis in conjunction with HTS enables billions of compounds to be assayed for biological activity against a single target.

Fragment-based compound screening approaches identify small molecular fragments that bind weakly to the target receptor, and covalently couple them to form larger compounds with hopefully much stronger binding affinity.[107,108] Multiple implementations of this approach have been reported. Wells *et al.* use mass spectrometry to detect fragments that bind to a target site on a protein and form a disulfide bond to a cysteine residue placed adjacent to the site.[109,110] Sharpless *et al.* perform *in situ* click chemistry that uses a protein binding site as a catalytic template to position fragments that independently bind to different regions of the site and are covalently coupled to each other.[111,112] While fragment-based methods are lower throughput than HTS and combinatorial chemistry methods, the combinatorial nature of these techniques allows them to explore virtually a large number of compounds that are known to make favorable interactions with the receptor, and they have demonstrated the ability to generate very high affinity ligands.

The major docking application used during this stage of drug design is virtual screening. Virtual screening uses fast docking methods to predict the relative binding affinities of a database of compounds. In order to enable the screening of large compound databases, faster but less accurate docking methods are used, which are generally unable to resolve differences between compounds that bind with similar energies. Therefore the predicted binding energies are useful for global ranking of a compound database (identifying the best 10% and the worst 10% of the compounds), but do not perform well for local ranking (*e.g.* exactly ranking the top 100 compounds). If accurate local ranking is required, a hierarchical docking strategy employing increasingly accurate energy scoring methods should be used.

Virtual screening can be used to augment HTS efforts in the design of targeted libraries enriched with compounds that are predicted to bind well to the target receptor.[113–117] Virtual screening can rank order an existing compound library in order to identify which subset of the library to use for HTS experiments, or to direct combinatorial synthesis by screening combinatorial compound libraries *in silico* and using the predicted binding energies to determine which scaffolds and substituents to use. An example of this is the CombiDOCK program[118] developed by Kuntz *et al.*, which docks combinatorial compound libraries very rapidly by assuming that the side chains of the molecule bind the receptor independently. CombiDOCK uses docking to position a scaffold molecule, and then sequentially iterates through a library of side chains at each substitution position on the scaffold, computing the binding energy that each side chain contributes. While this is an oversimplification of the receptor–ligand binding energy, it allows a large combinatorial library to be docked in linear time to a target.

It is also worth noting that many groups use virtual screening in lieu of HTS experiments, in many cases because they do not have access to HTS facilities. This approach demands the use of a hierarchical docking strategy in order to identify more accurately a sufficiently small number of compounds that can be synthesized or purchased, and then tested for biological activity with a conventional low throughput assay.

The last approach discussed in this section *is de novo* ligand design.[76,119–121] Although *de novo* design is not a docking technique *per se*, many *de novo* design algorithms make use of docking methods, which is why it is briefly described in this section. Receptor-based *de novo* design methods build ligands into a target site, relying upon the shape and chemical functionality of the site to guide the shape and chemistry of the ligand. These techniques first use docking methods to position small molecule fragments in a site, and then apply an algorithm, such as fragment linking or fragment growing, to build up the fragments into larger molecules.

For virtual screening applications, the most important factor influencing the selection of docking parameters is speed. Docking a molecule should require a few minutes of processor time in order for a virtual screen to feasibly process 10^5–10^6 compounds. In order to achieve this speed, the docking methods should consider full ligand flexibility, but should treat the receptor as rigid. Knowledge-based potentials are popular for use in virtual screening applications because of their speed and relative accuracy of the predicted binding energies. If a knowledge-based scoring function is used, it is important to compare the compounds in the training data set used to parameterize the scoring function with the compounds in the virtual screening library, to ensure that the binding energy predictions will be relevant for the compounds in the screening library. If an energy scoring function based on molecular mechanics is used, a number of options can greatly speed up the scoring function evaluation. A potential energy grid should be used to accelerate the calculation. The energy scoring function should include an intermolecular van der Waals potential term, an intermolecular Coulombic potential term with distance-dependent dielectric screening to approximate electrostatic solvation effects and an internal energy term for the ligand. The simplistic treatment of the electrostatic potential and electrostatic solvation effects is a source of error. If virtual screening is being performed in place of HTS, and only a small number of compounds will be assayed, a greater level of accuracy will be required. In this case, a hierarchical docking approach should be taken in which a second round of docking is performed on as many of the top ranked compounds as is computationally feasible, using a complete intramolecular molecular mechanics potential energy for both the ligand and the receptor and an intermolecular potential energy term using an implicit solvent model (GB/SA or PB/SA) to correct for electrostatic solvation effects. In summary, virtual screening should use a fast, inaccurate energy scoring function to approximately rank a large database of compounds and identify a subset that should be enriched with hits. This enables the design of targeted screening libraries that increase the efficiency of lead screening efforts.

9.3.3 Lead Compound Optimization

Once lead compounds have been identified for a drug target, they must be characterized and optimized with respect to a number of properties before they can be considered to be candidates for preclinical and clinical development.[122,123] In addition to the primary biological activity assay used during lead discovery, a number of secondary assays, often including tests of ADME/Tox properties[124,125] and biological selectivity, are used. The goal of lead optimization is to identify derivatives of the lead compound that have greater biological activity as well as favorable secondary properties.

While there are many different lead optimization strategies,[74,122] a fairly typical approach involves designing and synthesizing derivatives of the lead compound, which are then screened using both the primary biological activity assay and a number of secondary property assays. The assay results are used to generate a quantitative structure–activity relationship (QSAR) model[126,127] that can be used to deconvolute the contributions of lead compound modifications to biological activity, and these data are used to design another set of derivatives. This cycle is repeated until derivatives are identified that meet the criteria for preclinical development and clinical trials. The goal of this process is to minimize the amount of chemical synthesis that must be performed to identify candidate compounds for clinical development. It is therefore critical to decide whether to invest the effort to make a synthetically difficult modification when the biological effect of the modification is uncertain.

Docking can be used to guide the optimization of ligand–receptor binding, and presumably the biological activity, by predicting the binding energies of a library of lead compound derivatives. If the predicted energies are sufficiently accurate, then a great deal of chemical synthesis can be avoided by excluding derivatives that bind with lower affinity than the lead compound. One particular advantage of using docking for lead optimization is that the structural biology of lead compound binding is often known by this stage, which enables very accurate calculations of the free energy of binding using a molecular dynamics ensemble scoring method such as MM-PBSA or MM-GBSA.[73] The docking strategy is therefore to predict the ligand-binding pose as accurately as possible for each derivative in the library, and then perform a rescoring step to calculate the $\Delta\Delta G$ of binding with respect to the lead compound. These values are used to prioritize synthesis of the lead compound derivatives; for example, a very favorable predicted binding free energy might be used to justify performing a difficult synthesis.

For lead optimization, the accuracy of the docked ligand–receptor complex structures is the most important consideration when selecting docking parameters. The accuracy of molecular dynamics ensemble scoring methods is very sensitive to the receptor–ligand complex structure, and each calculation requires a good deal of computer time to complete. Therefore, docking errors in complex structure prediction will reduce the efficiency of this step. The docking method should consider full ligand flexibility, receptor side-chain flexibility and perhaps partial receptor backbone flexibility. The energy scoring function used should calculate an approximate free energy of binding for the ligand to the receptor. There should be a molecular mechanics internal energy term, a van der Waals potential term and an electrostatic potential term. The electrostatic potential term should account for electrostatic solvation effects through the use of an implicit solvation model, such as the GB or PB models, and there should be a term for the free energy of solvation of non-polar atoms. The theory and application of the MM-GBSA and MM-PBSA simulation methods, while outside the scope of this chapter, are discussed in the literature.[128–131]

9.4 Summary

The widespread availability of high-performance computers and the proliferation of high-resolution macromolecular structures have enabled the development and maturation of sophisticated docking methods. Docking programs are capable of predicting complex structures and binding energies with high accuracy, and they have been successfully incorporated into drug design research. As the field further matures, there is little doubt that these techniques will continue to play an important role in the discovery of new therapeutics.

References

1. H. M. Berman, J. Westbrook, Z. Feng, G. Gilliland, T. N. Bhat, H. Weissig, I. N. Shindyalov and P. E. Bourne, The Protein Data Bank, *Nucleic Acids Res.*, 2000, **28**(1), 235–42.
2. H. Berman, K. Henrick and H. Nakamura, Announcing the worldwide Protein Data Bank, *Nat. Struct. Biol.*, 2003, **10**(12), 980.
3. F. C. Bernstein, T. F. Koetzle, G. J. B. Williams, E. F. Meyer, M. D. Brice, J. R. Rodgers, O. Kennard, T Shimanouchi and M. Tasumi,

The Protein Data Bank: a computer-based archival file for macromolecular structures, *J. Mol. Biol.*, 1977, **112**(3), 535–42.
4. R. L. DesJarlais, R. P. Sheridan, G. L. Seibel, J. S. Dixon, I. D. Kuntz and R. Venkataraghavan, Using shape complementarity as an initial screen in designing ligands for a receptor binding site of known three-dimensional structure, *J. Med. Chem.*, 1988, **31**(4), 722–9.
5. T. J. Ewing and I. D. Kuntz, Critical evaluation of search algorithms for automated molecular docking and database screening, *J. Comput. Chem.*, 1997, **18**, 1175–89.
6. D. A. Gschwend and I. D. Kuntz, Orientational sampling and rigid-body minimization in molecular docking revisited: on-the-fly optimization and degeneracy removal, *J. Comput. Aided Mol. Des.*, 1996, **10**(2), 123–32.
7. I. D. Kuntz, J. M. Blaney, S. J. Oatley, R. Langridge and T. E. Ferrin, A geometric approach to macromolecule–ligand interactions, *J. Mol. Biol.*, 1982, **161**(2), 269–88.
8. E. C. Meng, B. K. Shoichet and I. D. Kuntz, Automated docking with grid-based energy evaluation, *J. Comput. Chem.*, 1992, **13**, 505–24.
9. E. C. Meng, D. A. Gschwend, J. M. Blaney and I. D. Kuntz, Orientational sampling and rigid-body minimization in molecular docking, *Proteins*, 1993, **17**(3), 266–78.
10. B. K. Shoichet and I. D. Kuntz, Matching chemistry and shape in molecular docking, *Protein Eng.*, 1993, **6**(7), 723–32.
11. D. B. Kitchen, H. Decornez, J. R. Furr and J. Bajorath, Docking and scoring in virtual screening for drug discovery: methods and applications, *Nat. Rev. Drug Discovery*, 2004, **3**(11), 935–49.
12. D. A. Gschwend, A. C. Good and I. D. Kuntz, Molecular docking towards drug discovery, *J. Mol. Recognit.*, 1996, **9**(2), 175–86.
13. R. D. Taylor, P. J. Jewsbury and J. W. Essex, A review of protein–small molecule docking methods, *J. Comput. Aided Mol. Des.*, 2002, **16**(3), 151–66.
14. N. Brooijmans and I. D. Kuntz, Molecular recognition and docking algorithms, *Annu. Rev. Biophys. Biomol. Struct.*, 2003, **32**, 335–73.
15. J. Krumrine, F. Raubacher, N. Brooijmans and I. Kuntz, Principles and methods of docking and ligand design, *Methods Biochem. Anal.*, 2003, **44**, 443–76.
16. I. Halperin, B. Ma, H. Wolfson and R. Nussinov, Principles of docking: An overview of search algorithms and a guide to scoring functions, *Proteins*, 2002, **47**(4), 409–43.
17. P. J. Gane and P. M. Dean, Recent advances in structure-based rational drug design, *Curr. Opin. Struct. Biol.*, 2000, **10**(4), 401–4.
18. T. J. Marrone, J. M. Briggs and J. A. McCammon, Structure-based drug design: computational advances, *Annu. Rev. Pharmacol. Toxicol.*, 1997, **37**, 71–90.
19. E. Garman and G. Laver, Controlling influenza by inhibiting the virus's neuraminidase, *Curr. Drug Targets*, 2004, **5**(2), 119–36.
20. M. von Itzstein, W. -Y. Wu, G. B. Kok, M. S. Pegg, J. C. Dyason, B. Jin, T. van Phan, M. L. Smythe, H. F. White, S. W. Oliver, P. M. Colman,

J. N. Varghese, D. M. Ryan, J. M. Woods, R. C. Bethell, V. J. Hotham, J. M. Cameron and C. R. Penn, Rational design of potent sialidase-based inhibitors of influenza virus replication, *Nature*, 1993, **363**(6428), 418–23.
21. S. W. Kaldor, V. J Kalish, J. F. Davies, B. V. Shetty, J. E. Fritz, K. Appelt, J. A. Burgess, K. M. Campanale, N. Y. Chirgadze, D. K. Clawson, B. A. Dressman, S. D. Hatch, D. A. Khalil, M. B. Kosa, P. P. Lubbehusen, M. A. Muesing, A. K. Patick, S. H. Reich, K. S. Su and J. H. Tatlock, Viracept (nelfinavir mesylate, AG1343): a potent, orally bioavailable inhibitor of HIV-1 protease, *J. Med. Chem.*, 1997, **40**(24), 3979–85.
22. Tripos Inc., 1699 South Hanley Road, St. Louis, MO 63144, USA.
23. Chemical Computing Group, 1010 Sherbrooke St. West, Suit 910, Montreal H3A 2R7, Canada.
24. Accelrys Inc., 9685 Scranton Road, San Diego, CA 92121-3752, USA.
25. OpenEye Scientific Software, 3600 Cerillos Road, Suite 1107, Santa Fe, NM 97507, USA.
26. C. I. Bayly, P. Cieplak, W. D. Cornell and P. Kollman, A well-behaved electrostatic potential based method using charge restraints for determining atom-centered charges: The RESP model, *J. Phys. Chem.*, 1993, **97**, 10269.
27. A. Jakalian, D. B. Jack and C. I. Bayly, Fast, efficient generation of high-quality atomic charges. AM1-BCC model: II. Parameterization and validation, *J. Comput. Chem.*, 2002, **23**(16), 1623–41.
28. Y. Pang, E. Perola, K. Xu and F. G. Prendergast, EUDOC: a computer program for identification of drug interaction sites in macromolecules and drug leads from chemical databases, *J. Comp. Chem.*, 2001, **22**, 1750–71.
29. M. R. McGann, H. R. Almond, A. Nicholls, J. A. Grant and F. K. M. Brown, Gaussian docking functions, *Biopolymers*, 2003, **68**(1), 76–90.
30. R. A. Friesner, J. L. Banks, R. B. Murphy, T. A. Halgren, J. J. Klicic, D. T. Mainz, M. P. Repasky, E. H. Knoll, M. Shelley, J. K. Perry, D. E. Shaw, P. Francis and P. S. Shenkin, Glide: a new approach for rapid accurate docking and scoring. 1. Method and assessment of docking accuracy, *J. Med. Chem.*, 2004, **47**(7), 1739–49.
31. T. A. Halgren, R. B. Murphy, R. A. Friesner, H. S. Beard, L. L. Frye, W. T. Pollard and J. L. Banks, Glide: a new approach for rapid, accurate docking and scoring.2. Enrichment factors in database screening, *J. Med. Chem.*, 2004, **47**(7), 1750–9.
32. H. A. Gabb, R. M. Jackson and M. J. Sternberg, Modelling protein docking using shape complementarity, electrostatics and biochemical information, *J. Mol. Biol.*, 1997, **272**(1), 106–20.
33. T. J. Ewing, S. Makino, A. G. Skillman and I. D. Kuntz, DOCK4.0: search strategies for automated molecular docking of flexible molecule databases, *J. Comput. Aided Mol. Des.*, 2001, **15**(5), 411–28.
34. D. Hoffmann, B. Kramer, T. Washio, T. Steinmetzer, M. Rarey and T. Lengauer, Two-stage method for protein–ligand docking, *J. Med. Chem.*, 1999, **42**(21), 4422–33.

35. M. D. Miller, S. K. Kearsley, D. J. Underwood and R. P. Sheridan, FLOG: a system to select 'quasi-flexible' ligands complementary to a receptor of known three-dimensional structure, *J. Comput. Aided Mol. Des.*, 1994, **8**(2), 153–74.
36. W. Welch, J. Ruppert and A.N. Jain, Hammerhead: fast, fully automated docking of flexible ligands to protein binding sites, *Chem. Biol.*, 1996, **3**(6), 449–62.
37. D. Joseph-McCarthy, B. E. T. Thomas, M. Belmarsh, D. Moustakas and J. C. Alvarez, Pharmacophore-based molecular docking to account for ligand flexibility, *Proteins*, 2003, **51**(2), 172–88.
38. D. Joseph-McCarthy and J. C. Alvarez, Automated generation of MCSS-derived pharmacophoric DOCK site points for searching multi-conformation databases, *Proteins*, 2003, **51**(2), 189–202.
39. D. M. Lorber and B. K. Shoichet, Flexible ligand docking using conformational ensembles, *Protein Sci.*, 1998, **7**(4), 938–50.
40. D. M. Lorber and B. K. Shoichet, Hierarchical docking of databases of multiple ligand conformations, *Curr. Top. Med. Chem.*, 2005, **5**(8), 739–49.
41. G. M. Morris, D. S. Goodsell, R. S. Halliday, R. Huey, W. E. Hart, R. K. Belew and A. J. Olson, Automated docking using a Lamarckian genetic algorithm and empirical binding free energy function, *J. Comput. Chem.*, 1998, **19**, 1639–62.
42. M. L. Verdonk, J. C. Cole, M. J. Hartshorn, C. W. Murray and R. D. Taylor, Improved protein–ligand docking using GOLD, *Proteins*, 2003, **52**(4), 609–23.
43. J. Fernandez-Recio, M. Totrov and R. Abagyan, ICM-DISCO docking by global energy optimization with fully flexible side-chains, *Proteins*, 2003, **52**(1), 113–7.
44. J.-Y. Trosset and H. A. Scheraga, Prodock: Software package for protein modeling and docking, *J. Comput. Chem.*, 1999, **20**(4), 412–27.
45. C. McMartin and R. S. Bohacek, QXP: powerful, rapid computer algorithms for structure-based drug design, *J. Comput. Aided Mol. Des.*, 1997, **11**(4), 333–44.
46. H. A. Carlson and J. A. McCammon, Accommodating protein flexibility in computational drug design, *Mol. Pharmacol*, 2000, **57**(2), 213–8.
47. H. A. Carlson, Protein flexibility and drug design: how to hit a moving target, *Curr. Opin. Chem. Biol.*, 2002, **6**(4), 447–52.
48. D. A. Case, T. E. Cheatham, T. Darden, H. Gohlke, R. Luo, K. M. Merz, A. Onufriev, C. Simmerling, B. Wang and R. J. Woods, The Amber biomolecular simulation programs, *J. Comput. Chem.*, 2005, **26**(16), 1668–88.
49. J. W. Ponder and D. A. Case, Force fields for protein simulations, *Adv. Protein Chem.*, 2003, **66**, 27–85.
50. B. R. Brooks, R. E. Bruccoleri, B. D. Olafson, D. J. States, S. Swaminathan and M. Karplus, CHARMM: A program for macromolecular energy, minimization, and dynamics calculations, *J. Comput. Chem.*, 1983, **4**(2), 187–217.

51. J. C. Phillips, R. Braun, W. Wang, J. Gumbart, E. Tajkhorshid, E. Villa, C. Chipot, R. D. Skeel, L. Kalé and K. Schulten, Scalable molecular dynamics with NAMD, *J. Comput. Chem.*, 2005, **26**(16), 1781–802.
52. W. Sherman, T. Day, M. P. Jacobson, R. A Friesner and R. Farid, Novel procedure for modeling ligand/receptor induced fit effects, *J. Med. Chem.*, 2006, **49**(2), 534–53.
53. A. D. MacKerell, B. Brooks, C. L. Brooks III, L. Nilsson, B. Roux, Y. Won and M. Karplus, CHARMM: The energy function and its parameterization with an overview of the program, in *The Encyclopedia of Computational Chemistry*, ed. P. v. R. Schleyer, N. L. Allinger, T. Clark, J. Gasteiger, P. A. Kollman, H. F. Schaefer III and P. R. Schreiner, John Wiley & Sons, Chichester, 1998, pp. 271–7.
54. T. A. Halgren and R. B. Nachbar, Merck molecular force field. IV. Conformational energies and geometries for MMFF94, *J. Comput. Chem.*, 1996, **17**(5–6), 587–615.
55. T. A. Halgren, MMFF VI. MMFF94s option for energy minimization studies, *J. Comput. Chem.*, 1999, **20**(7), 720–9.
56. T. E. Cheatham 3rd and M. A. Young, Molecular dynamics simulation of nucleic acids: successes, limitations, and promise, *Biopolymers*, 2000, **56**(4), 232–56.
57. W. C. Still, A. Tempczyk, R. C. Hawley and T. Hendrickson, Semianalytical treatment of solvation for molecular mechanics and dynamics, *J. Am. Chem. Soc.*, 1990, **112**, 6127–9.
58. G. D. Hawkins, C. J. Cramer and D. G. Truhlar, Parameterized models of aqueous free energies of solvation based on pairwise descreening of solute atomic charges from a dielectric medium, *J. Phys. Chem.*, 1996, **100**, 19824–39.
59. D. Sitkoff, K. A. Sharp and B. Honig, Accurate calculation of hydration free energies using macroscopic solvent models, *J. Phys. Chem.*, 1994, **98**, 1978–88.
60. M. D. Eldridge, C. W. Murray, T. R. Auton, G. V. Paolini and R. P. Mee, Empirical scoring functions: I. The development of a fast empirical scoring function to estimate the binding affinity of ligands in receptor complexes, *J. Comput. Aided Mol. Des.*, 1997, **11**, 425–45.
61. D. K. Gehlhaar, G. M. Verkhivker, P. A. Rejto, C. J. Sherman, D. B. Fogel, L. J. Fogel and S. T. Freer, Molecular recognition of the inhibitor AG-1343 by HIV-1 protease: conformationally flexible docking by evolutionary programming, *Chem. Biol.*, 1995, **2**(5), 317–24.
62. W. L. Jorgensen and J. Tirado-Rives, The OPLS [optimized potentials for liquid simulations] potential functions for proteins, energy minimizations for crystals of cyclic peptides and crambin, *J. Am. Chem. Soc.*, 1988, **110**(6), 1657–66.
63. W. L. Jorgensen, D. S. Maxwell and J. Tirado-Rives, Development and testing of the OPLS all-atom force field on conformational energetics and properties of organic liquids, *J. Am. Chem. Soc.*, 1996, **118**(45), 11225–36.

64. G. A. Kaminski, R. A. Friesner, J. Tirado-Rives and W. L. Jorgensen, Evaluation and reparametrization of the OPLS-AA force field for proteins via comparison with accurate quantum chemical calculations on peptides, *J. Phys. Chem. B*, 2001, **105**(28), 6474–87.
65. R. S. DeWitte and E. I. Shakhnovich, SMoG: *de novo* design method based on simple, fast, and accurate free energy estimates.1. Methodology and supporting evidence, *J. Am. Chem. Soc.*, 1996, **118**(47), 11733–44.
66. I. Muegge, PMF scoring revisited, *J. Med. Chem.*, 2005, **49**(20), 5895–902.
67. H. F. Velec, H. Gohlke and G. Klebe, DrugScore(CSD) – knowledge-based scoring function derived from small molecule crystal data with superior recognition rate of near-native ligand poses and better affinity prediction, *J. Med. Chem.*, 2005, **48**(20), 6296–303.
68. H. Gohlke and G. Klebe, Statistical potentials and scoring functions applied to protein–ligand binding, *Curr. Opin. Struct. Biol.*, 2001, **11**(2), 231–5.
69. H. Gohlke, M. Hendlich and G. Klebe, Knowledge-based scoring function to predict protein–ligand interactions, *J. Mol. Biol.*, 2000, **295**(2), 337–56.
70. H. J. Bohm, The development of a simple empirical scoring function to estimate the binding constant for a protein-ligand complex of known three-dimensional structure, *J. Comput. Aided Mol. Des.*, 1994, **8**(3), 243–56.
71. H. J. Bohm, Prediction of binding constants of protein ligands: a fast method for the prioritization of hits obtained from *de novo* design or 3D database search programs, *J. Comput. Aided Mol. Des.*, 1998, **12**(4), 309–23.
72. D. Kozakov, K. H. Clodfelter, S. Vajda and C. J. Camacho, Optimal clustering for detecting near-native conformations in protein docking, *Biophys. J.*, 2005, **89**(2), 867–75.
73. M. Feig and C. L. Brooks, Evaluating CASP4 predictions with physical energy functions, *Proteins: Struct., Funct., Genet.*, 2002, **49**(2), 232–45.
74. K. H. Bleicher, H. J. Bohm, K. Muller and A. I. Alanine, Hit and lead generation: beyond high-throughput screening, *Nat. Rev. Drug Discovery*, 2003, **2**(5), 369–78.
75. A. C. Anderson, The process of structure-based drug design, *Chem. Biol.*, 2003, **10**(9), 787–97.
76. D. Joseph-McCarthy, Computational approaches to structure-based ligand design, *Pharmacol. Ther.*, 1999, **84**(2), 179–91.
77. G. Klebe, Recent developments in structure-based drug design, *J. Mol. Med.*, 2000, **78**(5), 269–81.
78. I. D. Kuntz, Structure-based strategies for drug design and discovery, *Science*, 1992, **257**(5073), 1078–82.
79. J. Bajorath, Integration of virtual and high-throughput screening, *Nat. Rev. Drug Discovery*, 2002, **1**(11), 882–94.
80. M. A. Marti-Renom, A. C. Stuart, A. Fiser, R. Sanchez, F. Melo and A. Sali, Comparative protein structure modeling of genes and genomes, *Annu. Rev. Biophys. Biomol. Struct.*, 2000, **29**, 291–325.

81. D. S. Goodsell, G. M. Morris and A. J. Olson, Automated docking of flexible ligands: applications of AutoDock, *J. Mol. Recognit.*, 1996, **9**(1), 1–5.
82. Y. Z. Chen and D. G. Zhi, Ligand–protein inverse docking and its potential use in the computer search of protein targets of a small molecule, *Proteins*, 2001, **43**(2), 217–26.
83. G. R. Smith and M. J. Sternberg, Prediction of protein–protein interactions by docking methods, *Curr. Opin. Struct. Biol.*, 2002, **12**(1), 28–35.
84. K. E. Lind, Z. Du, K. Fujinaga, B. M. Peterlin and T. L. James, Structure-based computational database screening, *in vitro* assay, and NMR assessment of compounds that target TAR RNA, *Chem. Biol.*, 2002, **9**(2), 185–93.
85. I. I. Drews, Drug discovery today – and tomorrow, *Drug Discovery Today*, 2000, **5**(1), 2–4.
86. J. Drews, Drug discovery: a historical perspective, *Science*, 2000, **287**(5460), 1960–4.
87. J. Knowles and G. Gromo, A guide to drug discovery: Target selection in drug discovery, *Nat. Rev. Drug Discovery*, 2003, **2**(1), 63–9.
88. E. H. Ohlstein, A. G. Johnson, J. D. Elliott and A. M. Romanic, New strategies in drug discovery, *Methods Mol. Biol.*, 2006, **316**, 1–11.
89. E. H. Ohlstein, R. R. Ruffolo Jr. and J. D. Elliott, Drug discovery in the next millennium, *Annu. Rev. Pharmacol. Toxicol.*, 2000, **40**, 177–91.
90. G. C. Terstappen and A. Reggiani, *In silico* research in drug discovery, *Trends Pharmacol Sci*, 2001, **22**(1), 23–6.
91. T. Kortvelyesi, S. Dennis, M. Silberstein, L. Brown III and S. Vajda, Algorithms for computational solvent mapping of proteins, *Proteins*, 2003, **51**(3), 340–51.
92. S. H. Sheu, T. Kaya, D. J. Waxman and S. Vajda, Exploring the binding site structure of the PPAR gamma ligand-binding domain by computational solvent mapping, *Biochemistry*, 2005, **44**(4), 1193–209.
93. M. Silberstein, S. Dennis, L. Brown III, T. Kortvelyesi, K. Clodfelter and S. Vajda, Identification of substrate binding sites in enzymes by computational solvent mapping, *J. Mol. Biol.*, 2003, **332**(5), 1095–113.
94. I. Halperin, H. Wolfson and R. Nussinov, SiteLight: binding-site prediction using phage display libraries, *Protein Sci.*, 2003, **12**(7), 1344–59.
95. A. Miranker and M. Karplus, Functionality maps of binding sites: a multiple copy simultaneous search method, *Proteins*, 1991, **11**(1), 29–34.
96. E. Kellenberger, J. Rodrigo, P. Muller and D. Rognan, Comparative evaluation of eight docking tools for docking and virtual screening accuracy, *Proteins*, 2004, **57**(2), 225–42.
97. E. Perola, W. P. Walters and P. S. Charifson, A detailed comparison of current docking and scoring methods on systems of pharmaceutical relevance, *Proteins*, 2004, **56**(2), 235–49.
98. C. Bissantz, G. Folkers and D. Rognan, Protein-based virtual screening of chemical databases. 1. Evaluation of different docking/scoring combinations, *J. Med. Chem.*, 2000, **43**(25), 4759–67.

99. J. Janin, The targets of CAPRI rounds 3–5, *Proteins*, 2005, **60**(2), 170–5.
100. J. Janin, K. Henrick, J. Moult, M. J. Sternberg, S. Vajda, I. Vakser and S. J. Wodak, CAPRI: a critical assessment of predicted interactions, *Proteins*, 2003, **52**(1), 2–9.
101. J. Janin, Assessing predictions of protein–protein interaction: the CAPRI experiment, *Protein Sci.*, 2005, **14**(2), 278–83.
102. R. P. Hertzberg and A. J. Pope, High-throughput screening: new technology for the 21st century, *Curr. Opin. Chem. Biol.*, 2000, **4**(4), 445–51.
103. S. A. Sundberg, High-throughput and ultra-high-throughput screening: solution- and cell-based approaches, *Curr Opin Biotechnol*, 2000, **11**(1), 47–53.
104. H. J. Bohm and M. Stahl, Structure-based library design: molecular modelling merges with combinatorial chemistry, *Curr. Opin. Chem. Biol.*, 2000, **4**(3), 283–6.
105. W. P. Walters and M. Namchuk, Designing screens: how to make your hits a hit, *Nat. Rev. Drug Discovery*, 2003, **2**(4), 259–66.
106. S. R. Wilson and A. W. Czarnik, *Combinatorial Chemistry: Synthesis and Application*, Wiley-IEEE, Chichester, 1997.
107. R. A. Carr, M. Congreve, C. W. Murray and D. C. Rees, Fragment-based lead discovery: leads by design, *Drug Discovery Today*, 2005, **10**(14), 987–92.
108. D. C. Rees, M. Congreve, C. W. Murray and R. Carr, Fragment-based lead discovery, *Nat. Rev. Drug Discovery*, 2004, **3**(8), 660–72.
109. D. A. Erlanson, A. C. Braisted, D. R. Raphael, M. Randal, R. M. Stroud, E. M. Gordon and J. A. Wells, Site-directed ligand discovery, *Proc. Natl. Acad. Sci. U. S. A.*, 2000, **97**(17), 9367–72.
110. D. A. Erlanson and S. K. Hansen, Making drugs on proteins: site-directed ligand discovery for fragment-based lead assembly, *Curr. Opin. Chem. Biol.*, 2004, **8**(4), 399–406.
111. H. C. Kolb and K. B. Sharpless, The growing impact of click chemistry on drug discovery, *Drug Discovery Today*, 2003, **8**(24), 1128–37.
112. W. G. Lewis, L. G. Green, F. Grynszpan, Z. Radic, P. R. Carlier, P. Taylor, M. G. Finn and K. B. Sharpless, Click chemistry *in situ:* acetylcholinesterase as a reaction vessel for the selective assembly of a femtomolar inhibitor from an array of building blocks, *Angew Chem., Int. Ed.*, 2002, **41**(6), 1053–7.
113. P. D. Lyne, Structure-based virtual screening: an overview, *Drug Discovery Today*, 2002, **7**(20), 1047–55.
114. G. Schneider and H. J. Bohm, Virtual screening and fast automated docking methods, *Drug Discovery Today*, 2002, **7**(1), 64–70.
115. B. K. Shoichet, Virtual screening of chemical libraries, *Nature*, 2004, **432**(7019), 862–5.
116. J. C. Alvarez, High-throughput docking as a source of novel drug leads, *Curr. Opin. Chem. Biol.*, 2004, **8**(4), 365–70.
117. B. K. Shoichet, S. L. McGovern, B. I. Wei and J. J. Irwin, Lead discovery using molecular docking, *Curr. Opin. Chem. Biol.*, 2002, **6**(4), 439–46.

118. Y. Sun, T. J. A. Ewing, A. G. Skillman and I. D. Kuntz, CombiDOCK: structure-based combinatorial docking and library design, *J. Comput. Aided Mol. Des.*, 1998, **12**(6), 597–604.
119. J. Apostolakis and A. Caflisch, Computational ligand design, *Comb. Chem. High Throughput Screen*, 1999, **2**(2), 91–104.
120. A. Caflisch, Computational combinatorial ligand design: application to human alpha-thrombin, *J. Comput. Aided Mol. Des.*, 1996, **10**(5), 372–96.
121. H. J. Bohm, Current computational tools for *de novo* ligand design, *Curr. Opin. Biotechnol.*, 1996, **7**(4), 433–6.
122. G. W. Caldwell, D. M. Richie, J. A. Masucci, W. Hageman and Z. Yan, The new pre-preclinical paradigm: compound optimization in early and late phase drug discovery, *Curr. Top. Med. Chem.*, 2001, **1**(5), 353–66.
123. L. J. Lesko, M. Rowland, C. C. Peck and T. F. Blaschke, Optimizing the science of drug development: opportunities for better candidate selection and accelerated evaluation in humans, *J. Clin. Pharmacol.*, 2000, **40**(8), 803–14.
124. A. P. Li, Screening for human ADME/Tox drug properties in drug discovery, *Drug Discovery Today*, 2001, **6**(7), 357–66.
125. H. Yu and A. Adedoyin, ADME-Tox in drug discovery: integration of experimental and computational technologies, *Drug Discovery Today*, 2003, **8**(18), 852–61.
126. H. Kubinyi, QSAR and 3D QSAR in drug design. Part 2: applications and problems, *Drug Discovery Today*, 1997, **2**(12), 538–46.
127. H. Kubinyi, QSAR and 3D QSAR in drug design Part 1: methodology, *Drug Discovery Today*, 1997, **2**(11), 457–67.
128. J. M. Swanson, R. H. Henchman and J. A. McCammon, Revisiting free energy calculations: a theoretical connection to MM/PBSA and direct calculation of the association free energy, *Biophys. J.*, 2004, **86**(1 Pt 1), 67–74.
129. F. Fogolari, A. Brigo and H. Molinari, Protocol for MM/PBSA molecular dynamics simulations of proteins, *Biophys. J.*, 2003, **85**(1), 159–66.
130. J. Wang, P. Morin, W. Wang and P. A. Kollman, Use of MM-PBSA in reproducing the binding free energies to HIV-1 RT of TIBO derivatives and predicting the binding mode to HIV-1 RT of efavirenz by docking and MM-PBSA, *J. Am. Chem. Soc.*, 2001, **123**(22), 5221–30.
131. I. Massova and P. A. Kollman, Combined molecular mechanical and continuum solvent approach (MM-PBSA/GBSA) to predict ligand binding, *Perspect. Drug Discovery Des.*, 2000, **18**(1), 113–35.

CHAPTER 10

Strength in Flexibility: Modeling Side-Chain Conformational Change in Docking and Screening

LESLIE A. KUHN

Departments of Biochemistry & Molecular Biology, Computer Science & Engineering, and Physics & Astronomy, and the Quantitative Biology and Modeling Initiative, 502C Biochemistry Building, Michigan State University, East Lansing, Michigan, 48824, USA

10.1 Introduction

Modeling protein flexibility in structure-based drug design and virtual screening remains a strong challenge due to the number of degrees of freedom involved and the co-optimization of protein and ligand shape and chemistry. However, there is a growing trend towards incorporating some protein side-chain flexibility modeling in docking, which enables better ligand positioning and scoring, which in turn can enhance the success of virtual screening. Here, we present several methods used for side-chain flexibility modeling in docking and recent insights gained from analyzing conformational transitions between ligand-free and bound crystal structures.

10.2 Background

10.2.1 Improving Docking and Screening Through Side-chain Flexibility Modeling

Several studies have shown that better sampling of motions during docking, including sampling ligand orientations more finely and modeling induced fit between the protein and ligand, improves the ability of protein–ligand complementarity scoring functions to detect the most accurate docking.[1–4] For instance, when using the docking and screening tool SLIDE (Screening for Ligands with Induced-fit Docking, Efficiently) to dock 42 known thrombin ligands and 15 glutathione S-transferase (GST) ligands into the apo protein structures (reflecting the ligand-free binding site conformations), only nine of the 42 thrombin ligands and nine of the 15 GST ligands could be docked without modeling protein conformational change, even when the ligands were provided in their protein-bound conformations.[3] Modeling modest conformational change – by choosing the single bond(s) in the protein or ligand that could resolve the steric overlaps with the smallest rotation – allowed 86% of the thrombin ligands and 93% of the GST ligands to be docked accurately [to within 1.3 Å ligand root mean square deviation (RMSD), on average]. The same approach for modeling side-chain flexibility allowed SLIDE to identify nine out of 10 known thymidine kinase ligands within the 40 top-scoring compounds (and six in the top 25 compounds), when using the unbiased, ligand-free conformation of thymidine kinase as the screening template and a database of 80 000 conformers of ligand candidates (representing low-energy conformations of the known ligands mixed with a set of 1000 drug-like molecules).[5] In a virtual screening project to discover inhibitors for asparaginyl-tRNA synthetase from *Brugia malayi* (a human parasite causing elephantiasis), this approach resulted in a 15% hit rate; seven out of 45 compounds identified by SLIDE were experimentally confirmed to be micromolar inhibitors.[6]

The need for side-chain flexibility modeling is not so obvious from the numerous redocking studies that have been published, which tend to emphasize the ability of existing docking tools to predict ligand binding modes to within 2 Å RMSD across a range of protein structures and ligand chemical classes. However, even in this easy case of redocking, in which the protein and ligand structures are provided in their bound conformations, the best of these methods[7] currently fail to dock 45% of the ligands. The problem becomes much more complex in a predictive mode, in which virtual screening is used to identify new classes of ligands, given a protein binding site that is not pre-conformed to fit any of them. Thus, docking and screening studies using apo protein structures are likely to represent a much more realistic test. Fortunately, ligand conformations, at least for the cases with relatively few rotatable bonds (which tend to be favorable, in any case, due to the decreased entropic cost of binding), can be sampled reasonably well by existing Monte Carlo, genetic algorithm, and exhaustive torsional search methods in tools such as GOLD,[8] AutoDock,[9] FLEXX,[10] DOCK,[4,11] and Omega.[12–13]

10.2.2 Enhancing Target Specificity Through Flexibility Modeling

For many protein targets of interest, the druggable binding site (*e.g.* the ATP site in protein kinases) is highly conserved in homologous proteins. Inhibitors that bind to highly conserved sites present a risk of serious side effects or toxicity, which is typically evaluated through costly *in vitro* screening of compounds against a broad panel of homologous proteins to assess cross-reactivity, followed by extensive pharmacological testing. Many otherwise promising, high-affinity compounds discovered by screening and improved by structure-based design are lost in the process.

Structural plasticity among human protein kinases[14] and differences in flexibility and dynamics of bacterial thymidylate kinases[15] have been proposed as the basis for designing more specific inhibitors. For the folate biosynthetic enzyme and antibiotic drug target, 6-hydroxymethyl-7,8-dihydropterin pyrophosphokinase (HPPK), differences in active-site loop and side-chain conformations between three bacterial enzymes have been identified by crystallography,[16] framework dynamics,[17–19] and molecular dynamics (MD) analysis (L. Yao, M. Tonero, L. A. Kuhn and R. I. Cukier, unpublished results). We are now representing these conformational differences as a series of design templates to screen for species-selective inhibitors. A similar approach has elucidated the specificity of long side-chain pyrrolopyrimidines for asparaginyl-tRNA synthetase from *B. malayi* relative to the human enzyme. Their ligand-binding residues are absolutely conserved. However, a single side-chain difference (Thr to Ala) near the base of an active-site loop, and facing away from the site, apparently allows the loop to open more in the *Brugia* protein, allowing the ligand to bind preferentially to *Brugia*.[6] In general, we propose that low-energy protein conformations that differ from the closed, catalytic conformation are likely to present greater differences between species than the closed conformation. The existence of these unique, low-energy conformations can reflect sequence variation that occurs outside, but near, the binding site, and they are likely to be subject to decreased evolutionary selection relative to the catalytically productive, closed conformation. Thus, beyond improving docking and screening, the ability to accurately model side-chain (and main-chain) conformations in and around ligand binding sites is expected to open a range of new possibilities for gaining specificity between closely homologous enzymes.

10.3 Approaches

10.3.1 The State of the Art in Modeling Protein Side-chain Flexibility

We are fortunate that side-chain flexibility modeling can largely be decoupled from modeling main-chain flexibility (which involves many additional degrees of freedom). A study of almost 1000 pairs of ligand-free and bound protein structures found no correlation between the degree of main-chain and side-chain movement.[20] As a result, some groups have used rotamer libraries for protein side chains[21] to allow efficient sampling of their conformations in the ligand interface.[22–24] This is computationally feasible for docking, but not for screening. Additional drawbacks of rotamer sampling are discussed below.

Some methods for modeling side-chain flexibility effectively couple side-chain and main-chain motion by using as docking targets an ensemble of experimentally observed structures of the protein, often reflecting crystal structures solved in different space groups or with different ligands bound.[9,25–27] An advantage of this approach is that all the known, low-energy protein conformations can be considered. One of these methods considers combinations of conformations from different crystal structures that are mutually compatible,[26] but it is not clear whether this has an advantage over considering the different target conformations individually. Disadvantages of these approaches are that only existing protein conformations are sampled, and they do not reflect all the possible conformations, particularly when the protein binds to a substantially different class of ligands.

Soft docking is a more conservative approach which accepts that not all protein and ligand accommodations upon binding can be accurately predicted, and thus allows some degree of overlap between protein and ligand atoms during docking. Several methods either allow small van der Waals overlaps or dock the molecules using a smoothed representation of the protein surface.[28–30] This strategy can be combined with any of the others (*e.g.* side-chain sampling or docking into ensembles of structures), and often is. A complementary approach used at the end of docking is energy minimization to ensure that any van der Waals overlaps between protein and ligand atoms can be resolved. Energy minimization adjusts the atoms' positions to energetically improve interactions in the interface, but it does not attempt to overcome the energy barriers that would be involved in significant rearrangements; therefore, the motions are typically quite small. One successful application of energy minimization is GOLD, in which only polar, terminal hydrogen atoms on protein side chains are considered flexible, and the penultimate bond rotational angle is chosen to optimize hydrogen bond interactions.[9] However, as with other methods that perform detailed energy calculations in the course of optimization, this approach (which also includes genetic algorithm sampling of ligand conformations and orientations) proves too computationally intensive for large-scale high-throughput screening. At the far end of this spectrum, in terms of fineness of sampling and scoring, are Monte Carlo and MD techniques that consider all atoms as free to move within a force field including van der Waals, electrostatics, bond torsion and bending, and solvation energy terms.[27,31–35] These methods are generally appropriate for docking single protein–ligand complexes once there is a reasonably accurate initial placement of the ligand. However, MD simulations typically cannot surmount large conformational or orientational energy barriers within a reasonable timeframe.

SLIDE[3,28] represents an intermediate approach, in which all interfacial protein side chains and all single bonds in ligands are free to rotate during docking, but these motions are designed to remove van der Waals overlaps rather than thoroughly search the conformational space. As such, SLIDE's motions tend to be small, similar to those in energy minimization. However, because the bond angles to resolve collisions are calculated geometrically (Figure 10.1) rather than with respect to an energy function, the process is very fast. Protein side-chain and ligand flexibility have been modeled while screening and docking 150 000–800 000 compounds or 3D conformers using SLIDE,[6,28] with about 100 000 candidates screened per day on a two-processor

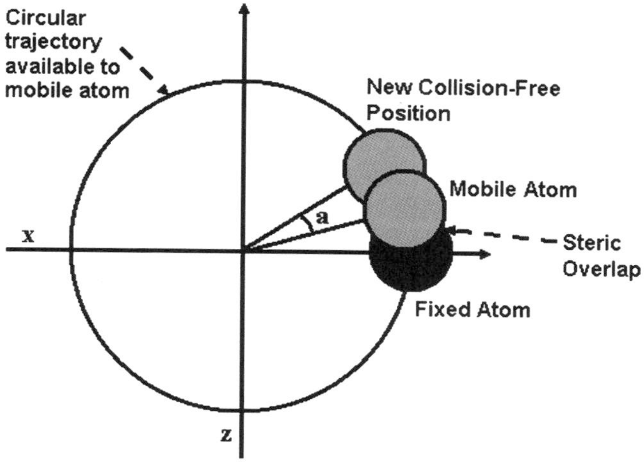

Figure 10.1 SLIDE performs directed rotations, calculated geometrically, to resolve protein–ligand van der Waals overlaps. Directed rotations are performed around a rotatable bond preceding the mobile atom. (Choice of the bond to be rotated, and therefore which atom is fixed versus mobile, is explained in Figure 10.2.) The rotatable bond is aligned along the *Y*-axis, with the hinge atom of the bond at $Y = 0$. The molecular system is rotated such that the fixed atom falls on the *X*-axis, in the positive quadrant of the X–Y plane. The rotation angle, **a**, is calculated such that the center of the mobile atom is displaced from the center of the fixed atom by the magnitude of their van der Waals overlap. This resolves the collision by performing a minimal rotation.

workstation. How SLIDE selects the bond(s) to rotate during flexibility modeling is described in Figure 10.2. The motions performed within SLIDE are typically disseminated throughout the interface, as shown for a protease–peptidyl ligand complex (with somewhat larger-than-typical motions; Figure 10.3). To complement its balanced protein and ligand flexibility modeling, SLIDE is typically combined with Omega to fully sample low-energy ligand conformations.[12,13] SLIDE can also be combined with ROCK (Ring-Optimized Conformational Kinetics[18,19]) to sample protein main-chain conformations that preserve the native non-covalent bond network, reflecting moderate- to large-scale motions that tend to be low in energy. Therefore, full protein and ligand flexibility can be modeled by providing a database of low-energy ligand conformations as input to SLIDE for screening against a panel of protein conformations sampled by ROCK, or by MD, crystallography, or nuclear magnetic resonance (NMR).[6,19]

10.3.2 Learning from Nature: Observing Side-chain Motions Upon Ligand Binding

An important question is: what kinds of motions do protein side-chains actually undergo when binding ligands? Are they minimal motions, rotameric

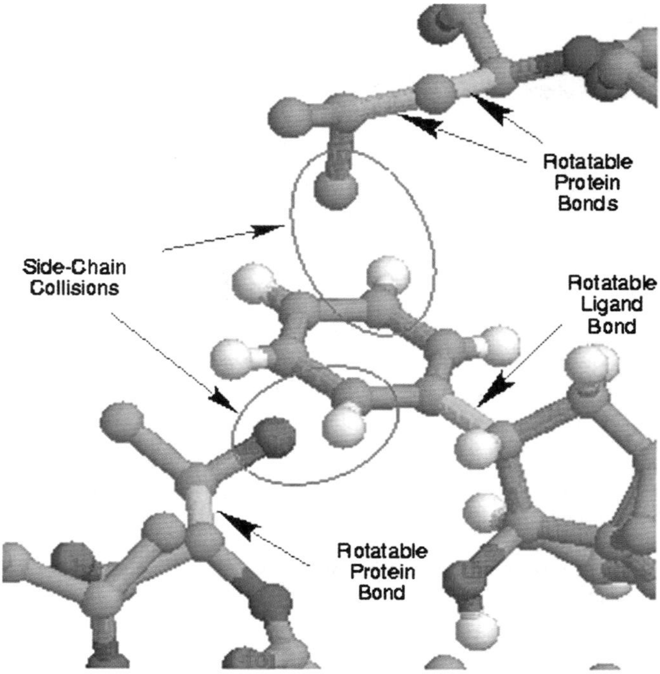

Figure 10.2 Intermolecular van der Waals collisions are resolved in SLIDE by directed rotations of single bonds in either the ligand or protein side chains. There are typically several possible rotations to resolve an intermolecular collision. An approach based on mean-field theory is used to decide which rotations are the most efficient for resolving one or more van der Waals overlaps in the current conformation of the complex. For all pair-wise intermolecular collisions, the bonds that can be rotated to resolve a particular collision are identified. They are stored in a matrix together with the corresponding rotation angle and the number of non-hydrogen atoms that will be displaced by the rotation. The product of the angle and the number of atoms (similar to a moment of inertia) provides a basis for the force that represents the cost of a rotation. A probability is associated with each rotation in the system. All rotations that can be used to resolve a particular collision are initialized with equal probabilities. During the optimization, the probabilities are updated to converge to an approximately optimal set of values, which assigns the highest probabilities to those rotations that solve the most collisions with the least overall cost.

transitions, or something more complex? Studies of conformational changes associated with protein–protein[36] and protein–ligand[20] binding in crystal structures show that even for proteins with conserved main-chain conformations upon ligand binding, there are side-chain conformational changes in at least 60% of the cases. However, side-chain conformations in these studies were considered to differ only if they reflected a rotameric transition, generally involving single-bond rotations of 60° or more. Other results, however, indicate

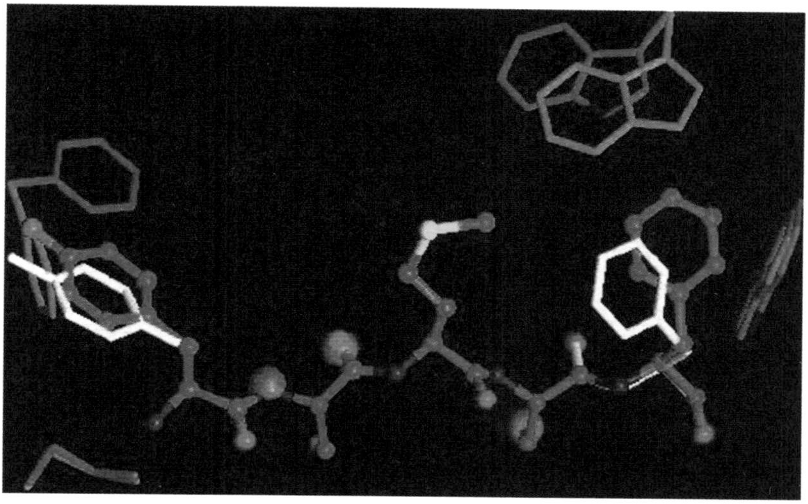

Figure 10.3 Conformational changes of the protein and the top peptidyl ligand (Tyr–Ser–Met–Ser–Phe) modeled by SLIDE after screening a set of 140 000 peptides [all five-residue peptide structures extracted from a low-homology subset of the Protein Data Bank (PDB)] against the ligand-free structure of the aspartic protease rhizopuspepsin (PDB 2apr). The docking of the ligand anchor fragment was based on a match of three polar ligand atoms onto the template points represented by spheres (hydrogen-bond acceptor, red; donor, blue). The native, ligand-free conformations of binding-site side chains are shown by blue tubes, and the initial conformation of the peptidyl ligand side chains are shown by white tubes; SLIDE's final conformation of the complex, involving induced fit of several aromatic residues in the protein and ligand, is shown in green.

that ligand binding induces strain or non-rotamericity in the preferred side-chain conformations.[37] To address this question without biasing towards a rotameric or non-rotameric interpretation, we analyzed protein side-chain rotations in 63 complexes; 32 of thrombin with different ligands (some highly flexible), 13 of GST with both hydrophobic (xenobiotic binding site) and polar (glutathione binding site) ligands, and 18 other, diverse complexes.[38] The goal was to observe protein side-chain motions in response to a variety of ligands, across a variety of proteins. 90% of interfacial protein side-chain rotations in the 63 structures were less than 45° upon ligand binding, and the remaining larger rotations were distributed broadly between 45° and 180°. Thus, most side-chain motions upon ligand binding are not rotameric transitions, instead reflecting small adjustments relative to existing rotamers and generating some degree of strain, consistent with the work of Heringa and Argos.[37] The same general trend holds for protein–protein interfaces (Zavodszky and Kuhn, unpublished results), although rotations of greater than 45° are found to occur more frequently (28% of the time). Ligand binding-site motions are similar to those elsewhere on protein surfaces, except that small-scale ($<15°$) side-chain

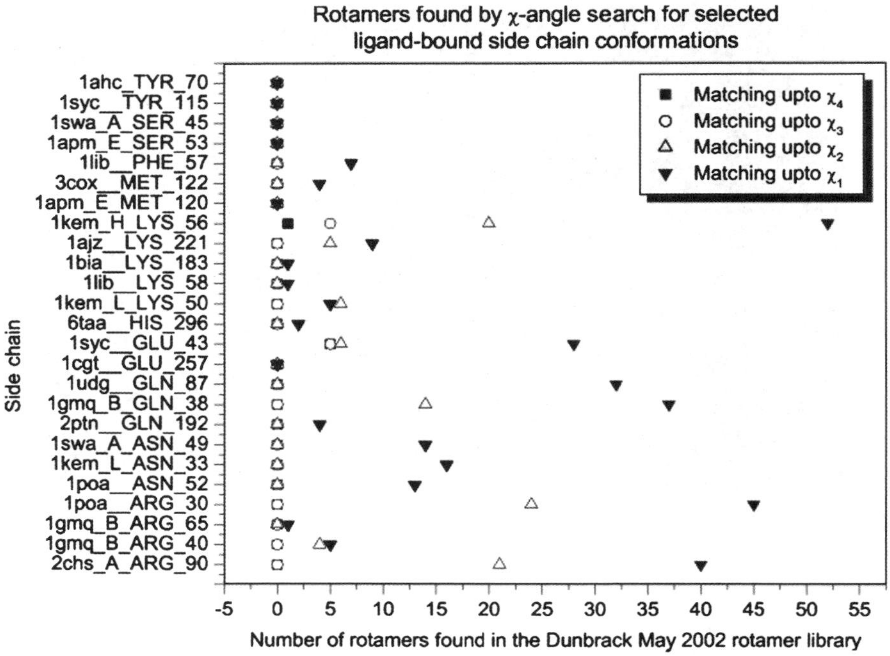

Figure 10.4 The number of rotamers approximating the dihedral angles of the ligand-bound conformations of 25 interfacial side chains undergoing large rotations (>60°) upon ligand binding. The May 2002 Dunbrack backbone-dependent rotamer library (http://dunbrack.fccc.edu/bbdep) was used to identify all rotamers matching the observed χ-values to within one standard deviation of the average value for that rotamer bin; all rotamer bins within ±10° of the interfacial residue's main-chain Φ and Ψ values were searched. Side-chain labels on the *Y*-axis include the PDB code, chain ID (if present), residue type, and residue number. Rotamer searches were done in incremental fashion, first comparing only χ_1, then χ_1 and χ_2, and so on, up to χ_4, depending on the number of rotatable bonds in the side chain. Only rotamers with probability values at least 5% as large as the probability of the most common rotamer within the same bin were considered. This helps exclude rotamers that are very rare, many of which may represent poorly resolved side-chain conformations in the PDB.

rotations are 20% more common in binding sites. These results indicate that side-chain motions upon ligand binding typically involve small-scale induced fit, which is found to be modeled appropriately by SLIDE.[38] The 10% of rotations that are larger cannot be classified simply as rotameric transitions, since they are distributed approximately evenly in dihedral angles between 45° and 180°.

We now understand that side-chain rotations upon ligand binding are not well represented by rotameric transitions or by energy minima within a rotameric state. Beyond not typically changing to a new rotamer, interfacial side chains observe favored dihedral angles only for χ_1 (C_α-C_β) and, to a lesser extent, χ_2 (C_β-C_γ) bond angles. Figure 10.4 shows the closeness with which even a very detailed main-chain-dependent rotamer library[39] can sample observed ligand-bound conformations for 25 cases in which side chains were observed to undergo large rotations.[40] In 17 out of the 25 cases there exist no reasonable rotamer matches, even when considering just the side chains' χ_1 and χ_2 angles. This is consistent with interfacial side chains adopting strained conformations due to packing in a tight interface and a new chemical environment.

10.4 The Future: Knowledge-based Modeling of Side-chain Motions

Given that we cannot afford to thoroughly sample low-energy side-chain conformations during docking and screening, how can we intelligently identify which side chains move significantly, and how to move them? Some guidance is provided by diagnosing why particular side chains experience large rotations. In one-third of the 25 cases mentioned above, large-scale rotations were needed to avoid steric collisions with the ligand.[40] In an additional 50% of the cases, the side chains apparently moved to satisfy hydrogen-bonding groups that could not be satisfied in the ligand-free orientation. This is a complementary picture to that derived from analyzing 30 protein–ligand complexes, indicating that a majority (75%) of interfacial, intra-protein hydrogen bonds are preserved upon ligand binding, and side chains involved in intra-protein hydrogen bonds tend to move very little.[40] Together, these results create a relatively tractable scenario in which minimal motion can be used to resolve steric overlaps between atoms during docking, then a conformational search can be performed to satisfy the hydrogen-bonding potential of buried polar side chains. As it turns out, most of the large-rotation cases actually involve small displacements (side-chain RMSDs of 1 Å or less), due to compensatory rotations in successive χ_i angles.[40] Thus, starting with the ligand-free conformation of each buried, unsatisfied polar side chain (of which there are typically just one or two per interface) and performing a local conformational search to optimize hydrogen bonding is expected to generate more accurate side-chain positions and allow better scoring of interactions during docking and screening.

Acknowledgements

This work was supported by NIH grants GM 67249 and AI 53877 to L. A. Kuhn and colleagues and reflects the thesis and postdoctoral work of several talented and enthusiastic researchers, particularly Volker Schnecke, Maria Zavodszky, and Sameer Arora. We appreciate the generosity of OpenEye Software (Santa Fe, NM) in providing Omega for our use.

References

1. M. Kontoyianni, G. S. Sokol and L. M. McClellan, *J. Comput. Chem.*, 2005, **26**, 11.
2. E. Perola, W. P. Walters and P. S. Charifson, *Proteins*, 2004, **56**, 235.
3. M. I. Zavodszky, P. C. Sanschagrin, R. S. Korde and L. A. Kuhn, *J. Comput. Aided Mol. Des.*, 2002, **16**, 883.
4. D. M. Lorber, M. K. Udo and B. K. Shoichet, *Protein Sci.*, 2002, **11**, 1393.
5. M. I. Zavodszky and L. A. Kuhn, unpublished results.
6. S. C. K. Sukuru, T. Crepin, Y. Milev, L. C. Marsh, J. B. Hill, R. J. Anderson, J. C. Morris, A. Rohatgi, G. O'Mahony, M. Grøtli, F. Danel, M. G. P. Page, M. Härtlein, S. Cusack, M. A. Kron and L. A. Kuhn, *J. Comput. Aided Mol. Des.*, 2006, **20**, 159.
7. E. Kellenberger, J. Rodrigo, P. Muller and D. Rognan, *Proteins*, 2004, **57**, 225.
8. G. Jones, P. Willett, R. C. Glen, A. R. Leach and R. Taylor, *J. Mol. Biol.*, 1997, **267**, 727.
9. F. Osterberg, G. M. Morris, M. F. Sanner, A. J. Olson and D. S. Goodsell, *Proteins*, 2002, **46**, 34.
10. B. Kramer, M. Rarey and T. Lengauer, *Proteins*, 1999, **37**, 228.
11. T. J. Ewing, S. Makino, A. G. Skillman and I. D. Kuntz, *J. Comput. Aided Mol. Des.*, 2001, **15**, 411.
12. OpenEye Software, Santa Fe, New Mexico: http://www.eyesopen.com/products/applications/omega.html.
13. J. Bostrom, P. O. Norrby and T. Liljefors, *J. Comput. Aided Mol. Des.*, 1998, **12**, 383.
14. M. Huse and J. Kuriyan, *Cell*, 2002, **109**, 275.
15. J. S. Finer-Moore, A. C. Anderson, R. H. O'Neil, M. P. Costi, S. Ferrari, J. Krucinski and R. M. Stroud, *Acta Crystallogr., Sect. D: Biol. Crystallogr.*, 2005, **61**, 1320.
16. B. Xiao, G. Shi, J. Gao, J. Blaszczyk, Q. Liu, X. Ji and H. Yan, *J. Biol. Chem.*, 2001, **276**, 40274.
17. D. J. Jacobs, A. J. Rader, L. A. Kuhn and M. F. Thorpe, *Proteins*, 2001, **44**, 150.
18. M. Lei, M. I. Zavodszky, L. A. Kuhn and M. F. Thorpe, *J. Comput. Chem.*, 2004, **25**, 1133.
19. M. I. Zavodszky, M. Lei, M. F. Thorpe, A. R. Day and L. A. Kuhn, *Proteins*, 2004, **57**, 243.
20. R. Najmanovich, J. Kuttner, V. Sobolev and M. Edelman, *Proteins*, 2000, **39**, 261.
21. R. L. Dunbrack Jr. and M. Karplus, *J. Mol. Biol.*, 1993, **230**, 543.
22. L. Schaffer and G. M. Verkhivker, *Proteins*, 1998, **33**, 295.
23. A. R. Leach and A. P. Lemon, *Proteins*, 1998, **33**, 227.
24. P. Kallblad and P. M. Dean, *J. Mol. Biol.*, 2003, **326**, 1651.
25. R. M. Knegtel, I. D. Kuntz and C. M. Oshiro, *J. Mol. Biol.*, 1997, **266**, 424.

26. H. Claussen, C. Buning, M. Rarey and T. Lengauer, *J. Mol. Biol.*, 2001, **308**, 377.
27. D. Bouzida, P. A. Rejto, S. Arthurs, A. B. Colson, S. T. Freer, D. K. Gehlhaar, V. Larson, B. A. Luty, P. W. Rose and V. M. Verkhivker, *Int. J. Quantum. Chem.*, 1999, **72**, 73.
28. V. Schnecke, C. A. Swanson, E. D. Getzoff, J. A. Tainer and L. A. Kuhn, *Proteins*, 1998, **33**, 74.
29. F. Jiang and S. H. Kim, *J. Mol. Biol.*, 1991, **219**, 79.
30. A. M. Ferrari, B. Q. Wei, L. Costantino and B. K. Shoichet, *J. Med. Chem.*, 2004, **47**, 5076.
31. M. L. Lamb and W. L. Jorgensen, *Curr. Opin. Chem. Biol.*, 1997, **1**, 449.
32. B. A. Luty, Z. R. Wasserman, P. F. W. Stouten, C. N. Hodge, M. Zacharias and J. A. McCammon, *J. Comp. Chem.*, 1995, **16**, 454.
33. H. A. Carlson, K. M. Masukawa, W. L. Jorgensen, R. D. Lins, J. M. Briggs and J. A. McCammon, *J. Med. Chem.*, 2000, **43**, 2100.
34. M. Totrov and R. Abagyan, *Proteins*, 1997, **27**(Suppl. 1), 215.
35. J.-H. Lin, A. L. Perryman, J. R. Schames and J. A. McCammon, *J. Am. Chem. Soc.*, 2002, **124**, 5632.
36. M. J. Betts and M. J. Sternberg, *Protein Eng.*, 1999, **12**, 271.
37. J. Heringa and P. Argos, *Proteins*, 1999, **37**, 44.
38. M. I. Zavodszky and L. A. Kuhn, *Protein Sci.*, 2005, **14**, 1104.
39. R. L. Dunbrack Jr., *Curr. Opin. Struct. Biol.*, 2002, **12**, 431.
40. S. Arora, *Optimizing Side-chain Interactions in Protein–Ligand Interfaces*, M.Sc. Thesis, Michigan State University, East Lansing, MI, 2005.

CHAPTER 11
Avoiding the Rigid Receptor: Side-Chain Rotamers

AMY C. ANDERSON

Department of Pharmaceutical Sciences, University of Connecticut, 69 N. Eagleville Rd., Storrs, CT 06269, USA

11.1 Introduction

Structural biology has opened many new avenues for drug-lead discovery by making available an increasing number of high-resolution structures of drug targets. In order to realize the drug discovery potential of these structures, it is important to accurately evaluate, *in silico*, the interactions of small organic molecules with these protein targets as they exist in solution. Many current docking programs[1,2] approach this goal by attempting to orient and score the interactions of a library of ligands, which may or may not be treated as flexible molecules, against a single, rigid target, often a high-resolution X-ray crystal structure. However, it has been established that many proteins in solution adapt to ligands after the initial binding event, often in an induced-fit manner in which both the structures of the protein and the ligand influence and conform to each other. There is increasing evidence that proteins exist as populations of conformations and present a variety of binding-site shapes and sizes to incoming ligands.[3] Binding a ligand shifts the equilibrium and population density of the ensemble of unbound conformations, resulting in a "binding funnel", similar to a protein-folding funnel, in which the lowest energy surface is rugged with an ensemble of complexed isomers and low-energy barriers.[4] Maintaining the protein as a rigid entity during docking forces one specific active-site shape and often introduces a bias towards ligands that resemble those with which the protein was originally crystallized. This bias has been most clearly demonstrated in cross-docking experiments.[5,6] Cross-docking experiments use two structures of the same protein target bound to different ligands; each ligand is removed from its complex and docked to the opposite protein target. Even ligands that are known to have high affinity against the protein often score poorly against structures of that protein derived from a different co-crystal complex. The poor score may be because even small steric interactions increase the docking score to the point that high-affinity ligands are prevented from ranking highly enough to be recognized within a library[6] or that favorable interactions are missed due to improper orientations of the protein and ligand. One serious consequence is that truly novel classes of ligands are often rejected during *in silico* screening in favor of ligands that resemble those from the co-crystal complex. The need to incorporate protein flexibility into docking algorithms in order to increase the accuracy of binding affinity prediction is critical and has been the subject of many excellent recent reviews.[7–9]

The ideal goal, then, is to dock a library of flexible ligands against all possible conformations of a protein target in order to maximize the correlation between high docking scores and experimentally determined activity. Although building flexibility into a ligand is routine, incorporating flexibility into the target can come at great computational cost since proteins can have thousands of degrees of freedom. Several approaches have been developed that attempt to find a compromise between allowing some degree of protein flexibility and managing computational time. The first of these approaches is called soft docking,[10,11] since it keeps the protein static while relaxing the high-energy penalty for the repulsive van der Waals force. The advantages of soft docking are that it can usually be easily incorporated into current docking algorithms and it does not incur additional computational cost. The disadvantage of soft docking is that it does not allow true flexibility, but simply reduces the penalty for slight steric interactions.

A second method to incorporate some degree of flexibility uses conformational sampling to generate a minimal ensemble of possible target conformations. Certain degrees of freedom can be selected for the protein target – a common method is to represent the side chains of the target with a rotamer library. The advantage of this method is that it incorporates a reasonable degree of flexibility without incurring large computational cost. The disadvantage of conformational sampling is the bias that is imposed on the ensemble: either specific residues must be identified as flexible, or the conformations that the residues assume must be discrete.

The third and most rigorous, but computationally demanding, method for incorporating target flexibility involves Monte Carlo or molecular dynamics (MD) simulations that model all degrees of flexibility, including the solvent.[12,13] While these simulations are accurate, they have the disadvantage of incurring enormous computational cost since even a 1 ns simulation can take several days with current computer processors. Optimizations to reduce this cost and maintain a high level of accuracy include generating a manifold of target conformations by choosing multiple snapshots from the MD simulation and varying the length of the MD simulation.[14]

This chapter explores the second approach, the role of conformational sampling in estimating protein flexibility and, specifically, the use of rotamer libraries to generate minimal ensembles of target conformations against which libraries of ligands can be docked and scored. A definition and brief discussion of rotamer libraries, validating the use of discrete conformations for residues, shows that the use of a minimal manifold of side-chain rotamers presents an efficient and reliable method for approximating protein flexibility. Several case studies illustrate the advantages of using rotamer libraries in the drug discovery process. In many of these examples, the conformational ensemble revealed an alternative target conformation that proved a successful route to new drug lead discovery.

11.2 Rotamer Libraries

Rotamer is short for "rotational isomer" and represents a single side chain conformation, defined by the dihedral angles of the side chain (Figure 11.1). A rotamer library is a collection of low-energy conformations for each residue type, tabulated with the frequency of observation.[15] Most rotamer libraries have been compiled by data mining from high-resolution X-ray structures in the Protein Database Bank (PDB). Lovell *et al.* compiled one of the most commonly used rotamer libraries.[16] They selected 240 X-ray crystal structures, determined with data to at least 1.7 Å resolution, deleted all the side chains with uncertain conformations, clustered the remaining side-chain conformations and reported conformation and frequency of observation. As one might expect, the number of rotamers increases as the length of the side chain increases, with arginine having 34 possible rotamers. Lovell's library accounts for >90% of the residues in the selected high-resolution structures, further suggesting that protein side chains only very rarely exhibit higher energy conformations. For example, Phe has four low-energy rotamers and there are almost no observations of other conformations of Phe.

Dunbrack and Cohen[17] present a backbone-dependent rotamer library based on an analysis of over 500 structures. They divided torsional space into bins and used Bayesian statistics to estimate the population in each bin. The statistical accuracy of their library is high and in populated areas of the Ramachandran plot, the agreement with data values is correlated. However,

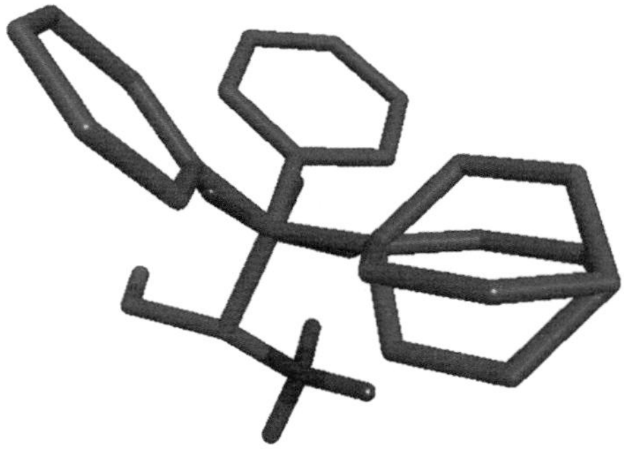

Figure 11.1 Four possible rotamers of phenylalanine.

rotamers that are defined in very sparsely populated regions of the Ramachandran plot may not be observed.

The evidence that proteins exist in conformational ensembles in solution and that side chains of amino acids exist primarily in rotameric states sets the stage for using minimal manifolds of protein conformations based on rotamer libraries. Najmanovich *et al.*[18] created a database of "paired" apo- (ligand-free) and holo- (ligand-bound) structures of the same protein and analyzed the observed induced fit in order to determine whether the complete complement of conformations of residues in the binding site should be calculated or whether a selection of a minimal set of flexible residues would suffice. From 221 different protein sequences, it was evident that in approximately 85% of cases, only three residues or fewer undergo conformational changes upon binding a ligand. These conformational changes were almost entirely due to rotations in side chains rather than translocations in the backbone.

11.3 Successful Applications of Rotamer Libraries in Drug Design

11.3.1 Aspartic Acid Protease Inhibitors

Novel, non-peptidic inhibitors of aspartic acid proteases such as human immunodeficiency virus (HIV) protease and renin, could be discovered only when the complete ensemble of conformations of the active site was considered.[19,20] Traditional drug design for aspartic acid proteases has focused on peptide-derived transition state analogs that bind the active site in an extended beta-strand conformation (Figure 11.2a). However, peptide-derived inhibitors have problems with bioavailability, including export and metabolism processes, often preventing them from being used successfully in the clinic. Crystal structures of chymosin, a closely related bovine aspartic peptidase, in a native form[21] and in a co-crystal complex with CP-113971[22] revealed a significant conformational change in the active site. Tyr 75 rotated almost 180 degrees in order to avoid a steric interaction with the inhibitor (Figure 11.2b). Roche scientists also discovered that a cascade of conformational changes, including rotameric shifts of Trp 39, Leu 73 and Tyr 75 in renin, allows piperidine-based inhibitors (Figure 11.2c) to stabilize an enzyme conformation different from the typical extended beta-strand binding site.[23,24] In order to overcome the bioavailability problems of the traditional inhibitors, Rich *et al.*[19] targeted a complete ensemble of available conformations of aspartic acid proteases. Using GrowMol,[25] a structure-generating program, they successfully predicted the novel binding mode of the piperidine-based inhibitors *in silico* and, in an extension, discovered several new aspartic acid protease inhibitors.

11.3.2 Matrix Metalloproteinase-1 Inhibitors

Kallblad and Dean[26] applied a rotamer-based approach to generate conformational ensembles of the S1′ pocket of matrix metalloproteinase-1 (MMP-1),

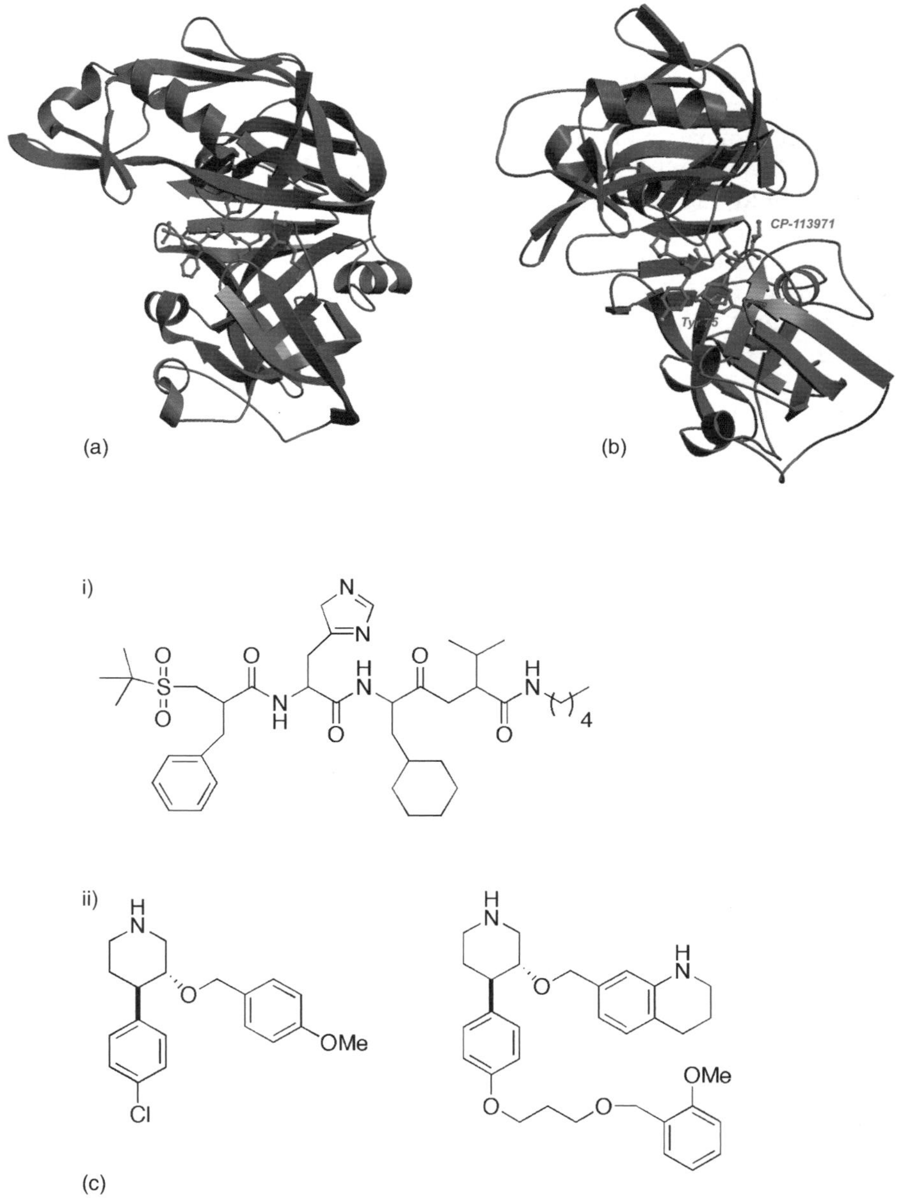

Figure 11.2 (a) Renin bound to a transition state analog, CGP 38'560 (green) in the traditional, beta-strand conformation (1RNE[37]). (b) The structure of native chymosin (1CMS[21]), an aspartic peptidase, was superimposed with chymosin bound to CP-113971 (1CZI[22]). For clarity, the structure of the protein from 1CZI was omitted. Tyr 75 (red) of native chymosin adopts a new rotamer to avoid a steric interaction with CP-113971 (green). (c) (i) CGP 38'560, a transition state analog. (ii) Roche piperidine-based inhibitors.

known to undergo conformational changes on binding certain inhibitors. First, the authors developed a method they call DYNASITE which, using a backbone-dependent rotamer library,[17] explores the side-chain conformations of a template-binding site. They chose a template structure for the S1′ pocket (2TCL[27]) and selected a set of six residues (Leu 181, Arg 214, Val 215, Ser 239, Tyr 240 and Phe 242) predicted to affect the shape and ligand-binding properties of the pocket (Figure 11.3a). They generated 2115 structures of the S1′ pocket, including a large number of alternative conformations not observed experimentally, but with relative energy lower than that of the minimized crystal structure. After minimization of each conformer, it was apparent that the low-energy conformers were formed with rotamers with higher probability. Clustering the conformers and selecting low-energy cluster representatives yielded 15 conformers with maximal diversity. The core ensemble exhibited much of the same flexibility observed in human MMP-1 entries in the PDB.

The MMP-1 inhibitor, RS-104966, is known to induce several conformational changes in the S1′ pocket, the most remarkable of which is the change associated with Arg 214 (Figure 11.3b). Kallblad and Dean[26] used GOLD[28] to dock RS-104966 to each of the 15 conformers in the core ensemble. They found that the RS-104966 binding mode was not reproducible using any of the template structures, but was reproducible using two members of the core ensemble. The main advantage to their method is that it samples a very wide range of conformational space and that a relatively low number of conformer representatives (15 in this case) can be used to represent conformational diversity.

In subsequent work, Kallblad *et al.*[29] compared virtual screening efforts using five target sets: a single crystal structure, 30 NMR apo-structures, 30 NMR holo-structures, the 15 conformers of the DYNASITE ensemble and a flexible crystal structure, generated by applying a rotamer library to the target within the *de novo* drug-design program, Skelgen.[30,31] The DYNASITE conformer ensemble showed significant variation in the S1′ cavity shapes and, furthermore, the conformational space of the DYNASITE ensemble overlapped the space from both the NMR apo- and holo-forms. Several compounds were designed in Skelgen against each of the five target sets. The designed compounds were compared to compounds with known inhibitory activity from the World Drug Index (WDI) and a set of compounds was discovered that strongly resembled highly potent MMP-1 inhibitors with K_i or IC_{50} values in the 8.2–90 nM range. This set of high-affinity compounds was successfully selected by the NMR models and the flexible models, but not by the static X-ray structure due to steric interactions. It is clear from these observations that the flexible-receptor approaches successfully identify highly potent molecules that would be missed with the use of a static receptor.

Yang *et al.*[32] found that ensembles generated from rotamer libraries reproduce experimentally observed structures as well as predict alternative ligand-binding conformations. These investigators chose a dataset of proteins whose structures had been determined by X-ray crystallography or NMR and for which multiple structures bound to different ligands were available. Using

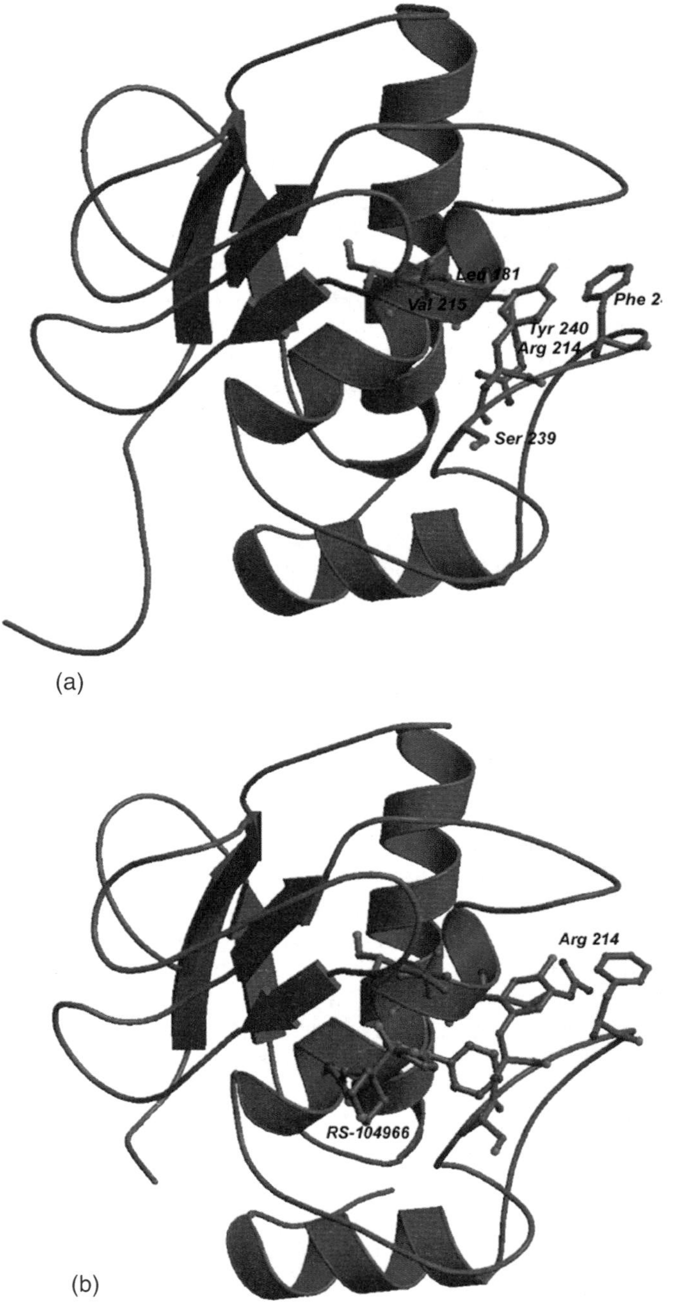

Figure 11.3 (a) The template structure for the S1′ pocket of MMP-1 (2TCL[27]) revealing the positions of Leu 181, Arg 214, Val 215, Ser 239, Tyr 240 and Phe 242. (b) The S1′ pocket of MMP-1 bound to RS-104966 (966C[38]). The shift for Arg 214 is noted.

DYNASITE,[26] they generated an exhaustive list of protein conformers and statistically clustered the list to choose representatives. In all cases except one they could reproduce the template structure with low root mean square deviation (RMSD). Additionally, in 22 out of 25 cases, using only one template structure, their method is able to reproduce an alternative binding site observed experimentally. The method fails only when ligand binding stabilizes a high-energy conformer.

11.3.3 Thymidylate Synthase Inhibitors

In another example of the use of rotamer libraries in drug design, Anderson *et al.* generated a minimal ensemble of conformations of thymidylate synthase, an enzyme known to undergo conformational changes on inhibitor binding[5] (Figure 11.4). The authors first demonstrated the failure of the rigid-receptor hypothesis by showing that each of two potent inhibitors, CB3717 ($K_i = 90$ nM) and 1843U89 ($K_i = 16$ nM), score well when docked into the structure from which they originated and score poorly when docked into the opposite structure. Furthermore, when the two inhibitors are included in the Available Chemicals Directory and the entire database was scored, neither inhibitor appeared in the list of top 500 compounds if the opposite structure was used as the target, whereas 1843U89 appears as the top-ranked compound if docked into its own structure and CB3717 appears in the top 40 compounds when docked into its own site. A set of flexible residues within the active site of

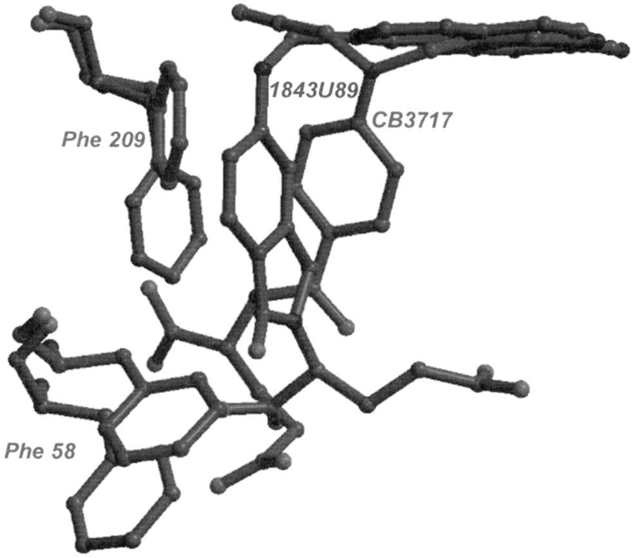

Figure 11.4 Thymidylate synthase from *Pneumocystis carinii* bound to CB3717 (green) (1CI7[39]) and 1843U89 (purple) (1F28[5]). Conformational changes for Phe 209 and Phe 58 are noted.

thymidylate synthase, comprising all hydrophobic residues and loop residues that allow room for conformational changes, as determined by assessing the hindrance of different rotamers of those residues, was selected. Once a minimal manifold of rotameric positions of the flexible side chains was imposed on the static structures, the two inhibitors received docking scores commensurate with their experimentally determined activity.

11.3.4 Protein Tyrosine Phosphatase 1B Inhibitors

Cross-docking studies with known high-affinity ligands against protein tyrosine phosphatase 1B failed when a ligand from one complex was docked to a protein-binding site derived from a complex with another ligand.[6] Frimurer *et al.* suggested that a single conformation of the binding site may not adequately represent the possible binding modes induced by the ligands.[6] The authors generated 96 different protein models from a single template structure, using a backbone-dependent rotamer library applied to four residues primarily involved in ligand binding in the active site (Asp 48, Lys 120, Asp 181 and Phe 182; Figure 11.5). They docked a set of three inhibitors, for which structures with the protein had been previously determined, to the parent structures as well as to the structures generated with the rotamer library. When the inhibitors were docked to the template structure, the correct binding mode

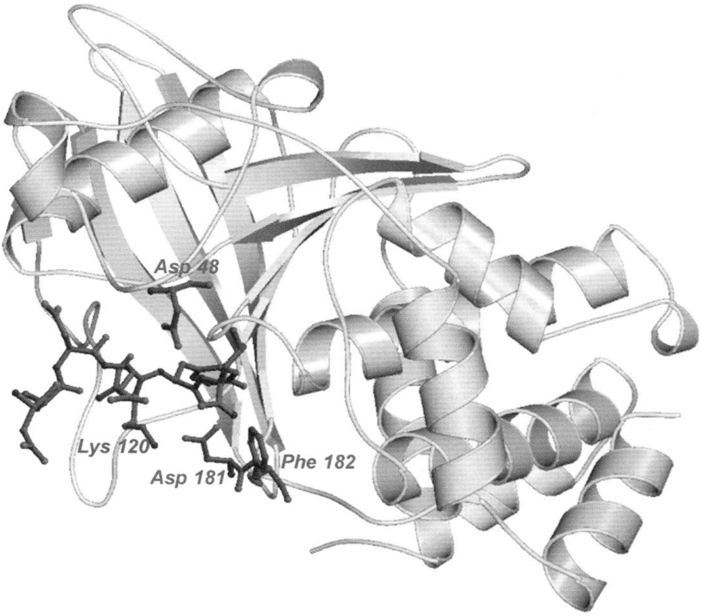

Figure 11.5 Protein tyrosine phosphatase 1B (1PTU[40]). Residues selected for flexibility are shown in green: Asp 48, Lys 120, Asp 181 and Phe 182. The protein is bound to a phosphopeptide (purple).

was not predicted. When the inhibitors were docked to the manifold of 96 protein structures generated with the rotamer library, the lowest ΔG_{bind} values obtained for each of the three inhibitors corresponded to the binding modes with the lowest RMS differences to the experimentally observed binding modes. In addition, the ranking of the estimated binding energies corresponded with the ranking from the experimentally determined binding affinities.

11.3.5 HIV Protease Drug-resistant Mutants Bound to Inhibitors

Schaffer and Verkhiver[33] present a method to predict the structures of two HIV protease drug-resistant mutants, V32I/I47V/V82I and V82A, bound to the ligands SB203386 and A77003, respectively. They used a hierarchical approach, combining flexible ligand docking, optimization of flexible protein side-chain conformations, minimizations and energy evaluation. First, they constructed an initial structure of the mutant protein from the wild-type protein, using a rotamer library of side chains. They then applied a variation of the dead-end elimination (DEE[34]) algorithm to eliminate rotamers that are not consistent with the global minimum energy conformation (GMEC), thereby reducing the conformational complexity. Finally, they performed a series of Monte Carlo simulated annealing procedures for ligand docking and minimized the final solutions. The predicted structures of the mutant complexes have less than 1.0 Å RMSD from the crystal structures of the mutant proteins bound to the ligands, thus demonstrating that the side-chain rotamer library adequately described the rotamers of the active-site residues.

11.3.6 Trypsin–benzamidine and Phosphocholine–McPC 603

One of the earliest examples of the use of side-chain rotamers determined all combinations of side-chain rotamers and ligand conformations within some defined energy relative to the global minimum energy. Leach[35] used DEE[34] to identify and eliminate all rotamers that are incompatible with the GMEC and then combined the results with the A* algorithm[36] to locate the least-cost path to the GMEC. The combined DEE–A* algorithm was used to generate all rotamer combinations of a defined set of residues near the active site that fell within 5 kcal mol^{-1} of the global energy minimum for the protein alone and for the protein–ligand complex. The algorithm was applied to dock benzamidine to trypsin and phosphocholine to the antibody McPC 603. For each acceptable conformation of the ligand (corresponding to the known crystal structure), the DEE–A* algorithm generated the rotameric states of 61 residues in the active site that gave the lowest energy structure. The lowest energy structures of the benzamidine–trypsin complex were identified. Leach also found that the algorithm predicted more conformational states in the complex than in the free protein and ascribes this to a modulation of the protein's energy surface by the ligand, leading to greater numbers of accessible conformational states.

11.4 Conclusions

It is evident that incorporating protein flexibility into docking and scoring algorithms leads to a more accurate depiction of the interaction of a protein and ligand in solution. The generation of conformational ensembles of a protein target using rotamer libraries of side chains achieves target flexibility while maintaining manageable computational cost. We have seen that the incorporation of protein flexibility leads to a better correlation between calculated and experimentally determined binding affinity. In addition, it can lead to the identification of alternative binding sites in target molecules, allowing an expansion towards novel classes of ligands. The identification of truly novel classes of compounds is critical to overcoming the resistance and bioavailability issues that plague us today.

Acknowledgements

The author acknowledges support from the National Science Foundation and the National Institutes of Health for her work investigating the use of rotamer libraries in drug design.

References

1. D. Lorber and B. Shoichet, *Protein Sci.*, 1998, **7**, 938.
2. B. Kramer, G. Metz, M. Rarey and T. Lengauer, *Med. Chem. Res.*, 1999, **9**, 463.
3. B. Ma, M. Shatsky, H. Wolfson and R. Nussinov, *Protein Sci.*, 2002, **11**, 184.
4. S. Kumar, B. Ma, C. Tsai, N. Sinha and R. Nussinov, *Protein Sci.*, 2000, **9**, 10.
5. A. Anderson, R. O'Neil, T. Surti and R. Stroud, *Chem. Biol.*, 2001, **8**, 445.
6. T. Frimurer, G. Peteres, L. Iversen, H. Andersen, N. Moller and O. Olsen, *Biophys. J.*, 2003, **84**, 2273.
7. H. Carlson and J. McCammon, *Mol. Pharm.*, 2000, **57**, 213.
8. H. Carlson, *Curr. Opin. Chem. Biol.*, 2002, **6**, 447.
9. M. Teodoro and L. Kavraki, *Curr. Pharm. Des.*, 2003, **9**, 1635.
10. K. Jiang, *J. Mol. Biol.*, 1991, **219**, 79.
11. V. Schnecke, C. Swanson, E. Getzoff, J. Tainer and L. Kuhn, *Proteins*, 1998, **33**, 74.
12. A. Mangoni, D. Roccatano and A. De Nola, *Proteins*, 1999, **35**, 153.
13. H. Carlson, K. Masukawa, W. Jorgensen, R. Lins, J. Briggs and J. McCammon, *J. Med. Chem.*, 2000, **43**, 2100.
14. K. Meagher and H. Carlson, *J. Am. Chem. Soc.*, 2004, **126**, 13276.
15. R. Dunbrack, *Curr. Opin. Struct. Biol.*, 2002, **12**, 431.
16. S. Lovell, J. Word, J. Richardson and D. Richardson, *Proteins*, 2000, **40**, 389.
17. R. Dunbrack and F. Cohen, *Protein Sci.*, 1997, **6**, 1661.

18. R. Najmanovich, J. Kuttner, V. Sobolev and M. Edelman, *Proteins*, 2000, **39**, 261.
19. D. Rich, M. Bursavich and M. Estiarte, *Biopolymers*, 2002, **66**, 115.
20. M. Bursavich and D. Rich, *J. Med. Chem.*, 2002, **45**, 541.
21. G. Gilliland, E. Winborne, J. Nachman and A. Wlodawer, *Proteins*, 1990, **8**, 82.
22. M. Groves, V. Dhanaraj, M. Badasso, P. Nugent, J. Pitts, D. Hoover and T. Blundell, *Protein Eng.*, 1998, **11**, 833.
23. C. Oefner, A. Binggeli, V. Breu, D. Bur, J. Clozel, A. D'Arcy, A. Dorn, W. Fischli, F. Gruninger, R. Guller, G. Hirth, H. Marki, S. Mathews, M. Muller, R. Ridley, H. Stadler, E. Vieira, M. Wilhelm, M. Winkler, F. Winkler and W. Wostl, *Chem. Biol.*, 1999, **6**, 127.
24. E. Vieira, A. Binggeli, V. Breu, D. Bur, W. Fischli, R. Guller, G. Hirth, H. Marki, M. Muller, C. Oefner, M. Scalone, H. Stadler, M. Wilhelm and W. Wostl, *Bioorg. Med. Chem. Lett.*, 1999, **9**, 1397.
25. R. Bohacek and C. McMartin, *J. Am. Chem. Soc.*, 1994, **116**, 5560.
26. P. Kallblad and P. Dean, *J. Mol. Biol.*, 2003, **326**, 1651.
27. N. Borkakoti, F. Winkler, D. Williams, A. D'Arcy, M. Broadhurst, R. Brown, W. Johnson and E. Murray, *Nat. Struct. Biol.*, 1994, **1**, 106.
28. G. Jones, P. Willet, R. Glen, A. Leach and R. Taylor, *J. Mol. Biol.*, 1997, **267**, 727.
29. P. Kallblad, N. Todorov, H. Willems and I. Alberts, *J. Med. Chem.*, 2004, **47**, 2761.
30. M. Stahl, N. Todorov, T. James, H. Mauser, H. Bohm and P. Dean, *J. Comput. Aided Mol. Des.*, 2002, **16**, 459.
31. N. Todorov and P. Dean, *J. Comput. Aided Mol. Des.*, 1997, **11**, 175.
32. A. Yang, P. Kallblad and R. Mancera, *J. Comput. Aided Mol. Des.*, 2004, **18**, 235.
33. L. Schaffer and G. Verkhivker, *Proteins*, 1998, **33**, 295.
34. J. Desmet, M. DeMaeyer, B. Hazes and I. Lasters, *Nature*, 1992, **356**, 539.
35. A. Leach, *J. Mol. Biol.*, 1994, **235**, 345.
36. A. Leach and K. Prout, *J. Comput. Chem.*, 1990, **11**, 1193.
37. J. Rahuel, J. Priestle and M. Grutter, *J. Struct. Biol.*, 1991, **107**, 227.
38. B. Lovejoy, A. Welch, S. Carr, C. Luong, C. Broka, R. Hendricks, J. Campbell, K. Walker, R. Martin, H. Van Wart and M. Browner, *Nat. Struct. Biol.*, 1999, **6**, 217.
39. A. Anderson, D. Freymann, K. Perry and R. Stroud, *J. Mol. Biol.*, 2000, **297**, 645.
40. Z. Jia, D. Barford, J. Flint and N. Tonks, *Science*, 1995, **268**, 1754.

Section 4
Screening

CHAPTER 12

Computational Prediction of Aqueous Solubility, Oral Bioavailability, P450 Activity and hERG Channel Blockade

DAVID E. CLARK

Argenta Discovery Ltd., 8/9 Spire Green Centre, Flex Meadow, Harlow, Essex, CM19 5TR, UK

12.1 Introduction

Over the past decade or so, the pharmaceutical industry has come under increasing pressure to make the drug-discovery process more efficient and thus more cost-effective. Partly in response to this, new technologies have been enthusiastically embraced: the "omics" (*e.g.* genomics, proteomics) promising larger numbers of validated disease targets, combinatorial chemistry promising orders of magnitude more compounds to screen against those targets, and high-throughput screening (HTS) promising to screen those compounds in a fraction of the time previously required. Arguably, these technologies have largely been victims of the hyperbole that surrounded their advents – expectations were perhaps (probably?) inflated and, consequently, the disappointment in their apparent failure to revolutionize drug discovery has been all the greater.[1]

However, another change has been going on in drug discovery in parallel to the adoption of the new technologies mentioned above; one that has been perhaps less trumpeted, but is possibly of greater significance. The change in question is the shift of what have traditionally been thought of as "Development" activities upstream to become part of "Research". Simply put, the historical focus of "Research" was to discover lead compounds and optimize their potency and selectivity for the desired biological target, with little concern for their physicochemical and ADMET (Absorption, Distribution, Metabolism, Excretion, Toxicity) properties. The optimization of these latter properties was deemed to be the responsibility of "Development". One of the consequences of this disconnect between R&D was that often Development was faced with very difficult, if not insoluble (pun intended!), problems to fix, with the inevitable result of increased time and cost.[2]

The paradigm shift of the past decade has been motivated by the realization that great savings can be made by the early elimination of drug candidates with poor physicochemical or ADMET properties ("fail fast, fail cheap"). Significant advances in the automation of physicochemical and *in vitro* ADMET assays have enabled the testing of many more compounds at a much earlier stage of drug discovery than previously.[3] In some cases, these properties can now be determined in synchrony with potency and selectivity information, giving rise to truly multi-objective (or parallel) optimization, which promises to deliver compounds for development with much-improved profiles and, thereby, much greater chance of successful transition to clinical candidates.

So far, so good. But what if a compound's properties could be reliably predicted solely from knowledge of its chemical structure, without the need even to synthesize and test it? This alluring, if not utopian, goal has been the spur in recent years for intense research into the computational prediction of physicochemical and ADMET properties. In this chapter, we review the progress of these "*in silico*" approaches with reference to four important properties of a chemical compound: aqueous solubility, oral bioavailability, cytochrome P450 (CYP450) activity and human ether-a-go-go related gene (hERG) channel blockade (a similarly focused review was recently published by Norinder[4]). This is not an exhaustive list – wider reviews are available[5–10]– but it is sufficient to permit an assessment of the state-of-the-art in this field and to identify some of the challenges that lie ahead on the road towards what has been termed "prediction paradise".[6]

12.2 Aqueous Solubility

Of all the physicochemical properties that a compound possesses, none has received such attention in recent years as aqueous solubility. This state of affairs is a reflection of the growing realization of the crucial role of aqueous solubility in a number of areas of drug discovery. The work of Lipinski and colleagues[11,12] has helped to highlight the fact that oral absorption is dependent on both solubility and permeability: after oral ingestion, a compound must both dissolve in the intestinal fluid and cross the gut wall if it is to reach the bloodstream. Lipinski[12] has presented a helpful chart in response to the frequently asked question "How soluble does a compound need to be to prevent poor absorption?" (Figure 12.1). The chart illustrates that, to achieve suitable systemic levels of an orally-dosed drug, higher solubilities are required for poorly permeable compounds than for moderately or highly permeable compounds. The solubility figures are calculated using the Maximum Absorbable Dose (*MAD*) equation originally derived by Johnson and Swindell[13] and illustrated by Curatolo[14]:

$$MAD = S \times K_a \times SIWV \times SITT \tag{12.1}$$

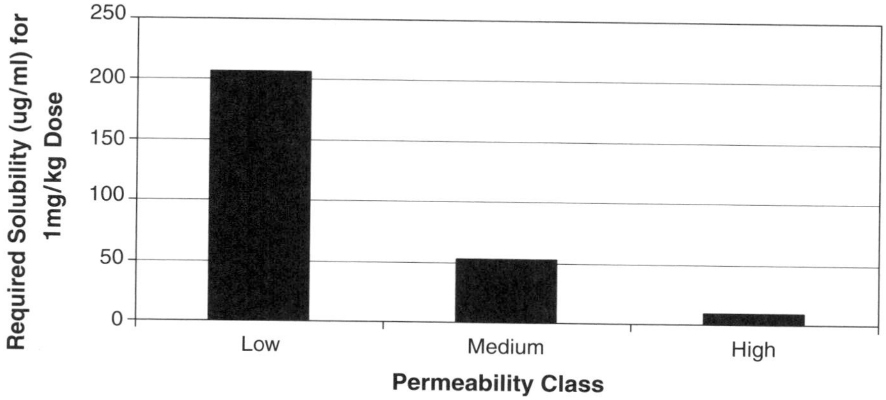

Figure 12.1 Chart showing the minimum solubility required for an oral dose of $1\,mg\,kg^{-1}$ for compounds of varying levels of permeability.[12]

where S is the aqueous solubility of the compound at pH 6.5 ($mg\,ml^{-1}$), K_a is the transintestinal absorption rate constant (min^{-1}), $SIWV$ is the small intestine water volume (assumed to be ~250 ml) and $SITT$ is the small intestine transit time (assumed to be 270 minutes).

Poor aqueous solubility is not only a hindrance to oral absorption, but often leads to difficulties in other aspects of drug discovery.[15] For instance, in biochemical and ADMET assay systems, which rely on an accurate knowledge of the concentration of compound present in the test solution, poorly soluble compounds tend to precipitate and this leads to uncertainties in experimental results. Later on in pharmaceutical development, poorly soluble compounds can give rise to significant, if not insuperable, challenges in formulation. For all these reasons, high-throughput systems for measuring thermodynamic and/or kinetic solubilities are now in place in many companies with the aim of identifying and addressing solubility problems as early as possible during lead optimization.

Unsurprisingly, recent years have also seen an increase in the intensity of research into computational methods for solubility prediction. Writing in 2003, Lobell and Sivarajah counted at least 17 different approaches reported in at least 30 publications, of which 27 have been published since 2000,[16] and there is no reason to believe this trend is likely to reverse in the near future. Reviews of the field have been published by Jorgensen and Duffy[17] and, more recently, by Delaney.[18] Despite all the research efforts, aqueous solubility is proving a very difficult property to predict accurately. Lipinski has remarked upon the complexity of the task given that the aqueous solubility of a compound is affected by its lipophilicity, hydrogen bonding to solvent, intramolecular hydrogen bonding, crystal packing and ionic charge status.[12] For charged compounds, solubility is affected additionally by the solution pH and the identity of the counter-ion. In general, the current generation of solubility models works best for neutral compounds and, as with many quantitative structure–activity relationship (QSAR) models, congeneric series of compounds.

Experimental aqueous solubility is often expressed as log S, this being the molar solubility of a compound. Experimental values range over 14 orders of magnitude, from −12 (extremely insoluble) to +2 (highly soluble). Typically, drug-like molecules fall in the narrower range of −5 to −1.[17] This reflects the need for drugs to be soluble (hence no lower than −5) and permeable (hence no higher than −1: very soluble compounds tend to be very polar and thus poor at membrane permeation). Probably the leading commercial software product for solubility prediction is that developed and marketed by ACD/Labs and this is quoted as being able to predict log S to within 1 log unit in 80% of cases.[19] The precise performance will, of course, vary depending on how well the structural class of interest is represented in the software's training set. However, the accuracy of prediction should improve in coming years as more solubility data are determined for drug-like compounds (much of the historical data that are publicly available comprise measurements on non-drug-like molecules), although the accuracy of predictions will never exceed the experimental uncertainty, which has been estimated to be of the order of 0.6 log units.[17]

To give just a single example of a QSAR model for the prediction of aqueous solubility, the ESOL model reported by Delaney[20] is shown in Equation (12.2):

$$\begin{aligned} \log S &= 0.16 - 0.63C\log P - 0.0062MWT + 0.066RB - 0.74AP \\ N &= 2874,\ r^2 = 0.72,\ s = 0.97,\ F = 1865 \end{aligned} \tag{12.2}$$

where N is the number of compounds used to derive the model, r^2 is the fraction of the variance explained by the model, s is the standard deviation of the regression and F is the Fisher value, a measure of statistical significance. As Equation (12.2) shows, the model depends on four variables that can be derived directly from the 2D chemical structure of a compound:

- $C\log P$ – logarithm of the octanol–water partition coefficient calculated, in this instance, by the ClogP program[21];
- MWT – the molecular weight;
- RB – number of rotatable bonds;
- AP – proportion of aromatic atoms.

The ESOL model was derived from solubility data for 2874 neutral compounds and showed reasonable predictive performance when applied to a set of 528 previously unseen compounds ($r^2 = 0.55$, $s = 0.96$). Compared to some computational solubility models, ESOL is attractive for at least three reasons:

- Simplicity – the equation is linear and comprises only a few variables.
- Interpretability – the variables are easy to understand and to relate to a compound's structure. Thus, it is simple to use the equation to suggest modifications to a compound that are predicted to improve its solubility (*e.g.* reduce molecular weight and/or $C\log P$).
- Speed – the variables are very rapidly computed and allow the profiling of large compound sets or interactive calculations *via* a corporate intranet.

The main drawback of the model – in common with many others – is that it cannot be used to predict solubility for charged compounds, which comprise a significant percentage of drug candidates.

12.3 Oral Bioavailability

For reasons of ease of dosing and patient compliance, the pharmaceutical industry seeks to develop drug compounds suitable for oral ingestion whenever possible. A compound that can successfully reach the desired biochemical target in the body at a concentration that is high enough to exert the intended therapeutic effect is said to be orally bioavailable. It is important to distinguish oral bioavailability (generally denoted as F) from two related terms: oral absorption and permeability.[22] Unfortunately, these three are often, and incorrectly, treated as if they were synonyms.

Following Lennernäs,[23] we use the following definitions: The *effective intestinal permeability*, P_{eff}, is expressed in cm s^{-1} and reflects the transport velocity across the apical (outer) membrane of the epithelial barrier of the intestine and into the cytosol.[24] Assuming no metabolism occurs at the intestinal lumen, the fraction of drug ($M_t/Dose$) that has disappeared from the intestine over a certain residence time, t, is given by:

$$M_t/Dose = \int \int_0^t \int AP_{eff}\, C_{lumen}\, \mathrm{d}A\mathrm{d}T \tag{12.3}$$

where A is the available surface area of the intestine, P_{eff} is the average value of the effective intestinal permeability of the drug along the region of the intestine where absorption occurs and C_{lumen} is the free reference concentration of the drug in the intestinal lumen.[23] From Equation (12.3), it is clear that P_{eff} is one of the key factors controlling the rate and extent of absorption. The *fraction absorbed* (f_a or, frequently, FA, expressed as percentage) includes all processes from the dissolution of the drug from its (solid) dosed form to its transport across the apical membrane of the epithelial barrier of the intestine. It is related to the *oral bioavailability* (F) by Equation (12.4)[23]:

$$F = f_a(1 - E_G)(1 - E_H) \tag{12.4}$$

where E_G represents the extent of first-pass metabolism in the gut and E_H represents first-pass metabolism and biliary excretion in the liver. It is important to note that bioavailability is a composite measure including contributions from absorption and first-pass metabolism. When metabolism is negligible, F approaches f_a, but more generally, $F < f_a$, and so the two should not be routinely equated.

Given the importance attached to the oral delivery of compounds, considerable effort has been applied in recent years to the prediction of permeability and oral absorption and several reviews have been published.[22,25–27] By contrast, the prediction of oral bioavailability has not received so much attention, probably because of the addition of first-pass metabolism makes it a more complex property. Some reviews summarize the progress made in this area.[26,28,29] An illustrative approach has been reported by Pintore *et al.*[30] who developed a classification model for the prediction of oral bioavailability. The model was trained using a test set of 352 compounds and a validation set of 80 compounds. An external test set of 75 compounds was used to test the predictive power of the model. Within these sets, the compounds were split into four classes according to their oral bioavailabilities:

- Class 1: $F \leq 20\%$;
- Class 2: $20\% < F < 50\%$;
- Class 3: $50\% \leq F < 80\%$;
- Class 4: $F > 80\%$.

Using a classification technique called Adaptive Fuzzy Partitioning (AFP), a nine-descriptor model was generated. The model included terms for log *P* and a count of hydrogen-bond donors, both of which have been shown to be important in other absorption-related models (*e.g.* the well-known "rule-of-5"[11]). Lipophilicity (represented here by log *P*) is also a key determinant of metabolism by the major CYP450s.[31] Given the complexity of the problem and the inherent variability in the data, the model's performance was quite respectable and certainly an improvement over previously published approaches. 70% of the training set was classified correctly and 69% of the validation set. For the previously unseen test set, 64% of compounds were assigned to the correct class.

It may be that in the future, as more oral bioavailability data become available, the prediction of this important quantity can be improved. Alternatively, a reductionist approach, in which solubility, permeability and first-pass metabolism are separately predicted and then combined, may prove to be a more fruitful avenue of research.

12.4 Cytochrome P450 Activity

The superfamily of CYP450 enzymes constitutes the major drug-metabolizing enzyme system in the human body.[32] The interaction of a (candidate) drug molecule with a particular CYP may result in the compound being metabolized or it may result in the CYP being inhibited. These events may be of concern for different reasons. If a compound is a substrate for a CYP, it may have a half-life that is too short to be compatible with a reasonable dosing regimen. Alternatively, if a compound inhibits a CYP, this can have serious implications for drug–drug interactions, leading to toxicity. For this reason, there has been much interest in trying to determine the interactions of compounds with the major CYPs (3A4, 2D6, 2C9, 2C19, 1A2) both experimentally and computationally.

From the latter perspective, two broad approaches have been adopted: structure-based and ligand-based.[33,34] The former makes use of the available X-ray structures of CYP450s either directly or to create homology models of

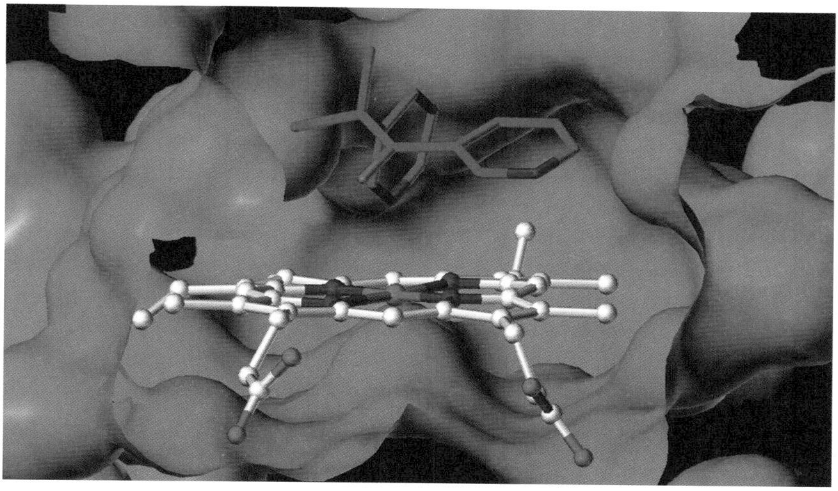

Figure 12.2 Close-up of CYP3A4 active site showing the heme group (white, ball-and-stick) and the inhibitor metapyrone (green, thick stick). (Created from RCSB entry 1W0G.)

other relevant CYP450s.[35,36] In either case, the aim is generally the same: to gain understanding of ligand-binding modes and suggest modifications to the ligand structures that may reduce affinity for the CYP450(s) in question. The structure-based approach has received a significant boost in recent years by the publication of structures of some key human drug-metabolizing CYPs: 3A4 (Figure 12.2), 2C9 and 2C8.[37–41] Previously, most models were built based on the X-ray structures of bacterial or rabbit enzymes.[35] By contrast, ligand-based approaches study known inhibitors and/or substrates in an attempt to generate pharmacophore or QSAR models.[42] However, the approaches should not be considered mutually exclusive, and are often best used in a complementary fashion.[43–45]

The solution of X-ray crystal structures of human CYP450s of relevance to drug discovery means that it is now possible to begin to consider rationally modifying candidate ligands to "design out" liabilities for particular cytochromes. For this to be possible, drug designers must be able to dock compounds into CYP450 active sites and be confident that the resulting binding modes are plausible. Kirton *et al.*[46] have recently reported tests of a popular docking program (GOLD) using a panel of 45 heme-containing protein–ligand co-crystal structures, including several CYP450s. Initially, GOLD was able to reproduce the binding modes [to within 2 Å root mean square deviation (RMSD)] of 64% of the ligands when the ChemScore scoring function was used. This prompted the re-engineering of the scoring function to improve the terms for acceptor–metal interactions and to account for the lipophilicity of planar nitrogen atoms in the heme environment. With the modified scoring function, the docking success rate increased to 73%. This is a little lower than the 79% success rate observed when docking into non-heme proteins, but is perhaps a reflection of the greater challenge posed by the CYP450s, which are notoriously promiscuous in their substrate recognition. Kirton *et al.*[46] suggest that for improved success rates, a better treatment of active-site flexibility may be required, as CYP450s are known to adopt different conformations when binding different ligands. As more CYP450 structures become available, this may be accomplished by docking ligands into multiple structures of the same enzyme. An additional consideration highlighted by de Graaf *et al.*[47] is the role of active-site water molecules in CYP450s. It is common practice in docking studies to delete water molecules from the active site of the protein in question; however, it was shown that ligand-binding modes were more reliably predicted when the crystallographic water molecules were retained during docking.

A perennial problem for ligand-based models has been the paucity of available CYP450 inhibition data. However, this is beginning to change as pharmaceutical companies routinely profile larger compound sets in panels of *in vitro* CYP450 assays. A good illustration of this trend is a recent paper in which QSAR models for CYP3A4 inhibition were developed from data on 930 structurally diverse, drug-like compounds.[48] Using D-optimal "onion" design, the data set was divided into a training set and a test set comprising 551 and 379 compounds, respectively. 315 descriptors were computed for each compound, ranging from simple descriptors, such as molecular weight, to more complex properties derived from quantum mechanical calculations. Various multivariate data analysis approaches were applied to relate the descriptors to the CYP3A4 inhibition activities. A partial least squares (PLS) model with four significant latent variables was able to predict the log IC_{50} values for the test set compounds with a root mean square error (RMSE) of 0.45 log units, which is at least four times smaller than the range of log IC_{50} values covered by the experimental data. The corresponding Q^2 value for the model was 0.6. PLS discriminant analysis was used to generate a model distinguishing between strong ($IC_{50} < 2\,\mu M$) and weak ($IC_{50} > 20\,\mu M$) inhibitors with a success rate of over 60% on the test-set compounds. An analysis of the models showed that high lipophilicity promotes CYP3A4 inhibition, as does increased polarizability and size. This observation is in keeping with what is known about the nature of the CYP3A4 binding site, which is large and comprises mainly hydrophobic residues.

Similarly, O'Brien and de Groot[49] have reported the development of classification models for CYP2D6 inhibition based on a training set of data from 1810 compounds. Compounds were designated as inhibitors ($IC_{50} < 3\,\mu M$) or non-inhibitors ($IC_{50} > 3\,\mu M$). Two different statistical methods – neural networks and Bayesian statistics – were used for model development and the best results were obtained by using a consensus approach that combined the output from both models. When the consensus model was applied to a test set of 600 previously unseen compounds, it provided extremely accurate predictions with >90% of compounds being classified correctly. This highly impressive performance is probably a reflection of, *inter alia*, the size of the training set and also illustrates the power of consensus models to yield improved predictions over single models (the individual neural network and Bayesian models typically showed >80% classification accuracy).

12.5 hERG Channel Blockade

In the latter years of the past decade, a number of high-profile ("blockbuster") drugs were withdrawn from the market or forced to carry strong "black box" warning labels because of sudden, cardiac-linked mortalities. The chemical structures of two such compounds, cisapride and terfenadine, are shown in Figure 12.3. Cisapride was marketed for the treatment of stomach acid reflux, while terfenadine was a widely used anti-allergy medicine. The problem with all these compounds was that, despite not being targeted at the heart, they induced cardiac arrhythmias; specifically, long QT syndrome (LQTS) with an increased propensity to develop the potentially fatal ventricular tachycardia known as Torsade de Pointes (TdP).

In almost all cases, the molecular basis for drug-induced LQTS is interaction with the cardiac ion channel known as hERG. Consequently, the early years of the twenty-first century have seen intense efforts by the pharmaceutical industry to develop methods that will allow the assessment of hERG channel blockade by potential drug compounds as early as possible in the drug-discovery process. Some excellent reviews, from different perspectives, of the progress made in recent years are available.[50–53]

As with the CYP450s, computational approaches towards understanding and predicting hERG blockade activity fall into two broad classes, structure-based and ligand-based.[54,55] In the former, attempts have been made to build models of the hERG channel based on the X-ray structure(s) of related ion channels and information derived from site-directed mutagenesis experiments. Models derived in this way can then be used to try to understand hERG structure–activity relationships and to suggest modifications to candidate compounds that show an undesirable level of hERG activity. The ligand-based approaches typically employ one of a variety of QSAR or classification techniques to derive predictive models that can provide an indication of a compound's hERG activity. In most cases, the ligand-based approaches are designed to be applied to large (virtual) compound collections and so need to be very computationally efficient.

A recent example of the structure-based approach has been reported by Rajamani *et al.*[56] Building on the information provided by the pioneering site-directed mutagenesis studies of Sanguinetti's group (recently summarized by Fernandez *et al.*[57]), two homology models of the hERG channel were

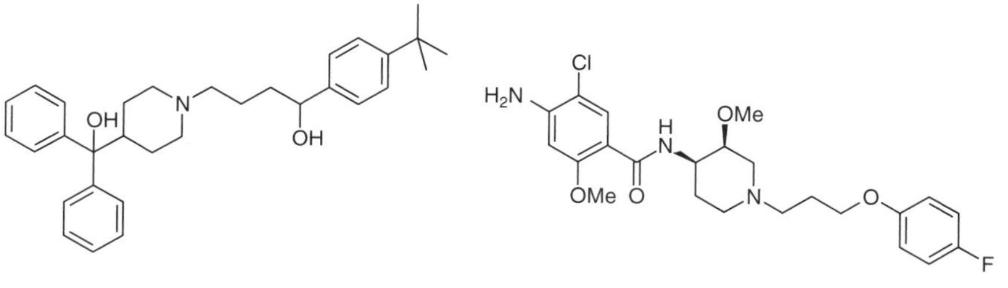

Figure 12.3 Terfenadine (left) and cisapride.

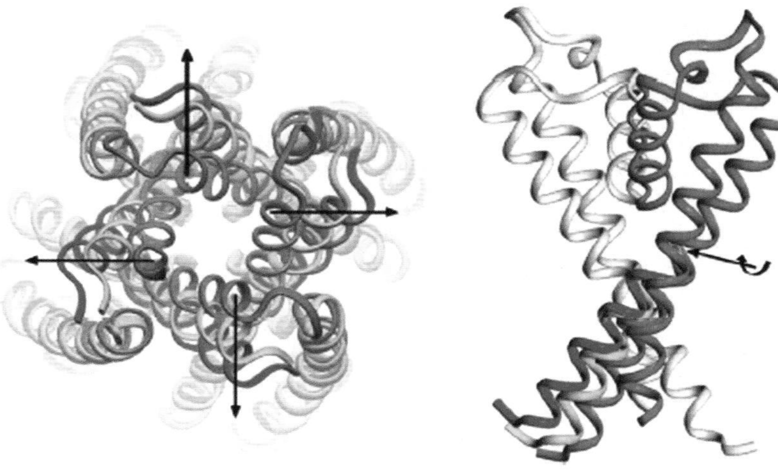

Figure 12.4 Representation of the direction of rotation and helix twist for channel pore opening. Color scheme: red, closed state derived from the reference KcsA structure; yellow, partially open structure represented by a 10 degree tilt; green, open state represented by 19 degree tilt. (Reprinted from *Bioorg. Med. Chem. Lett.*, **15**, R. Rajamani *et al.*, 1737–1741, Copyright 2005, with permission from Elsevier.)

constructed based on the X-ray structure of the potassium ion channel, KscA (Protein Data Bank code, 1K4C). One model was intended to represent the hERG channel in its fully open state, and the other in a partially open state (Figure 12.4). Using this "two-state" model, Rajamani *et al.*[56] hoped to incorporate to some degree the flexibility of the hERG channel and so obtain better predicted binding affinities for a diverse set of 32 known hERG ligands than could be obtained from a single model.

To test this hypothesis, the 32 hERG ligands were docked into the open and partially open models using the Glide program[58] and the best pose for each ligand was then optimized using the OPLS-AA force field in conjunction with the generalized Born/surface area (GB/SA) continuum solvent (water) model. During this optimization, all protein residues within 8 Å of the ligand were allowed to move. Following this, the final conformation of the ligand was extracted and minimized in isolation using the same conditions. From these calculations, the difference in computed electrostatic and van der Waals energies between the free and bound state of each ligand (Δele and Δvdw, respectively) was obtained. These two descriptors were then used to derive QSAR models predicting the ligands' experimental binding affinities for the hERG channel.

It was found that a good QSAR model could only be obtained when the Δele and Δvdw values for each ligand were those computed from the hERG model to which it preferred to bind. That is, if for a particular ligand, the sum of Δele and Δvdw was lowest (most favorable) for the open state, then the descriptors would be derived from that state rather than the partially open state, and *vice versa*. Of the 32 ligands, it was found that 21 preferred the open state and 11 the partially open state. The final QSAR equation, after the removal of five outliers, is shown in Equation (12.5):

$$pIC_{50}\ (\text{hERG}) = -0.163\ (\Delta vdw) + 0.0009(\Delta ele)$$
$$N = 27,\ r^2 = 0.82,\ rmsd = 0.56 \qquad (12.5)$$

where *rmsd* denotes the root-mean-squared difference between the pIC_{50} value predicted by Equation (12.5) and the experimental value. The equation suggests that one of the main contributions to hERG blockade activity is van der Waals interactions and this is in keeping with the presence of hydrophobicity terms in other hERG blockade models and also with what is known about the hERG active site.[57] No application of this equation to an external test set was reported and so its true predictive ability remains to be proven.

It was also shown that the binding modes of the compounds produced by the docking and optimization protocol were in general agreement with findings of site-directed mutagenesis experiments and with previously derived pharmacophore models. As an illustration, the docked conformation of cisapride in the partially open state is shown in Figure 12.5. The ability of the methods described by Rajamani *et al.*[56] to predict both affinity and binding mode for hERG-blocking compounds suggests that this approach could be useful for guiding the synthesis of compounds with reduced hERG affinity. Further validation by application to previously unseen compounds is awaited.

While structure-based approaches, such as that described above, can give powerful insights not merely into predicted affinity, but also into binding modes, they are, inevitably, time-consuming and thus unsuitable for high-throughput

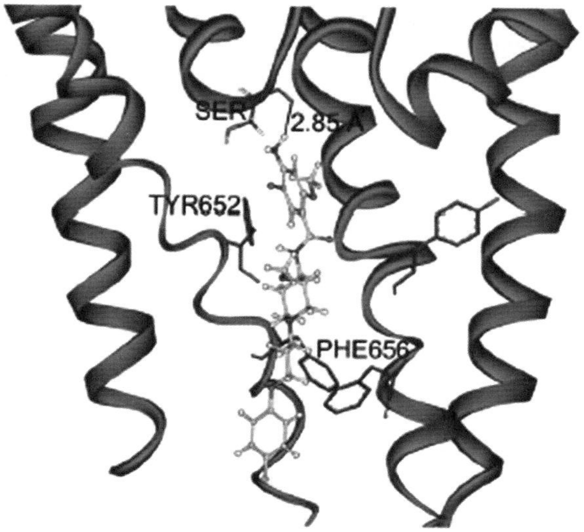

Figure 12.5 Cisapride docked into model of hERG channel in partially open state. (Reprinted from *Bioorg. Med. Chem. Lett.*, **15**, R. Rajamani *et al.*, 1737–1741, Copyright 2005, with permission from Elsevier.)

assessments of large sets of compounds. In such situations, ligand-based models are more likely to be of use.

The development of a ligand-based hERG-blockade prediction model has been reported by Aronov and Goldman.[59] The goal of this work was a classification-type model (*i.e.* predicting compounds as either active or inactive at the hERG channel) and so a set of 85 compounds, each with an IC_{50} value of 40 μM or lower, at the hERG channel was selected from the literature to form the "actives". 329 compounds with no reported hERG activity were chosen from a database of orally delivered drugs to comprise the "inactives". These sets were randomly split into training and test sets, with 80% of the data being used for the former and the remainder for the cross-validation tests.

Using the training set structures, several classification models based on molecular similarity were investigated. This kind of approach works by assessing the similarity of the molecule whose activity is to be predicted to the compounds in the training set, deciding whether it is most similar to the actives or to the inactives and classifying it accordingly. The best results were obtained with a "consensus" model combining two molecular similarity measures: an atom-pair-based measure founded on a compound's 2D structure and a 3D pharmacophore-based similarity. The outputs of these two approaches were combined in a "veto" approach; in other words, a compound was classed as being hERG active if *either* model predicted it as being active. Such consensus approaches have been applied for several years in virtual screening applications[60] and are becoming increasingly common in predictive ADMET models.[49]

In this kind of approach, it is crucial to minimize the number of hERG-inactive compounds that the model predicts to be hERG active (the false-positive rate) because this may result in promising compounds being erroneously excluded from further consideration in a drug-discovery project. While a somewhat higher false-negative rate can be tolerated, to be useful and credible a model needs to reduce this as much as possible. The overall classification

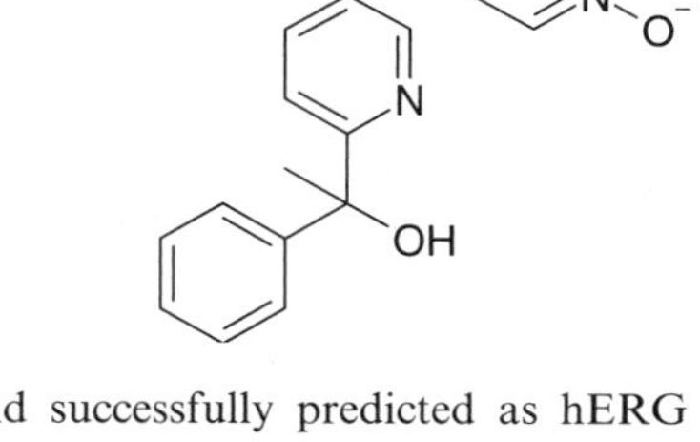

Figure 12.6 Compound successfully predicted as hERG active by a classification model.

accuracy of the model in this work was 82% and it exhibited a false-positive rate of 15% and a false-negative rate of 29%. The model was then applied to a previously unseen external test set comprising 15 compounds, eight of which were hERG actives. Of these eight, the model successfully predicted five (62.5%) to be active. While this, perhaps, indicates the model could be improved further by the inclusion of further compounds in its training set, the successful prediction of the compound in Figure 12.6 (hERG $IC_{50} = 1\ \mu M$) as a hERG active is notable, particularly given its lack of a strongly basic moiety characteristic of many hERG-blocking compounds.

12.6 Conclusions

The field of *in silico* ADMET prediction has received a significant investment of time and resources over the past decade. From the work described briefly above, it should be clear that some advances have been made during that period and models for various properties are now being applied in many drug-discovery programs, although still with an understandable degree of caution in many cases. In my opinion, the most realistic expectation of the current generation of ADMET models is that they can provide "enrichment" in the desired property or properties. For instance, if a solubility model is used to select a subset of compounds predicted to have good solubility from a larger set, then it should be expected that the chosen set will indeed exhibit, on average, better solubility than a set of the same size chosen at random.

Probably the most significant obstacle to continued progress is the relative paucity of data, particularly in the public domain. While it is clear that the larger pharmaceutical companies are now accumulating large data sets from *in vitro* ADMET assays (as was seen in the CYP450 section, for example), for most properties there are still only a few hundred data points at most that are generally available. This is small compared to the more than 10 000 data points available for training lipophilicity prediction software. In such a situation, it is tempting to collate data from multiple sources to increase the size of the training set, however it must be stressed that the *nature* as well as the *size* of a data set is a critical consideration.[61] Ideally, data sets should comprise molecules that are diverse in chemical structure and also in physicochemical properties. This will help to ensure that the resulting models will be as generally applicable as possible. In addition, it is important that the data cover the range of response as completely and evenly as possible. For example, many absorption data sets are heavily biased towards well-absorbed compounds and this obviously impacts the ability of any resulting models to predict poorly absorbed compounds. What this means in practice is that bespoke data generation will be required for modeling, and there is evidence that this is beginning to happen. Additionally, the challenges presented by the complexity of the fundamental processes should not be underestimated.[22] There is still much that remains to be learned and, for instance, new systems involved in active transport are being discovered on a regular basis.

The future of ADMET prediction is likely to benefit from the development of novel molecular descriptors and new statistical methods. In terms of the latter, support vector machines – a new generation of learning system based on recent advances in statistical learning theory – have been generating a great deal of interest in recent years and seem a promising technique.[62] Consensus scoring has been widely accepted in virtual screening[60] and it seems likely that the future will see an increased reliance on consensus models of ADMET properties. Early indications of this are beginning to appear, as seen in the CYP450 and hERG sections above. The first generation of *in silico* models rarely included an estimation of the likely error associated with the predicted values. In particular, there was often no warning given if a particular compound was remote from the training set used to develop the model. More attention is now being paid to these issues,[63–65] which are crucial if the credibility and utility of the next generation of models are to be enhanced. Finally, at present, some areas of ADMET remain almost untouched by modeling efforts. This is particularly true of excretion processes. As more data become available for these, and other currently poorly understood aspects of ADMET, it is likely that models will begin to emerge for their prediction.

References

1. P. Gribbon and A. Sewing, *Drug Discovery Today*, 2005, **10**, 17.
2. S. Venkatesh and R. A. Lipper, *J. Pharm. Sci.*, 2000, **89**, 145.
3. A. P. Li, *Curr. Top. Med. Chem.*, 2004, **4**, 701.
4. U. Norinder, *SAR QSAR Environ. Res.*, 2005, **16**, 1.
5. F. Lombardo, E. Gifford and M. Y. Shalaeva, *Mini Rev. Med. Chem.*, 2003, **3**, 861.
6. H. Van de Waterbeemd and E. Gifford, *Nat. Rev. Drug Discovery*, 2003, **2**, 192.
7. A. P. Beresford, M. Segall and M. H. Tarbit, *Curr. Opin. Drug Discovery Dev.*, 2004, **7**, 36.
8. J. E. Penzotti, G. A. Landrum and S. Putta, *Curr. Opin. Drug Discovery Dev.*, 2004, **7**, 49.
9. B. Pirard, *Comb. Chem. High Throughput Screening*, 2004, **7**, 271.
10. D. E. Clark, *Annu. Rep. Comput. Chem.*, 2005, **1**, 403.
11. C. A. Lipinski, F. Lombardo, B. W. Dominy and P. J. Feeney, *Adv. Drug Delivery Rev.*, 1997, **23**, 3.
12. C. A. Lipinski, *J. Pharmacol. Toxicol. Methods*, 2001, **44**, 235.
13. K. C. Johnson and A. C. Swindell, *Pharm. Res.*, 1996, **13**, 1795.
14. W. Curatolo, *Pharm. Sci. Technol. Today*, 1998, **1**, 387.
15. C. Lipinski, in *Methods and Principles in Medicinal Chemistry*, ed. H. van de Waterbeemd, H. Lennernäs and P. Artursson, Wiley-VCH, Weinheim, 2003, Vol. 18, p. 215.
16. M. Lobell and V. Sivarajah, *Mol. Div.*, 2003, **7**, 69.
17. W. L. Jorgensen and E. M. Duffy, *Adv. Drug Delivery Rev.*, 2002, **54**, 355.

18. J. S. Delaney, *Drug Discovery Today*, 2005, **10**, 289.
19. R. S. DeWitte, *personal communication*, 2004.
20. J. S. Delaney, *J. Chem. Inf. Comput. Sci.*, 2004, **44**, 1000.
21. The ClogP program is marketed by Daylight Chemical Information Systems Inc., 27401 Los Altos, Suite 360, Mission Viejo, CA 92691, USA; http://www.daylight.com and maintained and developed by Biobyte Corp., 201 W. 4th St., #204 Claremont, CA 91711, USA; http://www.biobyte.com.
22. P. S. Burton, J. T. Goodwin, T. J. Vidmar and B. M. Amore, *J. Pharmacol. Exp. Ther.*, 2002, **303**, 889.
23. H. Lennernäs, *J. Pharm. Pharmacol.*, 1997, **49**, 627.
24. H. Lennernäs, *J. Pharm. Sci.*, 1998, **87**, 403.
25. W. J. Egan and G. Lauri, *Adv. Drug Delivery Rev.*, 2002, **54**, 273.
26. H. van de Waterbeemd and B. C. Jones, in *Progress in Medicinal Chemistry*, ed. F. D. King, G. Lawton and A. W. Oxford, Elsevier, Amsterdam, 2003, Vol. 41, p. 1.
27. D. E. Clark, in *Encyclopedia of Computational Chemistry*, ed. P. v. R. Schleyer, W. L. Jorgensen, H. F. Schaefer III, P. R. Schreiner, W. Thiel, and R. Glen, John Wiley & Sons, Chichester, 2004 (http://www.mrw.interscience.wiley.com/ecc/articles/cu0049/frame.html).
28. R. D. Clark and P. R. Wolohan, *Curr. Top. Med. Chem.*, 2003, **3**, 1269.
29. A. K. Mandagere and B. C. Jones, in *Methods and Principles in Medicinal Chemistry*, ed. H. van de Waterbeemd, H. Lennernäs and P. Artursson, Wiley-VCH, Weinheim, 2003, Vol. 18, p. 444.
30. M. Pintore, H. van de Waterbeemd, N. Piclin and J. R. Chretien, *Eur. J. Med. Chem.*, 2003, **38**, 427.
31. H. van de Waterbeemd, D. A. Smith and B. C. Jones, *J. Comput. Aided Mol. Des.*, 2001, **15**, 273.
32. P. B. Danielson, *Curr. Drug Metab.*, 2002, **3**, 561.
33. M. J. de Groot, S. B. Kirton and M. J. Sutcliffe, *Curr. Top. Med. Chem.*, 2004, **4**, 1803.
34. C. de Graaf, N. P. E. Vermeulen and K. A. Feenstra, *J. Med. Chem.*, 2005, **48**, 2725.
35. C. A. Kemp, J. -D. Marechal and M. J. Sutcliffe, *Arch. Biochem. Biophys.*, 2005, **433**, 361.
36. J. Mestres, *Proteins*, 2005, **58**, 596.
37. P. A. Williams, J. Cosme, A. Ward, H. C. Angove, D. M. Vinkovic and H. Jhoti, *Nature*, 2003, **424**, 464.
38. P. A. Williams, J. Cosme, D. M. Vinkovic, A. Ward, H. C. Angove, P. J. Day, C. Vonrhein, I. J. Tickle and H. Jhoti, *Science*, 2004, **305**, 683.
39. J. K. Yano, M. R. Wester, G. A. Schoch, K. J. Griffin, C. D. Stout and E. F. Johnson, *J. Biol. Chem.*, 2004, **279**, 38091.
40. M. R. Wester, J. K. Yano, G. A. Schoch, C. Yang, K. J. Griffin, C. D. Stout and E. F. Johnson, *J. Biol. Chem.*, 2004, **279**, 35630.
41. G. A. Schoch, J. K. Yano, M. R. Wester, K. J. Griffin, C. D. Stout and E. F. Johnson, *J. Biol. Chem.*, 2004, **279**, 9497.

42. M. J. de Groot and S. Ekins, *Adv. Drug Delivery Rev.*, 2002, **54**, 367.
43. M. J. de Groot, A. A. Alex and B. C. Jones, *J. Med. Chem.*, 2002, **45**, 1983.
44. M. J. de Groot, M. J. Ackland, V. A. Horne, A. A. Alex and B. C. Jones, *J. Med. Chem.*, 1999, **42**, 1515.
45. M. J. de Groot, M. J. Ackland, V. A. Horne, A. A. Alex and B. C. Jones, *J. Med. Chem.*, 1999, **42**, 4062.
46. S. B. Kirton, C. W. Murray, M. L. Verdonk and R. D. Taylor, *Proteins*, 2005, **58**, 836.
47. C. de Graaf, P. Pospisil, W. Pos, G. Folkers and N. P. Vermeulen, *J. Med. Chem.*, 2005, **48**, 2308.
48. J. M. Kriegl, L. Eriksson, T. Arnhold, B. Beck, E. Johansson and T. Fox, *Eur. J. Pharm. Sci.*, 2005, **24**, 451.
49. S. E. O'Brien and M. J. de Groot, *J. Med. Chem.*, 2005, **48**, 1287.
50. B. Fermini and A. A. Fossa, *Nat. Rev. Drug Discovery*, 2003, **2**, 439.
51. K. Finlayson, H. J. Witchel, J. McCulloch and J. Sharkey, *Eur. J. Pharmacol.*, 2004, **500**, 129.
52. M. C. Sanguinetti and J. S. Mitcheson, *Trends Pharmacol. Sci.*, 2005, **26**, 119.
53. M. Recanatini, E. Poluzzi, M. Masetti, A. Cavalli and F. De Ponti, *Med. Res. Rev.*, 2005, **25**, 133.
54. R. Pearlstein, R. Vaz and D. Rampe, *J. Med. Chem.*, 2003, **46**, 2017.
55. A. M. Aronov, *Drug Discovery Today*, 2005, **10**, 149.
56. R. Rajamani, B. A. Tounge, J. Li and C. H. Reynolds, *Bioorg. Med. Chem. Lett.*, 2005, **15**, 1737.
57. D. Fernandez, A. Ghanta, G. W. Kauffman and M. C. Sanguinetti, *J. Biol. Chem.*, 2004, **279**, 10120.
58. The Glide program is developed and marketed by Schrödinger, Inc., Schrödinger, 1500 S.W. First Avenue, Suite 1180, Portland, OR 97201-5815, USA. http://www.schrodinger.com.
59. A. M. Aronov and B. B. Goldman, *Bioorg. Med. Chem.*, 2004, **12**, 2307.
60. P. S. Charifson, J. J. Corkery, M. A. Murcko and W. P. Walters, *J. Med. Chem.*, 1999, **42**, 5100.
61. T. R. Stouch, J. R. Kenyon, S. R. Johnson, X. Q. Chen, A. Doweyko and Y. Li, *J. Comput. Aided Mol. Des.*, 2003, **17**, 83.
62. M. W. B. Trotter and S. B. Holden, *QSAR Comb. Sci.*, 2003, **22**, 533.
63. A. M. Davis and R. J. Riley, *Curr. Opin. Chem. Biol.*, 2004, **8**, 378.
64. R. P. Sheridan, B. P. Feuston, V. N. Maiorov and S. K. Kearsley, *J. Chem. Inf. Comput. Sci.*, 2004, **44**, 1912.
65. R. Guha and P. C. Jurs, *J. Chem. Inf. Comput. Sci.*, 2005, **45**, 65.

CHAPTER 13
Shadows on Screens

BRIAN K. SHOICHET*, BRIAN Y. FENG AND KRISTIN E. D. COAN

Department of Pharmaceutical Chemistry & Graduate Group in Chemistry and Chemical Biology, University of California, San Francisco 1700 4th St., Byers Hall Room 508D, San Francisco, CA 94158, USA

13.1 Introduction

If high-throughput screening (HTS) has changed drug discovery, it has also introduced into it a bestiary of peculiar molecules. Some of these have turned out to be interesting and important, others have proven to be "nuisance compounds" with strange properties. Steep dose response curves, flat structure–activity relationships (SARs), and high sensitivity to assay conditions are rarely seen with classic, well-behaved drugs and reagents, but are common among nuisance hits. These are rarely suited to development, but much time and passion can be wasted chasing them before they are abandoned. Their prevalence has contributed to the evolution of screening practices over the past 15 years towards high-quality compound libraries, the maintenance of dry stocks of pure compounds, and ever-lower concentrations of compound in initial screens.

"Nonsense is always nonsense, but the study of nonsense can be scholarship," Saul Lieberman said of the Kabbalist Gershom Scholem. Much scholarly ink has been spilled on compounds in screening decks that are prone to artifactual inhibition. Lipinski's now famous rules focused on the physical properties of drugs, and were in reaction to an early tendency in HTS libraries towards large and hydrophobic molecules that made them unlikely to be orally bioavailable. Subsequent studies have focused on chemical reactivity,[1] assay interference,[2] high flexibility, oxidation potential,[3] formal molecular charge,[4] or liability to degradation and precipitation.[5] Doing so unambiguously has proven difficult; as one class of pathological inhibitor is identified, another emerges, like conversations at a tedious party from which one cannot escape. Partly this is a problem of the apparent "specificity" of nuisance compounds for the conditions of particular assays – a promiscuous hit in one assay can behave demurely in another, conferring on it a cruel imitation of fidelity. But there were also hits from screening that did not obviously manifest the nuisance properties identified in the initial studies. These molecules appeared chemically inert, seemed spectrally transparent, passed internal filters and Lipinski rules and, worst of all, had little in common other than their similar behavior in assays, including steep dose–response curves[2] (see below) and SARs. Such compounds were widely known among screeners, but their mechanism and the properties that related them remained obscure. They were not publicly discussed.

Here we consider a single mechanism to explain the behavior of many pathological hits, one that resolves apparent inconsistencies. At micromolar and sometimes submicromolar concentrations many organic molecules, even those considered "drug-like" by most metrics, aggregate into colloid-like particles in aqueous media. These aggregates can sequester protein targets, inhibiting them (Figure 13.1). Aggregating inhibitors are often unrelated chemically, though they often share certain physical properties. Like colloids and vesicles, they are sensitive to assay conditions; as particles present at low concentration, their level of inhibition depends on target concentration. Both properties contribute to their flickering, skittish behavior. From a classic perspective this behavior can be baffling, but viewed biophysically it is actually expected. Indeed, aggregation-based "promiscuous" inhibitors may be rapidly detected and controlled for based on these features. Here we summarize the range of molecules now known to behave this way, their mechanism of action, their rapid detection in a screening environment, and consider their possible effects in biological environments.

13.2 Phenomenology of Aggregation

We encountered nuisance compounds by accident, while looking for inhibitors of the enzyme β-lactamase. We had tested tens of compounds predicted by virtual and, eventually, HTS, finding many apparent inhibitors (Table 13.1). All had strange properties: they were non-competitive, time-dependent, and inhibited not only β-lactamase, but also dihydrofolate reductase (DHFR), chymotrypsin, β-galactosidase, and malate dehydrogenase (MDH).[6] They also had the odd, steep dose–response curves that were characteristic of nuisance

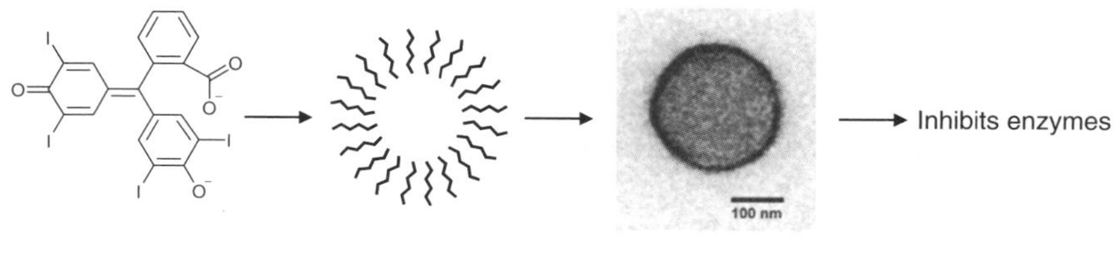

Figure 13.1 Large aggregates of organic molecules can sequester and inhibit proteins. (Adapted from the *Journal of Medicinal Chemistry*, 2002, **45**, 1712–1722, with permission. Copyright 2002 American Chemical Society.)

Table 13.1 Early inhibitors characterized as acting through aggregation. (Adapted from the *Journal of Medicinal Chemistry*, 2003, **45**, 1712–1722, with permission. Copyright American Chemical Society.)

Structure	*IC_{50} (μM)* *Original Target(s)*	*β-lactamase*	*Chymotrypsin*	*cDHFR*	*β-gal*
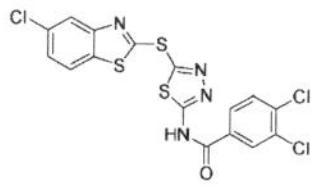	0.5 β-lactamase	0.5	2.5	5	15
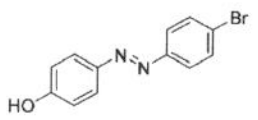	5 β-lactamase	5	25	35	90
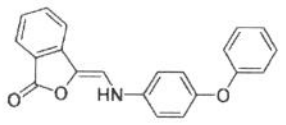	5 β-lactamase	5	15	N.D.	N.D.
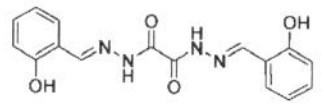	8 Malarial protease	10	55	70	180
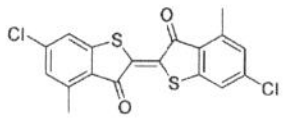	7 pDHFR	10	50	60	300
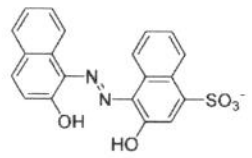	80 pDHFR	50	25	N.D.	600
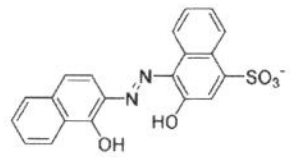	50 HIV Tar RNA	10	90	N.D.	600
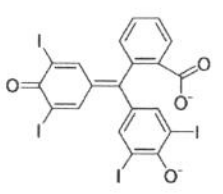	3 TS; 30 kinesin	3	11	20	200
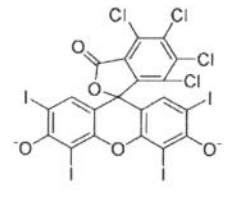	20 insulin receptor; 7.5 kinesin	16	50	N.D.	80

Table 13.1 (*continued*).

	IC_{50} (μM)					
Structure	*Original Target(s)*		*β-lactamase*	*Chymotrypsin*	*cDHFR*	*β-gal*
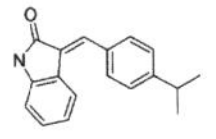	5.2 VEGF	10.0 IGF-1	6	30	30	55
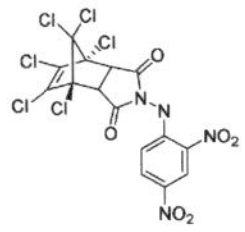	25 farnesyltransferase		3	9	25	150
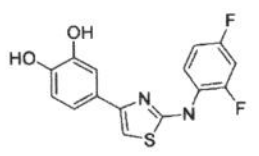	15 gyrase		18	100	150	320
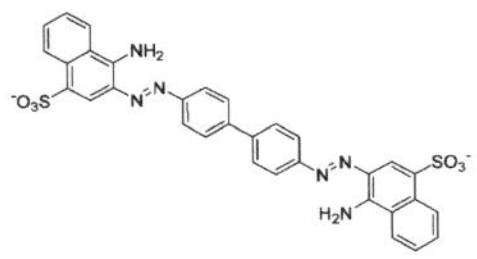	1 prion	30.4 TIM	3.9	40	0.4	100
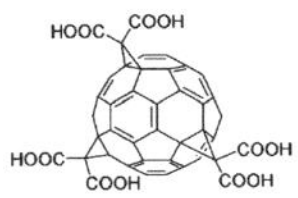	17 eNOS	24 nNOS	7	60	N.D.	N.D.
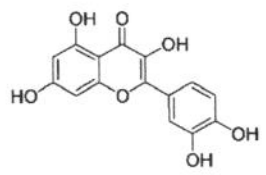	3.8 PI3K	11.0 integrase	4	100	N.D.	220

cDHFR, chicken DHFR; β-gal, β-galactosidase; pDHFR, *Pneumocystis carinii* DHFR; TS, thymidylate synthase; VEGF, vascular endothelial growth factor receptor tyrosine kinase; IGF-1, insulinlike growth factor receptor tyrosine kinase; TIM, triosephosphate isomerase; eNOS, endothelial nitric oxide synthase; nNOS, neuronal nitric oxide synthase; PI3K, phosphoinositide 3-kinase; N.D., not determined.

compounds (see below). We were initially concerned that these compounds were covalent inhibitors, but inhibition was reversible by dilution, inconsistent with such a mechanism. We then wondered if these inhibitors, so dissimilar structurally, were acting as denaturants. If this were true, we might expect inhibition to be increased by guanidinium, urea, or temperature, but the opposite occurred – it was attenuated. Intriguingly, the potency of these compounds was strangely sensitive to protein concentration, diminishing considerably on addition of large amounts of bovine serum albumin or even increased amounts of the target enzyme.[6] For instance, increasing the concentration of β-lactamase in the assay ten-fold diminished potency dramatically. Efficacy returned if the inhibitor concentration was raised by a similar amount. This was difficult to reconcile with any classic mechanism of enzyme inhibition with which we were then aware. Rather, it seemed to point to a stoichiometric mechanism of inhibition, except that the stoichiometries would not be 1:1 or even 10:1, but more like thousands of inhibitor molecules to one enzyme molecule. The only mechanism that we could think of with such baroque molar ratios was one where the inhibitors acted as colloid-like aggregates that somehow sequestered and inhibited enzyme targets without specificity.

A virtue of this hypothesis was that it was easily tested by direct methods. Both dynamic light scattering (DLS) and transmission electron microscopy (TEM) experiments clearly indicated the presence of particles (Figure 13.2). Both techniques agreed that the particles were huge, often 200 nm in diameter or more, five orders of magnitude larger than the enzymes that they were inhibiting. Consistent with these particles being colloid-like aggregates of the organic small molecules, they were sensitive to the ionic strength of the buffer. Thus, on moving to lower ionic strength buffers, particle size decreased but the number of particles appeared to increase and inhibition improved. At high ionic strength, particle size increased, the apparent number of particles diminished, as did inhibition.[6]

By 2001 it was clear that there was a chemically disparate class of nuisance compounds that inhibited enzymes not through a classic, single-molecule mechanism, but rather through sequestration. Because these compounds inhibited multiple enzymes, we began to call them "promiscuous" inhibitors, a term probably first coined by Mic Lajines at Pharmacia. What was uncertain at that time was the range of molecules that behaved this way, their relationship to the nuisance hits that had bedeviled HTS campaigns, and what their mechanism of action was.

13.3 What Sort of Compounds Aggregate?

Many of the compounds that had been shown to aggregate were conjugated, dye-like molecules. Indeed, such compounds continue to be discovered as aggregation-based inhibitors – see, for instance, the study by Tipton, Zou, and colleagues on the inhibition of phosphomannomutase/phosphoglucomutase by the dye disperse blue 56.[7] These compounds would probably have been detected by the computational filters implemented in Pharmacia to flag nuisance compounds. For instance, there is a high overlap between these "dye-like" molecules and those flagged by programs such as REOS[2] and the "frequent hitters" virtual screening program used at Roche.[8] We wanted to learn if more drug-like, arguably more pharmaceutically relevant molecules might also form aggregates.

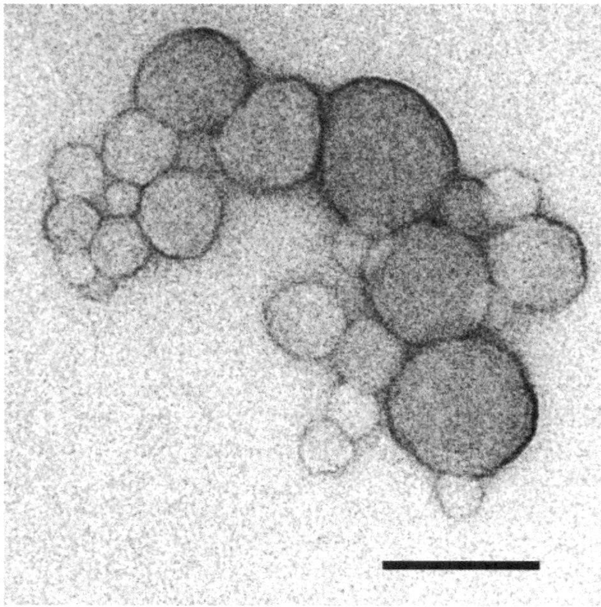

(A)

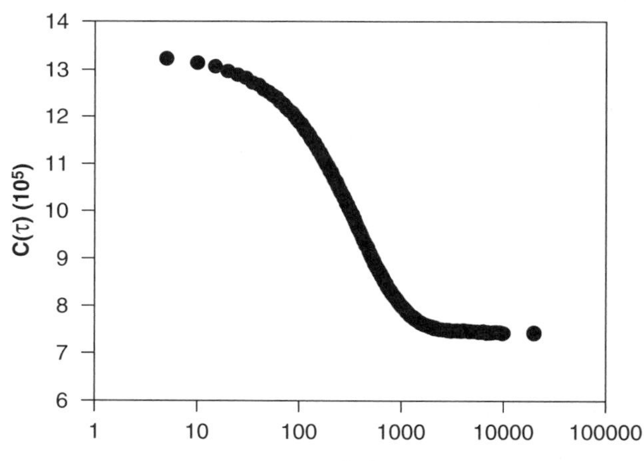

(B)

Figure 13.2 (**A**) Aggregates formed by tetra-iodophenolphthalein, a promiscuous inhibitor, visualized by transmission electron microscopy (Bar = 100 nm) and (**B**) by dynamic light scattering. (Reprinted with permission from the *Journal of Medicinal Chemistry*, 2002, **45**, 1712–1722. Copyright 2002 American Chemical Society.)

We first considered 30 compounds from an in-house Pharmacia screening deck. Of these, 20 inhibited three of our reporter enzymes, AmpC, DHFR, and chymotrypsin, non-competitively in a time-dependent manner at micromolar or tens-of-micromolar concentrations. The 20 also scattered light in a DLS assay. We concluded that these screening compounds were promiscuous aggregators, at least in biochemical buffers at micromolar concentrations.

We then investigated compounds that most would consider genuine leads for drug discovery. We picked 15 heavily studied kinase inhibitors, including quercetin, rottlerin, and bisindolylmaleimide, that Cohen and colleagues had shown to be promiscuous among kinase targets.[9] Of these 15 inhibitors, eight were active against three counter-screen enzymes (MDH, chymotrypsin, and β-lactamase) with what were by now tell-tale features, *i.e.* non-competitive, time-dependent inhibition, and steep dose–response curves (*e.g.* rottlerin's inhibition of β-lactamase, Figure 13.3). All eight formed particles in the hundreds of nanometer size range by DLS (Table 13.2).[10] We concluded that these eight, highly studied leads behaved as promiscuous, aggregation-based inhibitors in the micromolar and tens-of-micromolar concentration range. Subsequently, other investigators have found that other widely used biological reagents can behave this way. Flavenoid molecules seem particularly prone to hit in high-throughput screens, something alluded to by Rishton.[11] Recently, Zavodszky and Kuhn have found that fuchsin and morelloflavone will inhibit thrombin as promiscuous aggregator (Maria Zavodszky and Leslie Kuhn, personal communication). Similarly, several cholesterol-based hormones will aggregate and inhibit at micromolar concentrations.[12]

If "drug-like" molecules can behave as aggregation-based inhibitors, what about actual drugs? Our initial hope was actually that drugs would not

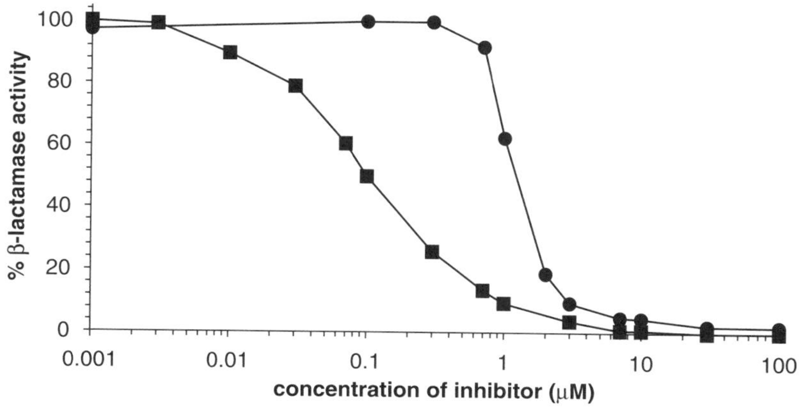

Figure 13.3 The dose–response curve of rottlerin, a promiscuous inhibitor, at micromolar concentrations (circles), and benzothiophene-2-boronic acid, a classic, competitive inhibitor (squares), with β-lactamase. (Reprinted with permission from the *Journal of Medicinal Chemistry*, 2003, **46**, 1478–1483. Copyright 2003 American Chemical Society.)

Table 13.2 Some kinase inhibitors inhibit non-kinases and form particles detectable by dynamic light scattering. Adapted from the *Journal of Medicinal Chemistry*, 2003, **46**, 1478–1483, with permission. Copyright 2003 American Chemical Society.

Compound	*IC_{50} vs. β-lactamase (μM)*	*DLS conc. (μM)*	*Intensity (kcps)*	*Diameter (nm)*
Rottlerin 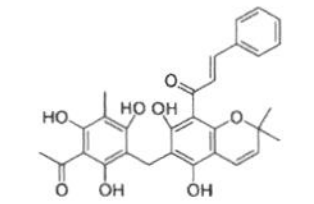	1.2	15	11.8 ± 1.2	99.0 ± 6.7
Quercetin 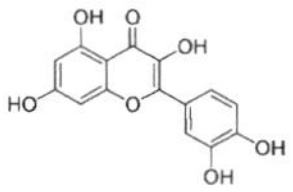	4	100	65.0 ± 11.1	>1000
K-252c 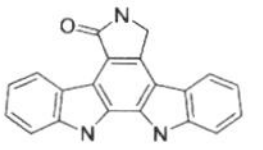	8	10	13.5 ± 1.7	780.9 ± 65.7
Bisindolylmaleimide IX 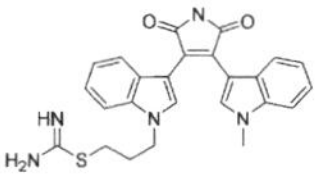	5	60	25.6 ± 3.0	578.6 ± 66.5
Bisindolylmaleimide I 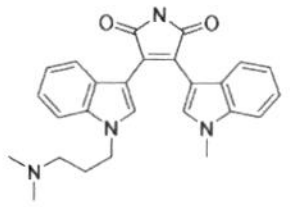	60	400	2.9 ± 0.5	287.1 ± 7.7
U0126 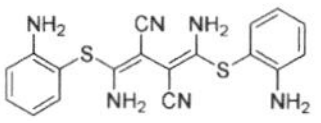	30	80	53.7 ± 9.9	432.2 ± 42.1
Indirubin 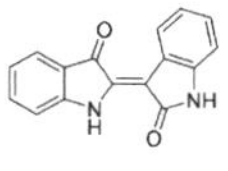	20	10	62.5 ± 6.9	>1000

Table 13.2 (*continued*).

Compound	*IC_{50} vs. β-lactamase (µM)*	*DLS conc. (µM)*	*Intensity (kcps)*	*Diameter (nm)*
Indigo 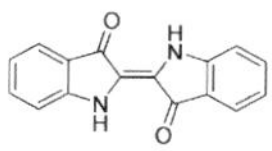	30	20	32.1 ± 1.2	>1000

kcps, kilocounts per second.

Table 13.3 Some drugs that inhibit promiscuously and that scatter light at micromolar and high micromolar concentrations. Adapted from the *Journal of Medicinal Chemistry*, 2003, **46**, 4477–4486, with permission. Copyright 2003 American Chemical Society.

Drug	*IC_{50} vs. β-lactamase (µM)*	*IC_{50} vs. Chymotrypsin (µM)*	*IC_{50} vs. MDH (µM)*	*DLS conc. (µM)*	*DLS scattering intensity (kcps)*	*Diameter of DLS particle (nm)*
clotrimazole	20	85	35	50	35.9 ± 12.6	323.2 ± 31.8
benzyl benzoate	90	250	125	250	30.5 ± 2.3	893.1 ± 56.4
nicardipine	20	175	50	60	23.8 ± 3.3	514.7 ± 67.8
delavirdine	90	225	85	100	43.3 ± 2.3	207.2 ± 15.2

aggregate, affording us a biophysical criterion for "drug-like." We tested over 50 drugs against a panel of enzyme counter-screens (MDH, chymotrypsin, and β-lactamase).[12] Most did not inhibit these enzymes, even up to 400 µM, consistent with expectations. Four drugs, however, inhibited all three enzymes in the ten to hundreds of micromolar concentration range, including the oral drugs delavirdine and nicardipine (Table 13.3). These four drugs showed all the characteristics of aggregation-based inhibition: non-competitive, time-dependent inhibition, high sensitivity to enzyme concentration and to detergent (see below), and formation of particles in the 200–400 nm size range by DLS.

Thus, whereas most drugs were well behaved, some clearly can form aggregates in aqueous buffers. How can this behavior be reconciled with their status as specific agents? One explanation is that aggregation was observed in biochemical buffers that lack the adjuvants, serum proteins, and bile salts that might disrupt or saturate aggregates in the body. Also, inhibition occurred at concentrations a thousand-fold greater than the IC_{50} values of the drugs for their targets. We thus do not consider the promiscuous inhibition in biochemical buffers at high concentrations to reflect the behavior of these drugs on the target in the body (for a possible *in vivo* effect of aggregation, see below). Still, drugs are molecules too; they are subject to the associations and equilibria that other organic molecules obey. Under screening conditions and at screening concentrations, drugs can aggregate and non-specifically inhibit, just like any other molecule.

13.4 Mechanism of Aggregation-based Inhibition

Whereas early studies suggested that there was an association between aggregates and inhibition,[6] the mechanistic link was unclear. After all, association with a solid support typically leaves an enzyme or a protein uninhibited, so why should association with an aggregate of organic molecule be so detrimental?

The first question was whether the aggregates directly associate with enzyme, something implied but never demonstrated by the initial studies. Three lines of evidence came to support this view. The simplest was co-precipitation of enzyme and aggregates followed by gel electrophoresis. Aggregates were incubated with enzyme and co-precipitated by centrifugation on a bench-top microfuge. The precipitant was run on an agarose gel and silver-stained. In the absence of enzyme, no protein band was observed, nor was one observed when the supernatant was run out on the gel. When the precipitant from the protein-aggregate solution was run, a clear enzyme band was visible, consistent with the idea that the aggregates, on pull-down, co-sequestered enzyme (Figure 13.4).[13] The second result supporting direct association was the observation that proteins such as β-galactosidase can be seen to associate directly with aggregates in TEM.[13] Lastly, when Green Fluorescent Protein (GFP) is mixed with aggregates, the formally uniform fluorescence of the protein becomes punctuate, consistent with direct sequestration of the GFP by the aggregate particles (Figure 13.4).[13]

If the aggregates inhibit by direct association and sequestration of proteins, how does this lead to inhibition? An initial hypothesis was that the aggregates were denaturing the enzymes, but this was not easily consistent with the retention of GFP fluorescence in association with aggregate. Also, no selectivity of aggregates for destabilized versus stabilized mutant β-lactamases was observed. If aggregates were working by protein unfolding, one would expect the destabilized β-lactamase to be more susceptible to inhibition than the up-stabilized enzyme, but this was not the case.[13] Additionally, aggregation-based inhibition was rapidly reversible by addition of detergent even after aggregate-based inhibition had reached saturation (Figure 13.5), implying that aggregation-based inhibition is rapidly reversible. Whereas these observations do not rule out some level of denaturation of the protein by the aggregate, they are inconsistent with gross structural changes.

Thus, aggregates associate with protein, but do not appear to globally denature them. How exactly this association takes place, what drives it, and why it leads to inhibition remain open questions. The nature of the interaction

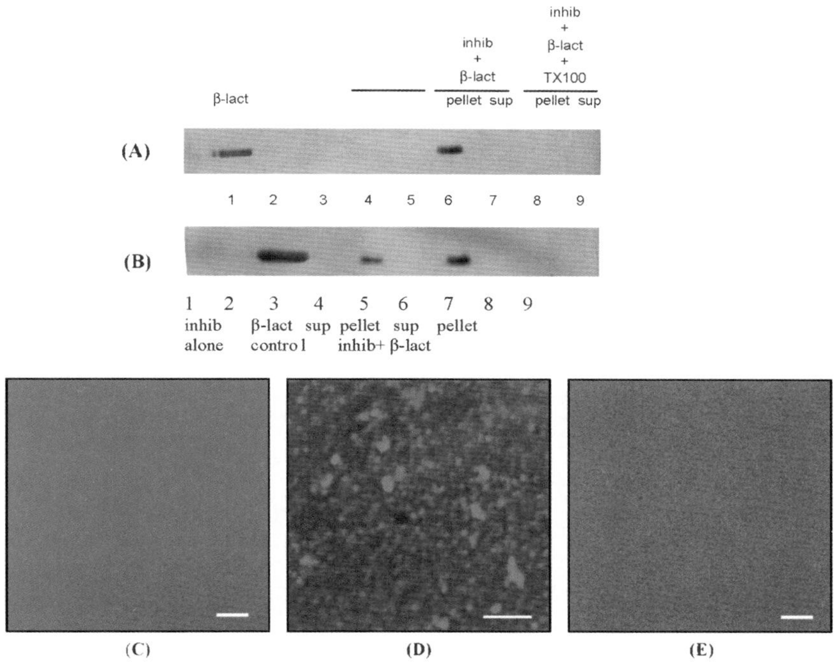

Figure 13.4 Direct association of enzyme by aggregates. (**A**) Co-precipitation of β-lactamase by tetra-iodophenolphthalein, a promiscuous aggregator, followed by gel electrophoresis. SDS-PAGE and silver-stain analysis of supernatants and pellets from centrifugation of β-lactamase in the presence or absence of inhibitor, with and without Triton X-100. (**B**) Same as in (**A**), except with the promiscuous inhibitor 4-bromophenylazo-(4')-phenol. (**C**) Fluorescence of GFP in the absence and (**D**) in the presence of the aggregator tetra-iodophenolphthalein. (**E**) Same as in (**C**), but with addition of the detergent Triton-X 100 at 0.01% concentration. Scale bar = 5 μm. (Adapted from the *Journal of Medicinal Chemistry*, 2003, **46**, 4265–4272, with permission. Copyright 2003 American Chemical Society.)

between aggregates and proteins also remains uncertain. It is tempting to suppose that association is driven by hydrophobicity, given its well-known lack of specificity, but other terms might also be important, including liquid-crystal-like stacking.

13.5 A Rapid Counter-screen for Aggregation-based Inhibitors

Two of the characteristics of aggregation-based inhibition lend themselves to rapid, large-scale detection: the high sensitivity of aggregates to non-ionic detergents[13,14] and the formation of large particles in solution. To explore how

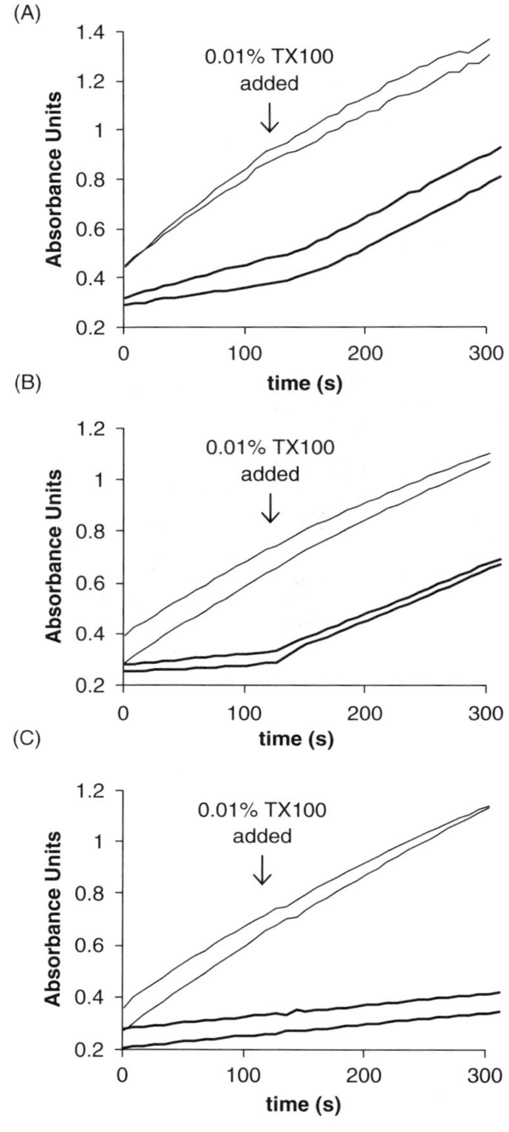

Figure 13.5 The effect of detergent, added partway through substrate hydrolysis by β-lactamase, on aggregation-based inhibition. (**A**) 10 μM tetra-iodophenolphthalein; (**B**) 5 μM rottlerin; (**C**) 0.6 μM BZBTH2B, a specific β-lactamase inhibitor.[26] Thick lines (—) denote inhibited reactions and thin lines (—) denote those with a dimethylsulfoxide control. (Reprinted with permission from the *Journal of Medicinal Chemistry*, 2003, **46**, 4265–4272. Copyright 2003 American Chemical Society.)

prevalent promiscuous aggregate-forming molecules might be in screening libraries, we examined 1030 diverse, Rule-of-Five-compliant[15] molecules from a widely used supplier of screening compounds. These molecules were tested in separate 96-well assays for detergent-sensitive inhibition of β-lactamase and light scattering in a DLS plate reader. The molecules may be divided into three groups: randomly selected, predicted aggregators, and predicted non-aggregators. We first consider the 298 randomly selected molecules, which may represent the behavior of an unbiased screening library. A startling 19% of these "drug-like" molecules inhibited β-lactamase in a detergent-dependent manner at 30 μM.[16] At 5 μM, 1.4% of these molecules continued to do so (Figure 13.6). To test the robustness of these results, a few of these molecules were further tested in a more precise, low-throughput version of the assay; all inhibition was reproducible. These molecules also inhibited chymotrypsin at 100 μM, consistent with the promiscuous nature of their inhibition.

Even more molecules formed particles in a plate-based DLS assay than inhibited β-lactamase.[16] Of the randomly selected molecules, 36% formed particles detectable at 30 μM, compared to only 19% displaying detergent-sensitive inhibition (Figure 13.7). This suggests that particle formation is a necessary-but-not-sufficient condition in the detection, and perhaps also the action, of promiscuous aggregates. Intriguingly, most compounds that visibly precipitated in the assay typically did not inhibit β-lactamase, suggesting that while precipitation and the formation of promiscuous aggregates are related,

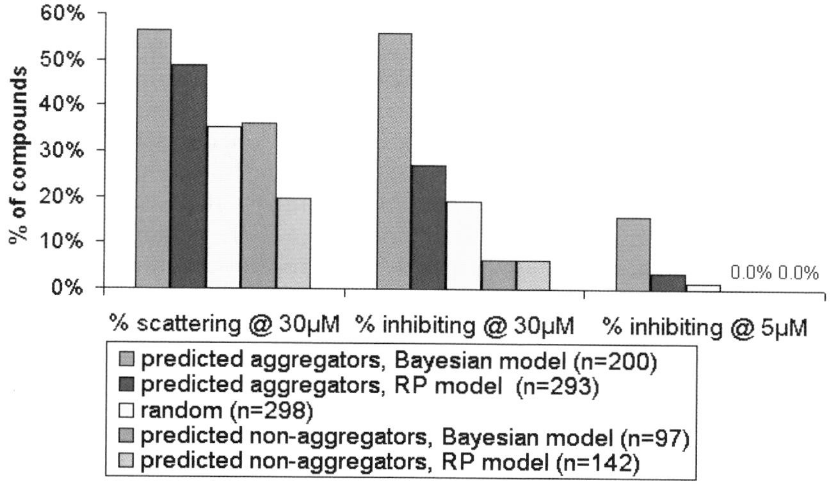

Figure 13.6 Percentage of molecules that scatter light by DLS and that inhibit in the absence, but not in the presence of detergent, using 96-well plate based assays. Molecules were tested in different sets: two sets of predicted aggregators and two sets of predicted non-aggregators, derived using different models,[14] and a fifth category of molecule chosen at random from a "drug-like" subset of the ChemDiv collection. (Reprinted with permission from *Nature Chemical Biology*, 2005, **1**, 146–148.)

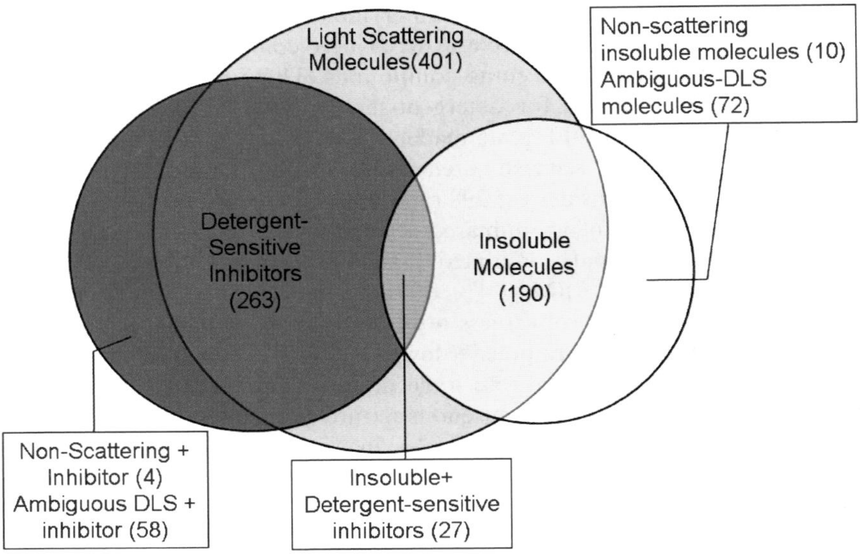

Figure 13.7 Comparison of the hits from the DLS screen and the detergent-sensitive enzyme inhibition screen. (Reprinted with permission from *Nature Chemical Biology*, 2005, **1**, 146–148.)

they are distinct phenomena (Figure 13.7). Both the tendency of DLS to conflate aggregation and precipitation and the long data-collection times on the instrument that we used limits the usefulness of the technique for rapid screening. This could change with advances in instrumentation.

A second purpose of these screens was to test the two different computational models for aggregation, both of which are based on the 2D-topology and physical properties of molecules as descriptors. The first model used a recursive partitioning algorithm[12–17] and the second used a naïve Bayesian model. Light scattering and detergent-sensitive inhibition were more common among the molecules computationally predicted to be aggregators than in the random subset, consistent with their predictions. Correspondingly, the predicted non-aggregators contained fewer hits from both assays. Both models predicted aggregators and non-aggregators with about the same success, though both had substantial numbers of false-positives and false-negatives. To improve the models, each was retrained on an expanded training set that included the experimentally characterized aggregators and non-aggregators from the predicted aggregator and non-aggregator subsets (*i.e.* about 750 molecules in addition to the original 110). The 298 molecules of the random subset were withheld as a final test set. This retraining substantially improved their performance; the misclassification rate (a combination of false-positives and false-negatives) of the initial Bayesian model fell from 26% to 20%, whereas the misclassification rate of the random partitioning model fell from 46%, to 11% on retraining and switching to the conceptually related random forest method.[18] These approaches thus hold some promise for prospective prediction of promiscuous aggregates, though they do not capture the steep concentration-dependence of this phenomenon and continue to mispredict molecules. When one is in doubt as to whether or not an inhibitor is acting through aggregation then experiment must be the final arbiter (see Shoichet[19] for a useful experimental protocol).

We next considered how the presence of multiple aggregators might affect the properties of a mixture of molecules. Screening such mixtures is a common way to minimize reagent consumption and to increase assay throughput. Whether the efficiency and speed gained is not undone by the increased chance of artifact has occasioned debate in the screening literature.[20–22] Different screening centers have adopted different positions on this question, with some using mixtures routinely and others adhering to one-well-one-compound. We wondered whether aggregates might not be more likely to form when mixtures are screened. Although mixtures are often screened at low concentrations of individual molecules (5 μM or less is common), the "chemical load," the overall concentration of all organic material in the assay, is often quite high. For instance, the chemical load of a mixture of ten molecules each at 5 μM would be 50 μM. Since aggregating molecules are unlikely to behave as perfect solutions, the presence of molecules with the potential to aggregate could affect mixture behavior in unexpected ways. We therefore assembled 80 mixtures of, on average, ten molecules, randomly chosen from among the 764 soluble molecules in the set of 1030 used in the previous study. The behavior of each mixture was analyzed as though it were a mixture of mutually exclusive inhibitors (*i.e.* the null hypothesis), with the individual behavior of each molecule, measured in the previous study, used to calculate the predicted activity of the mixtures using Equation (13.1):[23,24]

$$\text{Total \% Inhibition} = 100^* \frac{\sum_{i=1}^{n} \left(\frac{[I_i \cdot E]}{[E]}\right)}{\sum_{i=1}^{n} \left(\frac{[I_i \cdot E]}{[E]}\right) + 1} \tag{13.1}$$

Where E is the free enzyme and $I_i \cdot E$ is the enzyme bound to the i'th inhibitor in the mixture.

The actual experimental behavior of the mixtures often differed from the mutually exclusive, non-interacting model of Equation (13.1). Of the 80 mixtures, 45% were at least two-fold more potent than predicted and 16% were at least two-fold less potent. Almost one-quarter of the mixtures were greater than ten-fold more potent than predicted, whereas 10% of them were at least ten-fold less potent than expected. Overall, the deviation ranged from 200-fold less potent to 447-fold more potent than predicted (Figure 13.8). Both the synergistic and the antagonistic deviations were largely eliminated upon the addition of 0.01% Triton X-100 (Figure 13.8).

The results from both the single compound and mixture screens suggest that promiscuous aggregators are common enough in screening collections to cloud

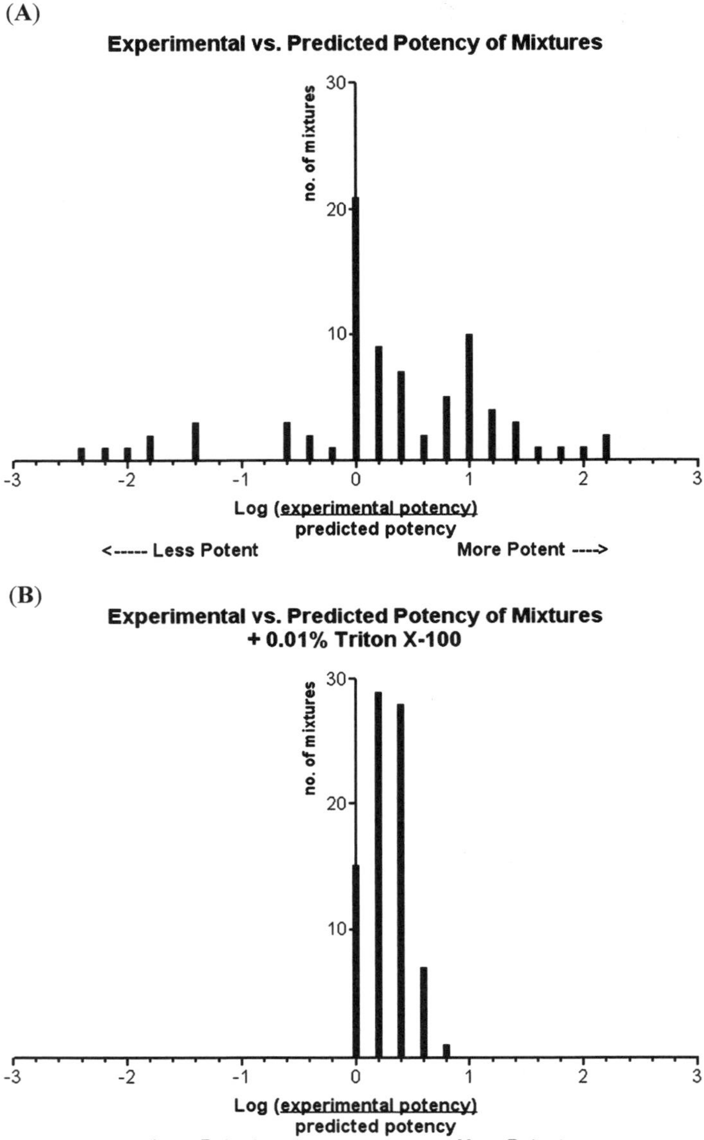

Figure 13.8 (**A**) 80 mixtures of molecules binned according to the deviation between the observed and predicted results. The deviation is a comparison of the predicted and observed potency, where the potency is calculated as a proxy for the IC_{50} of each mixture, and has the form ($[I_i \cdot E]/[E_0]$) or ([%inhibition]/[%activity]). (**B**) Identical analysis as in (**A**) for 80 mixtures of molecules screened in the presence of 0.01% Triton X-100. (Reprinted with permission from the *Journal of Medicinal Chemistry*, 2006, **49**, 2151–2154. Copyright 2006 American Chemical Society.)

the results of any screen that does not control for them. More heightened caution still is merited when interpreting the results of mixture-based HTS, which can behave in a non-additive, often synergistic manner when combined in aqueous media. Fortunately, there are simple techniques to resolve many of these ambiguities, the simplest of which is the addition of non-ionic detergent to assays (but see Shoichet[19] for a more extensive, step-wise protocol).

13.6 Biological Implications?

Until now, we have emphasized the role of aggregation in biochemical screening; it is here that we have definitive evidence for its existence and actions. Is it conceivable that aggregation affects compound behavior in cell-based assays or in whole animals? There is little evidence to support or falsify activity in cell-based assays at present. Weighing against such activity, the aggregate might have to cross the cellular membrane for the effects to be observed; this is possible, owing to dynamic equilibria between aggregated and non-aggregated compounds, but the plausibility of this mechanism is unknown. Aggregates could also affect cellular behavior indirectly, either through extra-cellular receptors or by membrane disruption. Consistent with the latter, many aggregates are hemolytic (Christian Parker, personal communication).

Recently, Arnold and colleagues have mooted a role for aggregation in whole-body pharmacology, through effects on absorption.[25] Several human immunodeficiency virus (HIV) non-nucleoside reverse transcriptase inhibitors (NNRTIs) were shown to form aggregation-based particles in the 90 nm size range under pH and buffer conditions similar to that experienced in the gut. The *in vivo* efficacies of these NNRTIs are often better than their IC_{50} values against isolated enzyme. Arnold and colleagues propose that, after an oral dose, these inhibitors aggregate in the gut and are absorbed by specialized, particle-recognizing cells in the gastrointestinal tract that feed into the lymphatic system.[25] This would deliver the drugs at unusually high concentrations to immune T-cells, the preferential target of HIV, explaining their unusual IC_{50} to effective dose ratios. This hypothesis of aggregation in the stomach and gut of some oral drugs may explain the relatively high bioavailability of some drugs that, based on their high hydrophobicity, might otherwise be predicted to have low bioavailability. This intriguing hypothesis remains to be tested directly by full animal pharmacology.

13.7 The Spirit-haunted World of Screening

High throughput screening is the first and most mechanized step in the drug-discovery pipeline. It is thus unnerving that its results can be among the most ambiguous. This is no trivial fault of instrumentation, assays, or analysis, but reflects the breathtaking ambition to rapidly screen libraries composed of 10^5–10^6 organic molecules. Each of these molecules has idiosyncratic physical and chemical properties affecting solubility, reactivity, and stability; that many will behave poorly in any given assay is unsurprising. Multiple mechanisms contribute to such misbehavior: some, like precipitation[5] or instability, contribute to false-negatives in screening; others, like chemical reactivity,[11] optical opacity,[2] and oxidation,[3] contribute to false-positives. It is the latter that are the most costly in time spent, if not opportunity. If their identification continues to be researched it is because molecules that behave badly under one condition might behave well under another. Knowledge-based tools to identify "nuisance" compounds thus must contend with a context dependence that confers on the molecules a maddening inconsistency.

Few screening phenomena are more prone to erratic behavior than colloidal aggregation, the observation of which depends on the physical properties of the assays, the presence or absence of adjuvants such as detergents or serum proteins, and the concentration of the target. It is thus unsettling that so many "drug-like" organic molecules have this property at micromolar concentrations. On the other hand, one might take encouragement from the physical basis of colloid-like particle formation, whose careful study long predates HTS. In this sense, one of the more shadowy pathologies of organic molecules in HTS is also among the easiest to detect and control for. Doing so early in the discovery process will save investigators much time and aggravation and diminish the number of reports that are based on artifact.

Acknowledgements

A shorter version of this chapter has appeared in *Drug Discovery Today*.[19] We thank our colleagues who have collaborated on this project, especially Susan McGovern, and also James Seidler, Tom Doman, Brian Helfand, Anang Shelate, and Kip Guy. We thank Mic Lajiness, Gerry Maggiora, Mike Schneider, Sophia Ribero, Jeff Alberts, and Clay Scott for many interesting conversations. This work is supported by NIH grant GM71630.

References

1. G. M Rishton, *Drug Discovery Today*, 1997, **2**, 382.
2. W. Walters and M. Namchuk, *Nat. Rev. Drug Discovery*, 2003, **2**, 259.
3. P. J. Hajduk, J. R. Huth and S. W. Fesik, *J. Med. Chem.*, 2005, **48**, 2518.
4. Y. C. Martin, *J. Med. Chem.*, 2005, **48**, 3164.
5. K. Oldenburg, D. Pooler, K. Scudder, C. Lipinski and M. Kelly, *Comb. Chem. High Throughput Screen*, 2005, **8**, 499.
6. S. L. McGovern, E. Caselli, N. Grigorieff and B. K. Shoichet, *J. Med. Chem.*, 2002, **45**, 1712.
7. H. Y. Liu, Z. Wang, C. Regni, X. Zou and P. A. Tipton, *Biochemistry*, 2004, **43**, 8662.
8. O. Roche, P. Schneider, J. Zuegge, W. Guba, M. Kansy, A. Alanine, K. Bleicher, F. Danel, E. M. Gutknecht, M. Rogers-Evans, W. Neidhart, H. Stalder, M. Dillon, E. Sjogren, N. Fotouhi, P. Gillespie, R. Goodnow,

W. Harris, P. Jones, M. Taniguchi, S. Tsujii, W. von Der Saal, G. Zimmermann and G. Schneider, *J. Med. Chem.*, 2002, **45**, 137.
9. S. P. Davies, H. Reddy, M. Caivano and P. Cohen, *Biochem. J.*, 2000, **351**, 95.
10. S. L. McGovern and B. K. Shoichet, *J. Med. Chem.*, 2003, **46**, 1478.
11. G. M. Rishton, *Drug Discovery Today*, 2003, **8**, 86.
12. J. Seidler, S. L. McGovern, T. N. Doman and B. K. Shoichet, *J. Med. Chem.*, 2003, **46**, 4477.
13. S. L. McGovern, B. Helfand, B. Feng and B. K. Shoichet, *J. Med. Chem.*, 2003, **46**, 4265.
14. A. J. Ryan, N. M. Gray, P. N. Lowe and C. Chung, *J. Med. Chem.*, 2003, **46**, 3448.
15. C. A. Lipinski, F. Lombardo, B. W. Dominy and P. J. Feeney, *Adv. Drug Delivery Rev.*, 1997, **23**, 3.
16. B. Y. Feng, A. Shelat, T. N. Doman, R. K. Guy and B. K. Shoichet, *Nat. Chem. Biol.*, 2005, **1**, 146.
17. D. M. Hawkins, S. S. Young and A. I. Rusinko, *QSAR*, 1997, **16**, 296.
18. L. Breiman, *Machine Learning*, 2001, **45**, 5.
19. B. K. Shoichet, *Drug Discovery Today*, 2006, **11**, 607.
20. M. Snider, *J. Biomol. Screening*, 1998, **3**, 169.
21. T. Chung, *J. Biomol. Screening*, 1998, **3**, 171.
22. L. Wilson-Lingardo, P. W. Davis, D. J. Ecker, N. Hebert, O. Acevedo, K. Sprankle, T. Brennan, L. Schwarcz, S. M. Freier and J. R. Wyatt, *J. Med. Chem.*, 1996, **39**, 2720.
23. T. C. Chou and P. Talalay, *Eur. J. Biochem.*, 1981, **115**, 207.
24. T. C. Chou and P. Talalay, *Adv. Enzyme Regul.*, 1984, **22**, 27.
25. Y. V. Frenkel, A. D. Clark Jr., K. Das, Y. H. Wang, P. J. Lewi, P. A. Janssen and E. Arnold, *J. Med. Chem.*, 2005, **48**, 1974.
26. G. S. Weston, J. Blazquez, F. Baquero and B. K. Shoichet, *J. Med. Chem.*, 1998, **41**, 4577.

CHAPTER 14

Iterative Docking Strategies for Virtual Ligand Screening

ALBERT E. BEUSCHER IV AND ARTHUR J. OLSON*

Molecular Graphics Laboratory, Department of Molecular Biology, The Scripps Research Institute, 10550 North Torrey Pines Road, MB-05, La Jolla, CA 92037-1000, USA

14.1 Introduction

Virtual ligand screening (VLS) is an evolving computational method for screening chemical libraries *in silico* to predict compounds that will best bind to a macromolecular structure. This approach is gaining increased acceptance in industry and a growing number of successful virtual screening experiments have recently been reported in the literature. In a typical VLS approach, a large chemical database (> 100 000 compounds) is reduced by drug-likeness, pharmacophore-matching or other desired chemical properties to a smaller size (~ 10 000 compounds) than can be screened with a docking program. This strategy works efficiently because it applies the computational methods in the order of their speed, with the fastest method first. Usually the docking step is the slowest, and last, computational method applied in a VLS study. Several docking programs have been introduced within the past few years to address virtual screening. AutoDock was one of the early pioneering docking programs and has been designed to be accessible to a wide scientific audience. The most recent version of AutoDock, AutoDock 3.0.5, has been licensed at no cost to over 3600 laboratories and cited over 750 times since its introduction in 1998.[1] AutoDock was designed to be a flexible molecular modeling tool rather than a specialized screening program, so it has not been used in reported VLS studies until recently, primarily due to concerns about its computational speed. However, progressive advances in other technologies have made virtual screening accessible through AutoDock. First, computer processors have become cheaper and more powerful, leading to the rise of fast Linux farms of hundreds of central processor units (CPUs), which are particularly suited for "embarrassingly parallel" applications such as AutoDock, which have no requirements for inter-processor communication. Additionally, chemoinformatics tools are becoming more widely available to the broad scientific community, which makes it easier to use a small number of docking results to find a larger number of compounds with similar structures and activities. Two recent reports demonstrate that AutoDock can be used for VLS experiments with large compound databases.[2,3] Both studies share a common strategy in finding inhibitors. They start with a relatively small, chemically diverse compound library that is screened with AutoDock. The top-ranked compounds from AutoDock are assayed and any verified inhibitors are used as leads for similarity and substructure searches against a larger (~100 000 compound) library. The compounds gathered from the database searches are screened with AutoDock and the top-ranked compounds are again assayed. This iterative process of similarity searches, VLS screens and experimental assays is repeated as often as new inhibitors are discovered. Both of these AutoDock VLS studies successfully employed this iterative strategy to identify micromolar potency inhibitors or better.

This review briefly describes the computational tools used to set up AutoDock virtual screening. The iterative diversity-based VLS studies using AutoDock are reviewed and compared with hierarchical VLS strategies. An additional AutoDock VLS study that used a chemically focused library, rather than a diverse library, is also reviewed, as well as a fourth AutoDock study that uses a more straightforward, monolithic approach to VLS. In addition, other issues related to a diversity-based strategy and AutoDock VLS are discussed.

14.2 AutoDock Background

14.2.1 Scoring Function

Docking algorithms can be divided into search and scoring functions. AutoDock uses a physics-based scoring function,[4] Equation (14.1), that includes energy terms for a 12-6 Lennard-Jones potential (E_{vdw}), a 12-10 hydrogen bonding potential (E_{hbon}), an electrostatic potential (E_{elec}) with a distance-dependent dielectric, an entropic penalty (E_{ent}) proportional to the number of rotatable bonds and a desolvation penalty (E_{desol}) that is only applied to aliphatic and aromatic carbons.

$$\Delta E = E_{vdw} + E_{hbond} + E_{elec} + E_{ent} + E_{desol} \qquad (14.1)$$

The intermolecular interactions between the ligand and the macromolecule are calculated utilizing three-dimensional affinity maps, generated by AutoGrid, for each atom type in the ligand. These pre-calculated energy maps greatly increase the speed of the docking calculations. The ligand internal energies are evaluated in a pairwise fashion that considers the van der Waals energies of atoms separated by rotatable bonds. Internal electrostatic energy evaluations may also be included. The sum of the AutoGrid-based intermolecular energies and the ligand internal energies is referred to as the docking energy and is used by AutoDock to rank the best ligand conformation. The sum of the AutoGrid-based intermolecular energies and the torsional penalty, which is a constant value for each ligand depending on the number of rotatable bonds, is known as the "free energy of binding", or binding energy for short. The binding energy has been successful at predicting, with an accuracy of 2.0 kcal mol^{-1}, the experimental binding energies for a set of human immunodeficiency virus (HIV) protease inhibitors.[1] The various weights assigned to the van der Waals and electrostatic terms are adjustable and can be tailored for different ligand or protein systems, such as the carbohydrate-based energy function developed by Laederach and Reilly.[5] Also, improvements have been made to the hydrogen bonding terms after the publication of AutoDock 3.0 that more accurately models the hydrogen bonding of oxygens, nitrogens and spatially adjacent hydrogens.[6]

14.2.2 Search Function

The AutoDock 3.0 search function uses a modified genetic algorithm, based on the work of Rik Belew and William Hart,[1,7] which represents the ligand translation, orientation and internal torsional rotations as state variables that form the ligand genotype. The size of this search space is $6+n$, where n is the number of free torsions, and thus an exhaustive systematic search is typically not feasible. AutoDock starts with a random population of genotypes that are selected for by phenotype fitness, which is the docking energy. The fittest individuals propagate to the next generation by random mutation and recombination. A localized variation is applied with a stochastic probability to the phenotype for local optimization that can be propagated back to the genotype, resulting in a search method termed Lamarckian Genetic Algorithm. This process is continued over several generations until a threshold number of phenotype evaluations have been calculated. The population size of each generation, the recombination crossover probability, local search probability, random mutation rate and evaluation threshold cutoff are all adjustable terms. Genetic algorithms are stochastic processes, so different results can be achieved for different runs using the same macromolecule and ligand. To improve confidence in the docking results, several docking runs are calculated; in several published cases, over a hundred times per ligand.[8–10]

14.2.3 AutoDockTools

The use of AutoDock is greatly facilitated by the AutoDockTools (ADT) suite of programs that run within the Python Molecular Viewer (PMV) software package. These tools simplify and enhance the preparation of the macromolecule and ligand files for docking and provide powerful visualization tools for examining the docking results. An excellent ADT Tutorial has been written by Ruth Huey and Garrett Morris and is available on the AutoDock website (http://www.scripps.edu/mb/olson/doc/autodock/). All of the steps necessary to start AutoGrid and AutoDock calculations, starting with the macromolecule and ligand structure files, are available within ADT, including steps for adding partial charge and solvation parameters, setting ligand rotatable bonds and

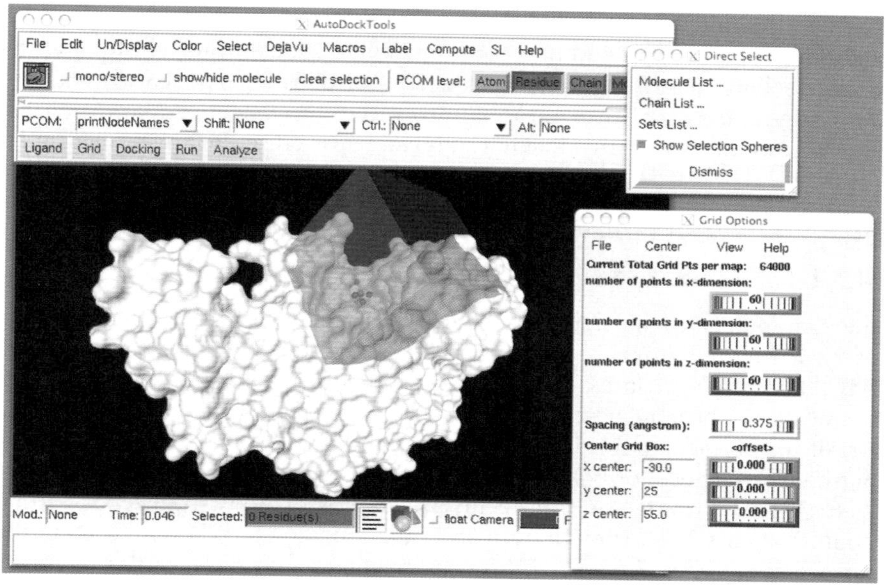

Figure 14.1 "Set Grid Options", an example of the AutoDockTools GUI.

grid box parameters, and launching the AutoGrid and AutoDock jobs. These commands are scriptable, so for a given application that will be applied many times, such as the conversion of a database of ligand.pdb files for virtual screening, all of the necessary commands can be placed into a script file and run for each molecule. Particularly useful are the functions that allow the user to make docking parameter decisions using the ADT molecular graphics, such as the "Set Rotatable Bonds" command, which allows the user to select which ligand bonds will be rotatable during docking, and the "Set Grid" command (Figure 14.1), which allows the user to visually adjust the docking volume around the protein binding site.

14.2.4 AutoDockTools Analysis

A separate set of commands within ADT are available to visualize the AutoDock results from the docking log files (.dlg files), which contain all of the results from an AutoDock calculation. Once a docking log file is read into ADT, the different conformations from each docking run can be visualized. Each conformation is shown with its docking energy and binding energy, and the atoms of each conformation can be colored by atom type, intermolecular electrostatic energy or van der Waals energy. Another powerful analysis tool within ADT is the ability to cluster the conformations from different runs based on different root mean square deviation (RMSD) thresholds. The clustering procedure groups the different conformations based on RMSD. Each cluster is seeded by a low-energy conformation that differs from the other seeds by the

RMSD threshold or more. The clustering results can be visualized in ADT as a histogram showing the total number of docking results in each cluster and the lowest docking energy within each cluster. By selecting a particular cluster from this histogram, each of the conformations within the cluster can be visualized. Thus the results from many docking runs can be visualized relatively efficiently.

14.3 Diversity-based Virtual Screening Studies

14.3.1 AICAR Transformylase

The first published diversity-based virtual screening study with AutoDock searched for inhibitors to the aminoimidazole-4-carboxamide ribonucleotide (AICAR) transformylase domain of the bifunctional AICAR transformylase/inosine monophosphate cyclohydrolase (ATIC) enzyme.[2] The ATIC enzyme is part of the purine nucleotide biosynthetic pathway, which is a 10-step pathway that converts phosphoribosylpyrophosphate (PRPP) into inosine monophosphate (IMP). The AICAR transformylase activity of ATIC transfers a formyl group from the N_{10}-formyl-tetrahydrofolate cofactor to AICAR to form the stable 5-formyl-AICAR (FAICAR) intermediate. Enzymes in this pathway have been targeted for cancer therapeutics because rapidly dividing cells utilize the *de novo* purine biosynthesis pathway rather than the competing purine salvage pathway. Crystal structures of AICAR transformylase in complex with various inhibitors have been determined,[11–13] including human AICAR transformylase in complex with AICAR and a folate cofactor analog. Compounds from the National Cancer Institute (NCI) Diversity Set were docked using AutoDock 3.0.5 to AutoGrid maps based on the crystal structure of the human enzyme. Based on these AutoDock calculations, the top 44 compounds were obtained from the NCI. Of these, 28 had solubility issues that precluded assays, leaving 16 compounds for experimental testing. Half of these compounds showed measurable AICAR transformylase inhibition (4.1–231 μM) (Table 14.1 and Figure 14.2). Four of these hits were used as leads for similarity searches, using a 70% Tanimoto similarity threshold. A particularly potent inhibitor discovered from the similarity searches was used in an additional round of similarity searching and screening. In total, 130 similar compounds were found from the 250 251 compounds in the open NCI database.[14] These compounds were docked and scored by AutoDock and the top-ranked compounds from each similarity search were ordered. Of the compounds discovered by similarity searches, 12 were tested and 11 were found to be inhibitors (Table 14.2), including an inhibitor with an IC_{50} of 600 nM, which was an almost 10-fold improvement over the initial NCI Diversity Set round.

14.3.2 Protein Phosphatase 2C

A second virtual screening study used AutoDock 3.0.5 and the NCI Diversity Set to search for inhibitors against the Ser/Thr protein phosphatase 2C (PP2C).[3] PP2C is a member of the PPM Ser/Thr protein phosphatase

Table 14.1 Experimentally validated inhibitors of AICAR transformylase discovered from AutoDock screening of the NCI Diversity Set.[2]

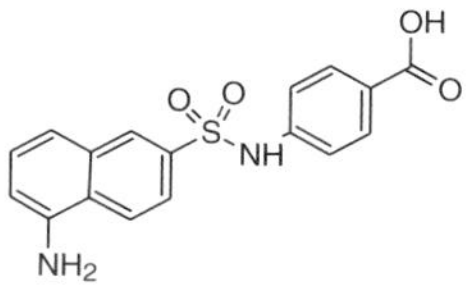

37173

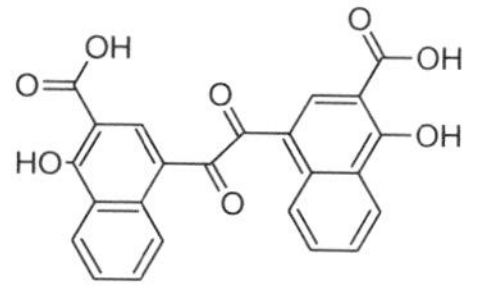

292213

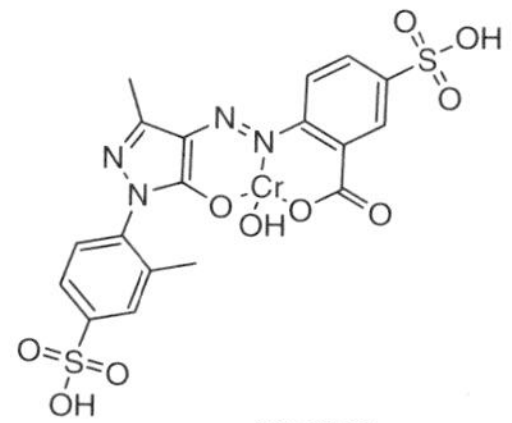

326203

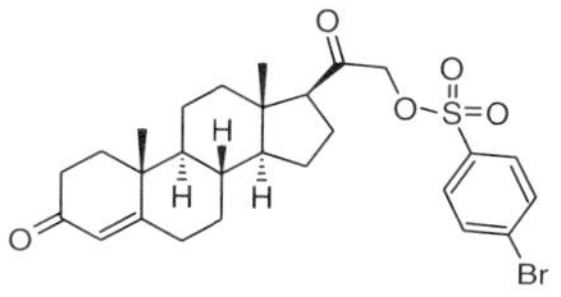

88915

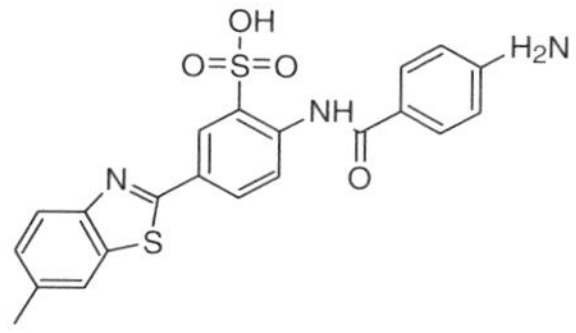

26699

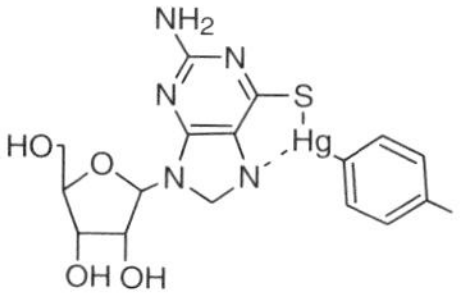

321237

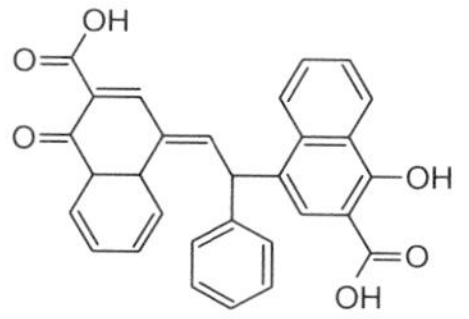

326211

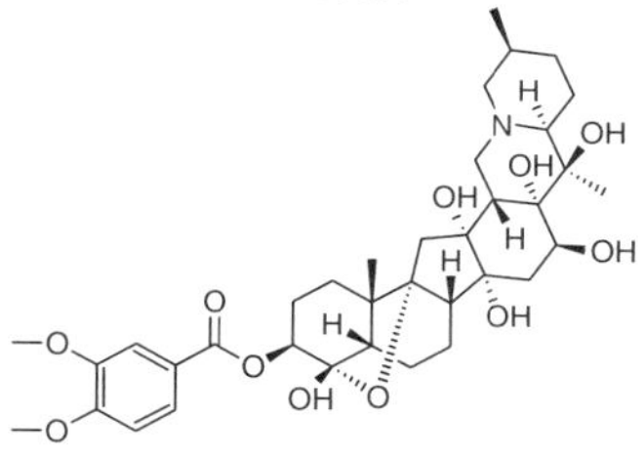

7524

NSC Number	*IC50 (μM)*	*AutoDock Binding Energy*	*MW*	*Rotatable Bonds*
37173	4.1	−12.5	342	5
292213	8.8	−13.5	430	7
326203	11.6	−11.9	607	7
88915	13.9	−14.4	550	5
26699	16.9	−13.0	440	6
321237	105.9	−10.9	591	8
326211	203.7	−13.7	508	7
7524	231.3	−10.6	674	13

family. Members from this protein family show little inhibition from potent inhibitors to the better studied PPP Ser/Thr protein phosphatase family, which includes PP1, PP2A and PP2B.[15] There is a crystal structure of a eukaryotic PP2C protein, human PP2C alpha, which has a bi-metal active site, putatively either Mg^{2+} or Mn^{2+}, and a free phosphate from the crystallization solution complexed to the metal-bound waters.[16] The initial docking model of PP2C included both metals, which were assigned values for Mg^{2+}, as well as

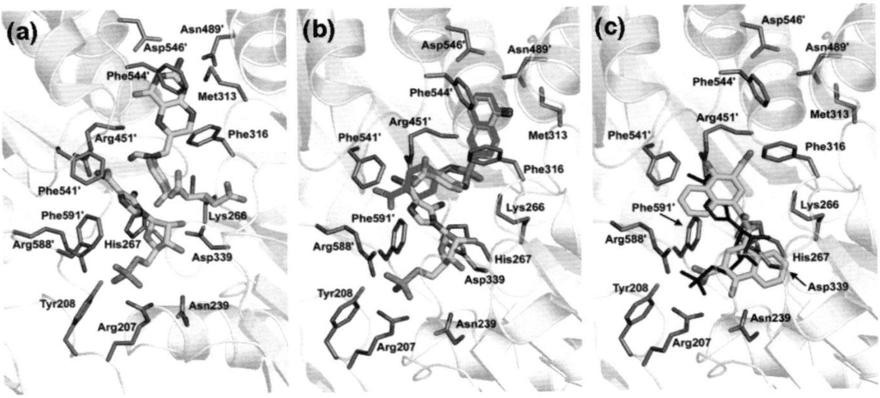

Figure 14.2 AICAR transferase (Tfase) active site shown with cofactor and selected inhibitors. (a) AICAR Tfase active site with substrate (lower ligand) and AutoDock-predicted position of the folate cofactor (upper ligand). (b) Predicted binding position of NSC37173 with substrate present (yellow) and without substrate (green). The substrate is shown as lower ligand. (c) Predicted binding position of NSC292213 overlapped with substrate (black). (Reproduced with permission from *J. Med. Chem.*, 2004, **47**(27), 6681–6690. Copyright 2004 American Chemical Society.)

the six metal-bound waters found in the crystal structure. An initial set of "pseudo-substrates" – phosphothreonine (pThr) and phosphoserine (pSer), as well as *para*-nitrophenol phosphate (PNPP) – were docked to PP2C grid maps using AutoDock, to test the validity of the docking predictions with the initial docking model. The phosphate groups from the docked compounds matched within 1 Å to the free phosphate position found in the crystal structure. The agreement between the docking and the crystallographic results gave some confidence that at least the phosphate recognition elements within the PP2C active site were modeled accurately.

The NCI Diversity Set was computationally screened against PP2C using AutoDock. From the top 40 compounds, 37 were tested, although several of these compounds had solubility problems under the assay conditions. Future compounds for testing were screened with the XLOGP program to remove compounds with low aqueous partition coefficients.[17] Three compounds from this first set of 37 showed significant PP2C inhibition at 100 μM inhibitor concentration (Table 14.3 and Figure 14.3). An additional 31 compounds were obtained and tested from the remainder of the 100 top-ranked NCI Diversity Set compounds, which led to the discovery of one additional inhibitor. The two best inhibitors from the initial screen were found to have IC_{50} values in the range of 5–10 μM and 20–30 μM. Similarity and compound substructure searches based on these initial four leads against the Open NCI Database resulted in several sets of chemically focused chemicals that totaled over 6000 compounds. The results from each database search were docked and ranked. The top compounds from each search were tested and, including the NCI

Table 14.2 Experimentally validated inhibitors of AICAR transformylase discovered from AutoDock screening selected compounds from the Open NCI Database.[2]

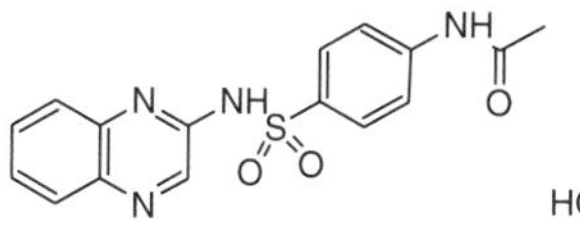

41806

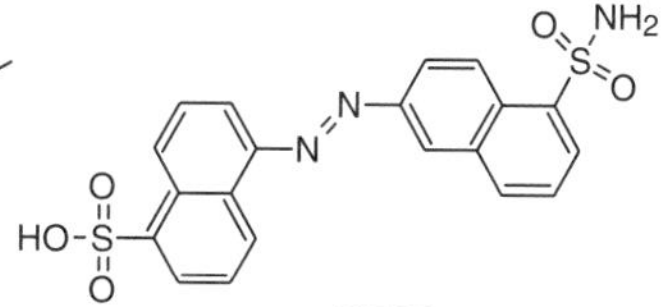

30171

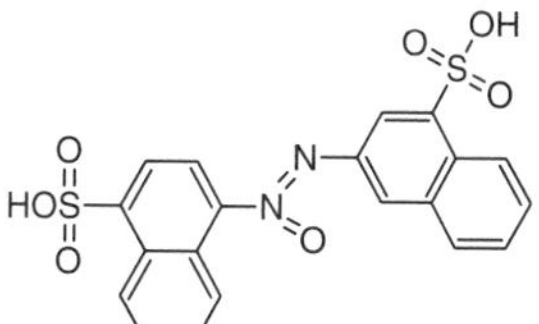

37031

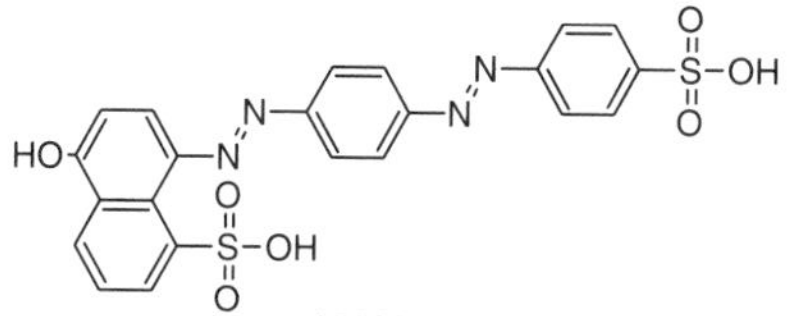

45592

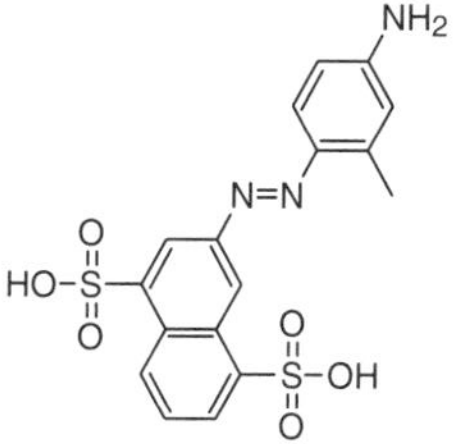

58046

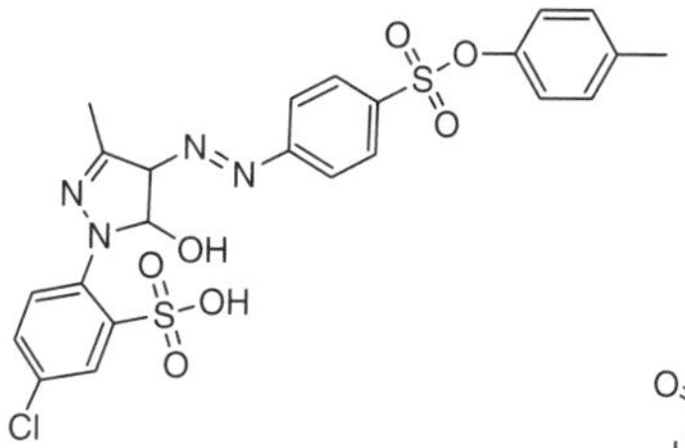

47729

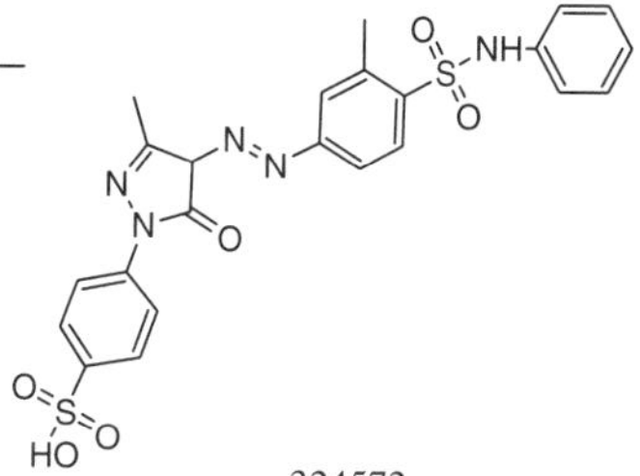

324572

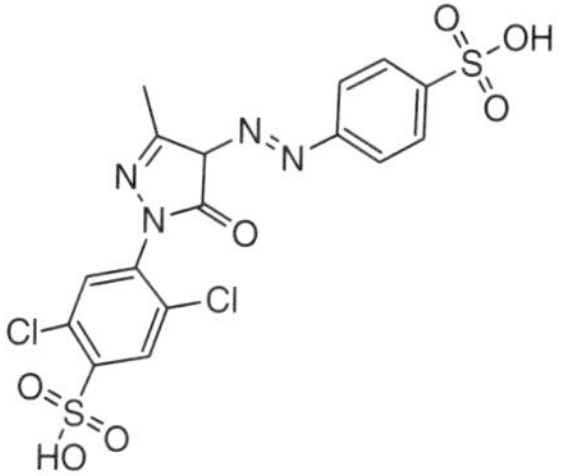

324981

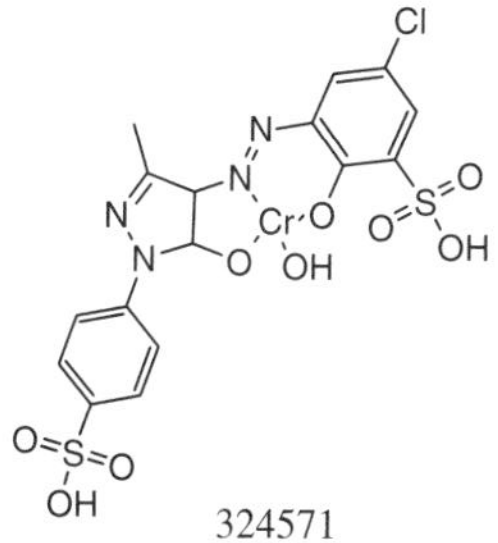

324571

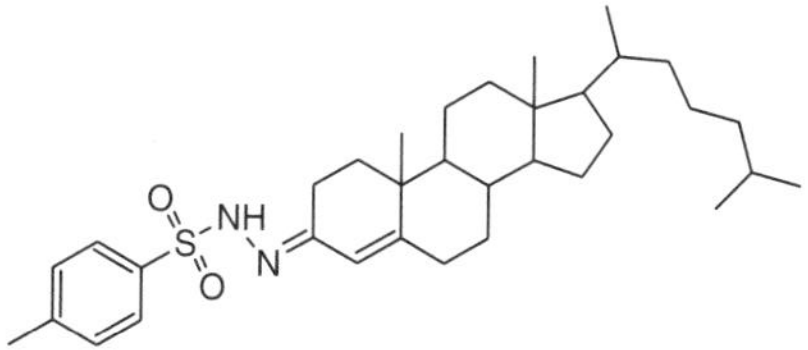

126445

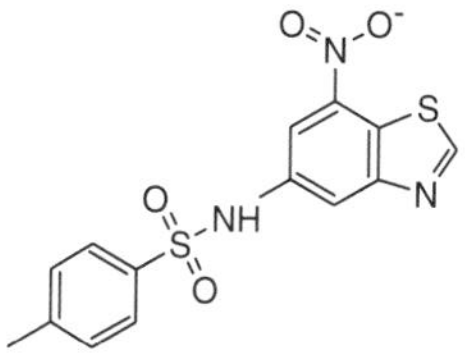

170645

(*Continued*)

Table 14.2 *(Continued)*.

NSC Number	*IC50 (μM)*	*AutoDock Binding Energy (kcal/mol)*	*MW*	*Rotatable Bonds*	*Similar To*
41806	54.7	−10.7	342	5	37173
30171	0.6	−11.7	441	5	37173
37031	7.8	−12.4	458	4	30171
45592	11.6	−12.3	559	7	30171
58046	7.6	−12.0	467	5	30171
47729	3.3	−12.8	586	7	326203
324572	11.6	−12.5	551	7	326203
324981	20.1	−12.2	553	5	326203
324571	20.1	−12.1	600	6	326203
126445	9.6	−15.1	553	8	88915
170645	36.1	−10.3	349	4	170645

Table 14.3 Experimentally validated inhibitors of protein phosphatase 2C discovered from AutoDock screening of the NCI Diversity Set and selected compounds from the Open NCI Database.[3]

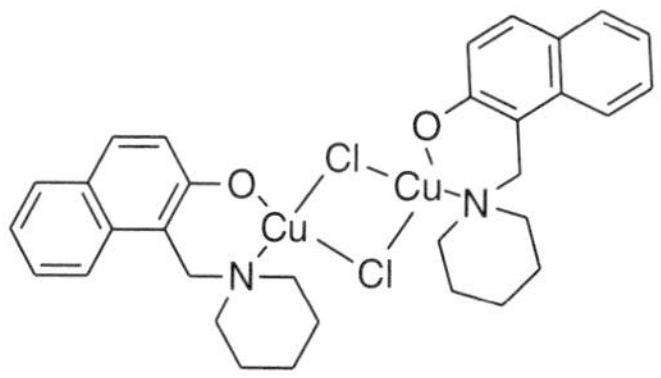

109268

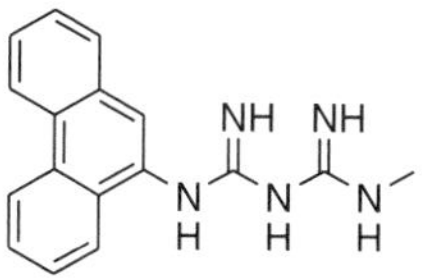

401366

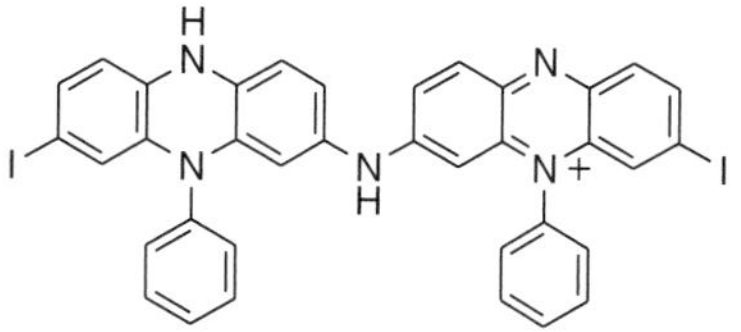

402959

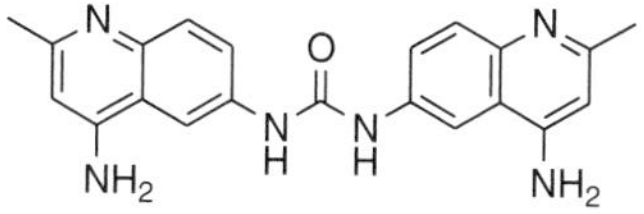

12155

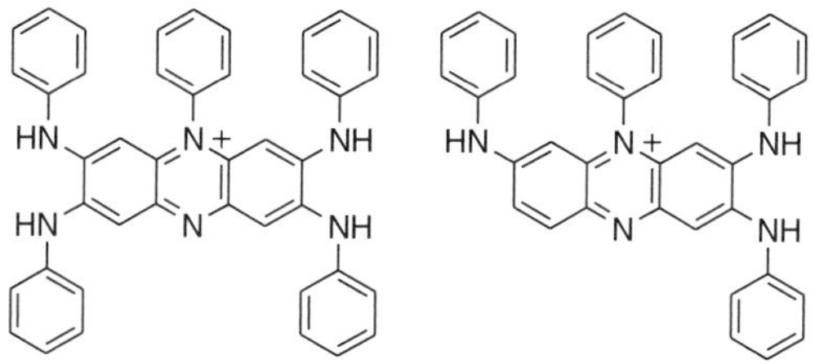

5020

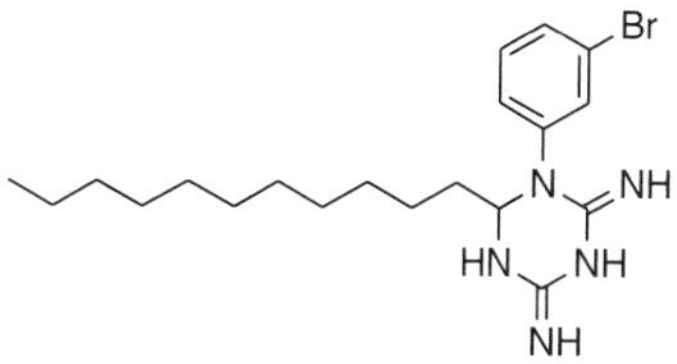

5206

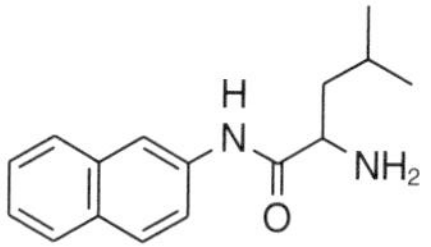

83633

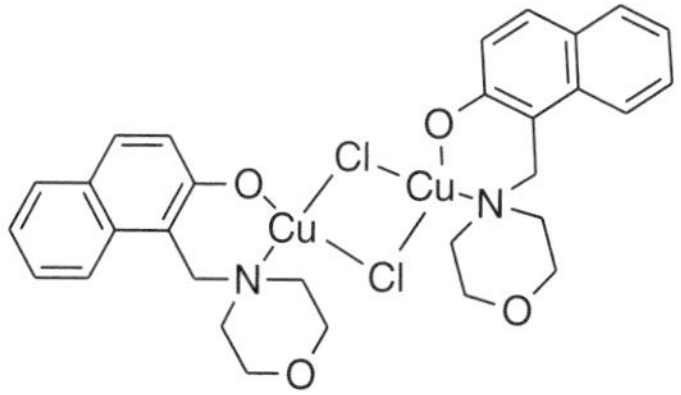

109272

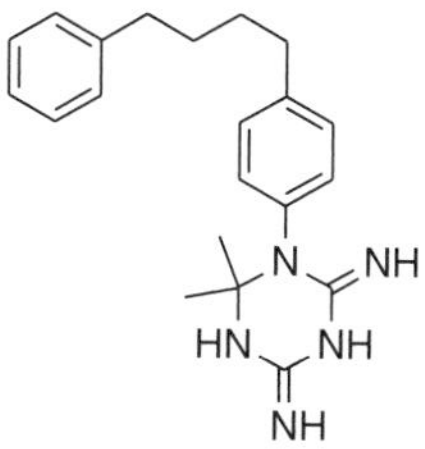

127153

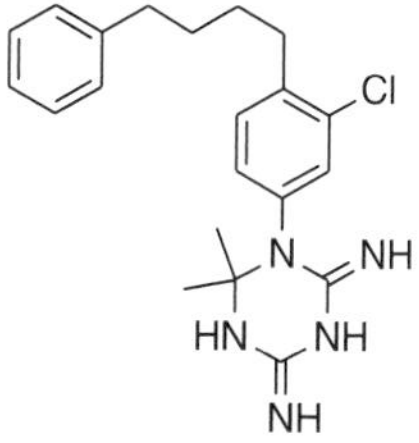

128184

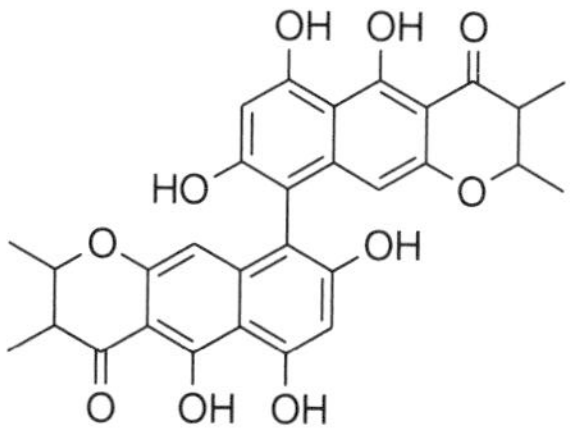

345647

(Continued)

Table 14.3 (*Continued*).

NSC Number	*Fraction of PP2C activity*	*AutoDock Binding Energy (kcal/mol)*	*MW*	*Rotatable Bonds*
109268	0.18	−11.79	679	0
401366	0.35	−9.24	328	5
402959	0.32	−10.33	780	4
12155	0.61	−9.98	372	3
5020	0.04	−8.61, −9.97	622, 530	9, 7
5206	0.06	−9.47	459	11
83633	0.64	−8.68	256	4
109272	0.22	−13.61	684	0
127153	0.51	−9.89	349	6
128184	0.54	−10.06	384	6
345647	0.38	−10.01	547	1

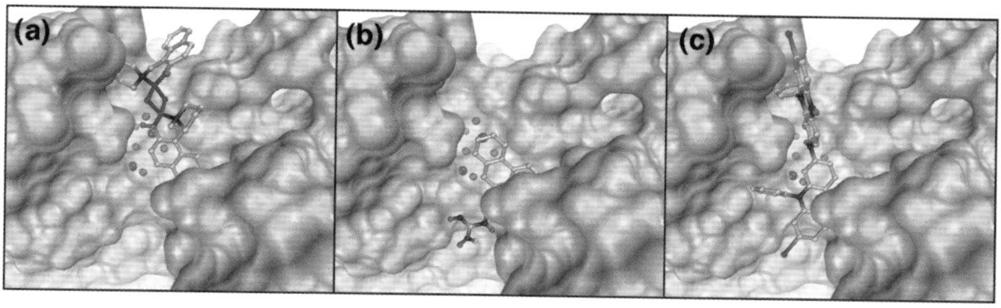

Figure 14.3 Predicted protein phosphatase 2C binding to (a) 109268, (b) 401366 and (c) 402959. Shown as spheres are Mg^{2+} (green) and waters (red) used in the docking model. (Reproduced with permission from *J. Med. Chem.*, 2006, **49**, 1658. Copyright 2006 American Chemical Society.)

Diversity Set screening, 222 compounds were tested. From this second round of screening, seven new inhibitors were found, although none of these surpassed the potency of the best inhibitor found from the NCI Diversity Set. The compounds discovered in this study were the first non-phosphate-containing inhibitors of PP2C.

14.4 Comparison with Existing VLS Strategies

A table of published VLS work is compiled in Table 14.4 to compare the iterative diversity-based VLS results to results from other VLS studies. VLS studies that did not include sufficient information about the hit rate for the docked results or those based on "decoy" compound libraries used for methods validation were excluded from the table. The previously discussed iterative VLS studies with AutoDock produced comparable results to those shown in Table 14.4 in terms of inhibitor affinity and hit rate, and accomplished these results while docking fewer compounds than most VLS studies. These comparisons are only a broad generalization, however. Both the hit rate and inhibitor potency are highly dependent on the nature of the protein target and available compound library, probably more so than the particular VLS method used. However, the comparison does show that the VLS approach with AutoDock is competitive with more specialized docking programs.

Aside from benchmarking the AutoDock VLS results, Table 14.4 demonstrates the wide variability of approaches used by VLS users. The starting size of the compound database varied from 50 000 to 1.2 million compounds, but there was little apparent correlation between this size and the resulting hit rates or inhibitor potencies. Some groups started with over 100 000 compounds in the initial database, but ultimately docked only as few as 66 of these compounds to discover an inhibitor. Despite the overall variability in the results or the numbers of compounds, most of the studies fell within one or two VLS strategies. We discuss these strategies and compare them to the iterative diversity-based approach.

Table 14.4 Results from published virtual ligand screening studies.

System	*Hits*[¶]	*Best Hit*	*Hit Rate*	*Software*	*DB Size*	*Docked*	*Tested*	*Citation*
Protein Phosphatase 2C	**11**	**5 μM** [b]	**5%**	**AutoDock3.0.5**	**250,000**	**8,000**	**222**	**Rogers *et al.*[3]**
Dvl PDZ domain	1	237 μM [d]	11%	Unity/FlexX	250,000	108	9	Shan *et al.*[61]
Protein kinaseB	5	1.1 μM [a]	0.8%	Gold, Flexx, CSCORE	50,000	50,000	600	Forino *et al.*[62]
Cyclooxygenase	4	low micromolar	25%	DOCK4.0	13,711	13,711	12	Mozziconacci *et al.*[39]
Estrogen receptor	7	18 nM [b]	58%	GOLD/Affinity	25,000	25,000	12	Zhao and Brinton[63]
Tyr. Kinase receptor	12	3.3 μM [a]	2.5%	GOLD	50,462	50,462	468	Toledo-Sherman *et al.*[64]
PeptidaseIV	62	micromolar	1.6%	Glide	500,000	40,000	4,000	Ward *et al.*[65]
Trypanothione reductase	2	2 uM [a]	8%	4Scan/FlexX, Propose	1,000,000	§ 5,000	25	Meiering *et al.*[66]
β-secretase	§ 12	50 μM [b]	17%	FFLD/SEED	35,000	35,000	72	Huang *et al.*[67]
ChK1 kinase	36	110 nM [b]	40%	FlexX-Pharm	560,000	200,000	103	Lyne *et al.*[34]
Cysteine protease	24	1 μM [b]	29%	GOLD	241,000	60,000	84	Desai *et al.*[22]
SirT2 deacetylase	5	56.7 μM [b]	33%	Unity/GOLD	§ 50,000	66	15	Tervo *et al.*[68]
Peptide deformylase	§ 7	890 nM [b]	0.2%	ICM	528,000	528,000	3169	Howard *et al.*[35]
Acetylcholine esterase	35	0.5 μM [b]	31%	ADAM & EVE	160,000	160,000	114	Mizutani and Itai [69]
AICAR Tfase	**19**	**600 nM** [b]	**68%**	**AutoDock3.0.5**	**250,000**	**2,100**	**28**	**Li *et al.*[2]**
NK1 receptor	1	0.25 μM [e]	14%	Unity/FlexX-Pharm	827,000	11,109	7	Evers and Klebe[55]
Aldose reductase	6	2.4 μM [b]	66%	Unity/FlexX	260,000	1,261	9	Kraemer *et al.*[70]
Rac GTPase	1	~50 μM	6.6%	Unity/FlexX	140,000	NR	15	Gao *et al.*[71]
CDK2	25	2.2 μM [b]	17%	LIDAEUS	50,000	50,000	148	Wu *et al.*[72]
Protein kinase CK2	4	80 nM [b]	33%	DOCK4.0	400,000	400,000	12	Vangrevelinghe *et al.*[73]
Thymidylate synthase	1	100 μM [b]	7.1%	DOCK3.5	152,000	152,00	14	Atreya *et al.*[32]
tRNA transglycosylase	9	250 nM [a]	100%	Unity/FlexX	800,000	856	9	Brenk *et al.*[74]

Carbonic anhydrase	11	0.6 nM [b]	85%	FlexX	90,000	3,314	13	Grüneberg *et al.*[18]
Aldose reductase	10	4.3 μM [b]	28%	ADAM&EVE	120,000	120,000	36	Iwata *et al.*[75]
Kinesin	22	2.5 μM [d]	21%	DOCK4.0	200,000	110,000	102	Hopkins *et al.*[40]
HIV TAR RNA	2	Low micromolar [c]	25%	DOCK3.5/ICM	152,000	30,000	8	Filikov *et al.*[24]
Farnyltransferase	12	25 μM [b]	57%	EUDOC	219,000	68,000	21	Perola *et al.*[76]
Retinoic acid receptor	2	micromolar	6%	ICM	153,000	153,000	32	Schapira *et al.*[36]
Thymidylate synthase	§ 3**	65 μM [a]	27%	DOCK3.5	153,000	153,000	11	Tondi *et al.*[33]
HGXPRTase	2	10 μM [a]	11%	DOCK3.5	§ 150,000	§ 150,000	18	Somoza *et al.*[4]

¶The criteria for a reported "hit" in a report varied widely; §, best guess based on report; NR, data not reported,
[a] K_i,
[b] IC_{50},
[c] CD_{50},
[d] K_d,
[e] – whole-cell affinity,
**this number is probably an underestimate.

14.4.1 Hierarchical VLS

A common strategy is apparent for VLS studies that started with a larger number of compounds than were actually docked (Table 14.4). Instead of docking all of the compounds, one or more fast computational steps are used to reduce the number of compounds prior to docking. We term this approach "Hierarchical VLS", where the fast and simple computational methods are applied early within the workflow and the slower and more complex computational methods are applied later. Ideally, these initial filtering steps enrich the pool of compounds for active ligands, although little work has been done to quantify the number of active ligands that are lost at various stages.

A good example of this hierarchical strategy is Grüneberg *et al.*[18] which led to the discovery of subnanomolar affinity inhibitors towards human carbonic anhydrase II. The study began with 98 850 compounds from the LeadQuest and Maybridge databases that had been filtered by Lipinski-type rules for drug-likeness.[19] These 98 850 compounds were filtered down to 5904 compounds based on a UNITY pharmacophore search (Tripos Inc.) for zinc-binding chemical properties. A three-dimensional five-point pharmacophore, based on the energy landscape of the protein binding pocket as evaluated by several different ligand scoring functions, was used in UNITY to further reduce the number of ligands to 3314 compounds. The FlexS program[20] was used to score the remaining ligands based on a superposition to the structures of two carbonic anhydrase II complexes. The top 100 compounds from FlexS were docked and ranked based on expected binding affinity by FlexX.[21] Based on the docking results and visual inspection, 13 compounds were experimentally tested. Three compounds inhibited at subnanomolar concentrations, one at nanomolar and seven at micromolar concentration, based on IC_{50} measurements. These results were extremely good and set the upper limits on hit rate and inhibitor potency of the VLS studies surveyed in Table 14.4.

Many of the VLS studies in Table 14.4 use variations of the hierarchical approach, such as Desai *et al.*[22] which applied a drug-likeness filter to 241 000 compounds from the Chembridge database as an initial step, then in subsequent steps used the GOLD docking program[23] with various "speed-up" factors to filter the database down to 1500 compounds, which were then docked with GOLD using standard parameters. Filikov *et al.*[24] used DOCK[25,26] as the initial screen for 143 000 compounds from the Available Chemicals Directory (ACD) and then docked the best 30 000 with ICM[27] using steps of increased conformational sampling. Also, AutoDock has been used in a hierarchical VLS strategy, where DOCK was used as an initial docking filter from the ACD and then the top results were re-ranked using AutoDock.[28] In fact, the docking calculation may not necessarily be the ultimate computational step in the hierarchical strategy, as a VLS validation report was recently published that followed the docking calculations with a Molecular Mechanics–Poisson Boltzmann Surface Area (MM/PBSA) calculation as the final computational step.[29]

Both the hierarchical and iterative diversity-based strategies use high-accuracy docking routines that are unable to search a 100 000 compound

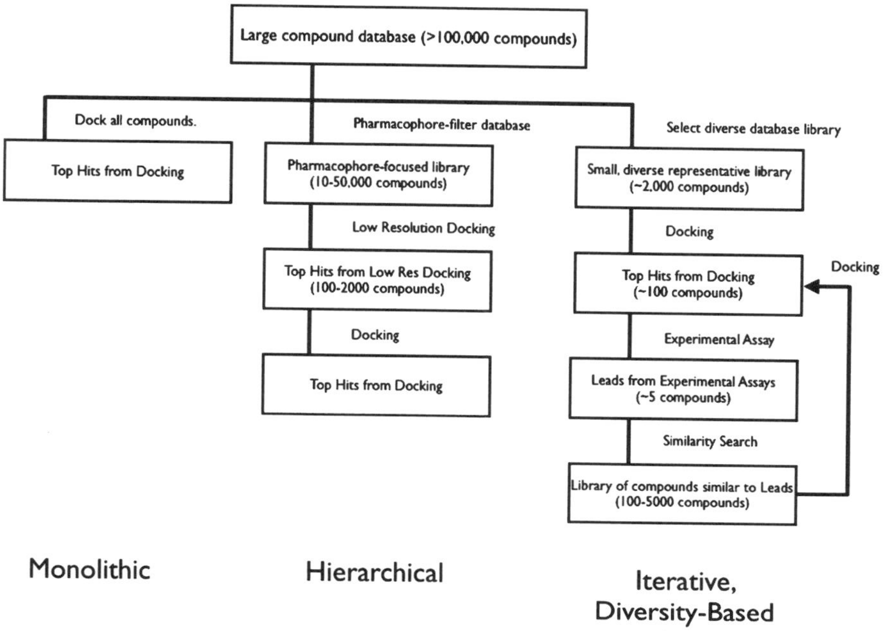

Figure 14.4 A comparison of three general workflows used in virtual ligand screening.

database in a satisfactory time frame, so they employ strategies to reduce the number of compounds to be docked. The hierarchical VLS approach filters away most of the compounds with pharmacophore filters and fast, lower accuracy docking routines, while the iterative diversity-based VLS filters away compounds based on chemical similarity to the initial VLS hits. A comparison of the two strategies is shown in Figure 14.4. These approaches have their benefits and they are not necessarily mutually exclusive. The use of the pharmacophore filter often improves the VLS hit rate, but reduces the range of chemistry that is explored. The pharmacophore step requires that the target has a set of known bioactive molecules and that the user has sufficient expertise and the appropriate software to develop a useful pharmacophore model. The development of the pharmacophore can often be a time-limiting step in a VLS study. By comparison, the diversity-based approach seeks to quickly complete a set of docking calculations as the initial step and then to get experimental feedback. The superposition of these compounds from the initial docking calculations can be used later in the development of a pharmacophore model.[30,31]

Conversely, the small, diverse compound library used in iterative diversity-based VLS permits a relatively rapid screening calculation, but it requires an accurate docking program because a false-negative result in the initial diversity screening may mean that a fertile region of chemical space is not searched in subsequent screening rounds. Thus some chemical feature redundancy in the initial diversity library is desirable. Also, the diversity-based VLS strategy is an iterative process that often requires frequent communication between computational and experimental workers, assuming these tasks are split to different people or groups. The repeated shift of workload between these two areas may increase the length of the project time wise, particularly if the experimental portion is a bottleneck. While the iterative approach may proceed without the experimental assay step by simply using top docking hits and similarity searches, utilizing experimental assays to verify each docking step reduces the uncertainty in the intermediate results.

The hierarchical and iterative VLS approaches are certainly not mutually exclusive. The docking step from the iterative, diversity-based strategy could be replaced with a set of hierarchical filtering and docking steps. The combination of the diversity sampling, similarity searches and iterative workflow with the speed of the hierarchical calculations could result in a software approach with the capacity for screening very large chemical databases.

14.4.2 Monolithic VLS Strategy

Another VLS strategy apparent from Table 14.4 is simply to dock every compound in a chemical database, a strategy most commonly seen with the docking program DOCK,[32,33] but other docking programs have been used to dock over 100 000 compounds.[34–36] The monolithic method is a diversity-based VLS approach, similar to the iterative VLS approach described earlier, where the full range of chemistry within the database is sampled. However the iterative diversity-based approach docks a small, representative sample of a database, while the monolithic approach docks the entire database. Thus the monolithic VLS method requires a particularly fast docking algorithm to screen large databases. DOCK[25,26] uses a geometric matching algorithm for searching the protein-binding site that is very fast but comes at the expense of scoring accuracy.[37,38] A similar tradeoff for speed versus accuracy can be made with many docking programs to increase the number of compounds that can be screened. If a docking program can process one compound per minute, on a 100-processor Linux cluster it can process 184 320 compounds per day, while a smaller and cheaper 10-processor Linux cluster can screen over 100 000 compounds per week. Large-scale docking methods are not confined to the monolithic VLS approach, but have also been used in both hierarchical and iterative VLS studies. In particular, DOCK has been used successfully as an initial step before either using more detailed docking methods on the top-ranked DOCK results[24,28] or post-processing the top results with new scoring functions.[39] Also DOCK has been used in an iterative fashion to screen a large database, and then to use the subsequent experimental assay hits for re-screening the database by similarity search.[4,40] In these iterative VLS studies with DOCK, the similarity search was applied to the same compound database that had been computationally screened with DOCK. The similarity searches picked up compounds with similar structure to the DOCK hits that did not score well in the docking calculations, some of which were found to have inhibitory activity. In this way, the iterative VLS strategy was able to augment the results from DOCK.

14.5 Other AutoDock VLS Studies

14.5.1 Acetylcholine Esterase Peripheral Anionic Site

A third virtual screening search for inhibitors using AutoDock was targeted against the acetylcholine esterase (AChE) peripheral anionic site (PAS).[41] This study differed from the previously mentioned AutoDock VLS studies in that it used a focused library approach, based on a combinatorial library with a common naphthalene scaffold, rather than a diversity-based approach. The AChE VLS was specifically focused towards finding inhibitors that bound to the PAS, which is a negatively charged binding pocket outside the deeply buried catalytic triad active site, rather than the active site itself. The PAS was targeted because PAS inhibitors, but not active site inhibitors, have been shown to inhibit amyloid β (Aβ) fibril formation. The Aβ peptide has been shown to interact with the AChE PAS, and this interaction accelerates the formation of Aβ fibrils, which have been linked to amyloid plaque formation and Alzheimer's disease. Separate crystal structures of mouse AChE bound to the PAS inhibitors propidium and decidium were available from the Protein Data Bank, and the propidium-complex structure was used as a docking model. The protein region chosen for docking included both the PAS and the catalytic active site, which made it possible to predict which compounds would bind to the PAS rather than the active site. Initial positive control dockings were done with propidium and decidium. When these compounds were first docked, they did not match well with the crystal structures and the RMSD differences were over 4.7 Å for both docked compounds compared to their crystallographic positions. Two crystallographic waters found in the PAS were added to the docking model. These waters were chosen because they appeared to mediate polar interactions between the inhibitors and the protein and they occupied a protein pocket that the inhibitors were incorrectly placed into by AutoDock. The new docking model reproduced the crystallographic positions much better, with RMSD values of 1.5 and 1.8 Å for decidium and propidium, respectively.

Known PAS inhibitors generally have a fused ring structure that stacks with the Trp286 side chain found in the PAS. Based on this observation, it was speculated that a chemical library of 271 compounds based on a flat, rigid naphthalene chemical scaffold (a so-called "credit card" library) would contain compounds that bound well to the PAS. All of the compounds in the chemical library were screened with AutoDock. Due to the high costs of purchasing the necessary reagents to assay for fibril formation, it was desirable to test as few compounds as possible. Ultimately, 10 compounds were chosen for biochemical assaying. The top-scoring compound was predicted to bind to the AChE active site and was not pursued. However the next 10 top-scoring compounds were all predicted to bind within the PAS, as well as the binding tunnel leading to the active site, and were screened with a colorimetric assay. All were found to have some inhibitory activity and the five that gave greater than 50% inhibition at 5 μM compounds were further studied and showed K_i values between 1.9 and 14.2 μM, which is comparable to other PAS inhibitors, such as propidium. All five compounds were found to inhibit Aβ fibril formation at a similar level to propidium, while the active-site inhibitor tacrine showed no inhibition. Still further experiments showed that the active credit card inhibitors displaced propidium and also blocked access to the active site of AChE to known active-site inhibitors. This inhibitory behavior is unusual for a PAS inhibitor, but was anticipated by the docking predictions.

14.5.2 Human P2Y$_1$ Receptor

Another recent AutoDock VLS study is the work of Hiramoto *et al.*[42] where AutoDock was used to screen compounds for binding to a homology model of the human P2Y$_1$ receptor, a G protein-coupled receptor (GPCR) protein activated by adenine and uridine nucleotides. This approach did not use a diversity-based VLS strategy, but instead screened a small compound library of 500 animal metabolites from the KEGG database plus four known agonists and four known antagonists to P2Y$_1$. A comparison of the docking energies of the known P2Y$_1$ ligands against neurotransmitters that are inactive towards P2Y$_1$ substantially favored the known ligands (more than 6.0 kcal mol^{-1} lower predicted binding energy). The ligand discrimination appears largely to be due to charge differences: highly negatively charged phosphate-containing compounds are favored as P2Y$_1$ ligands, while compounds with smaller electrostatic charges, such as gamma-aminobutyric acid (GABA) and acetylcholine, are not. All of the eight known P2Y$_1$ ligands were ranked within the top 30 of the VLS results. Based on the virtual screening results, 21 compounds within the top 30 results were experimentally tested for both agonist and antagonist activity against P2Y$_1$. The 21 compounds included the eight known P2Y$_1$ ligands so, in effect, 13 compounds with unknown activity were tested. Of these 13 compounds, three were found to be agonists and one was found to be a weak antagonist, giving a 47% hit rate. The best compound, 5-phosphoribosyl-1-pyrophosphate, activated P2Y$_1$ at a 50% effective dosage (ED$_{50}$) of 15 nM. Also, the predicted binding energy for the agonists had a surprisingly high 63% ($p = 0.05$) correlation with the measured ED$_{50}$ values.

14.6 Diversity-based *vs.* Issues

14.6.1 Library Choice

The diversity of the initial compound library should be representative of a much larger set of available compounds, which allows the VLS results with the diverse library to be used to focus efforts towards specific chemical regions of the larger chemical library. Several approaches have been developed for constructing a diverse chemical library.[43–46] The easiest is to select compounds randomly from the larger library, which can be effective depending on how evenly the molecular properties of the library are distributed. More rigorous approaches towards getting a diverse, representative library are to use clustering approaches, which group compounds based on their chemical features, or descriptors.[46] Another approach is to use dissimilarity-based methods, where compounds are added to the diverse library based on how dissimilar they are chemically from the compounds that have already been selected.[47] Thus a key attribute to consider when constructing a diverse database is the set of chemical descriptors used to describe the database compounds (see discussion Section 14.6.2). A great deal of work has been put into developing new molecular descriptors primarily for quantitative structure–activity relationship (QSAR) approaches, but that are also applicable to chemical similarity and diversity algorithms.[48] In the case of the NCI Diversity Set, which was used in the AutoDock diversity-based VLS studies, the molecular descriptors were based on three-point Chem-X pharmacophore representations, which were applied by researchers working with the NCI Developmental Therapeutics Program to 71 156 compounds available in gram quantities at the NCI. From this set of compounds, a dissimilarity filter was used to build the 1990 compound NCI Diversity Set. Another example of diverse library design is the Annotated Chemical Library,[49] which was designed from FDA-approved drugs and compounds with biological activity, but was also shown to be chemically diverse, particularly in comparison to 10-fold larger, commercially available compound libraries.

Some chemical properties are problematic for drug discovery or docking calculations. If the compounds will ultimately be tested *in vivo*, the compound library should be filtered to remove non-drug-like properties. These properties include the well-known "Lipinski Rule of Five", which states that most approved drugs fall within certain ranges of molecular weight, log P and number of hydrogen-bond acceptors and hydrogen-bond donors.[19] Some chemical properties are difficult for docking programs to describe or explore. These may include the presence of rare elements or chemical bonds that do not have standard atom types, macrocycles and other cyclic substructures (which are not treated as fully flexible ligands by most docking programs), and large numbers of rotatable bonds, which are difficult to fully conformationally sample. These properties may lead to inaccurately estimated docking energies, and subsequently to false-positive or false-negative results. The presence of one of these properties in a compound does not necessarily prevent it from being docked, but it does require additional treatment for the compound, which can be time consuming. For example, in the case of AutoDock, a macrocycle ring structure is treated as a rigid ring, but low-energy macrocycle conformations can be modeled with a secondary conformational sampling program and then docked individually.

14.6.2 Similarity Search

The goal of diversity-based virtual screening is to discover initial compound hits that are pursuable, that is, that they lead to related compounds that have better activities. The connection between the initial hits and the chemically related compounds is made by a similarity search. Excellent reviews on similarity are available elsewhere,[50] but a simple description is given here. To calculate the similarity between two compounds, their chemical properties, also termed molecular descriptors, must be calculated. Continuous chemical properties are commonly enumerated as binary values, *e.g.* a molecular weight descriptor might divide the possible range of molecular weights into bins and then give a compound a "1" if its molecular weight falls within a particular bin, a "0" if not. In this way, the molecular descriptors can be written as binary strings, which can be rapidly compared. A number of formulae may be used to quantify the distance between two molecular descriptors; the most widespread is the Tanimoto coefficient, Equation (14.2):[50]

$$\text{Tanimoto similarity} = c/(a + b - c) \tag{14.2}$$

where a is the number of "on" bits in molecule 1, b is the number of "on" bits in molecule 2 and c is the number of shared "on" bits.

In the case of AICAR[2], chemical similarity searches were conducted using a Tanimoto similarity threshold of 70%, which gathered a total of 130 compounds from five searches. Ultimately 12 compounds were tested and 11 compounds were verified inhibitors, a rate of 92%, compared with the rate of 50% from the initial diversity library screenings. The dockings from the similarity search resulted in a 600 nM potency inhibitor, which was better than the 4 μM inhibitor found from the diversity library screenings. In the case of PP2C[3], the chemical searches based on the NCI Diversity Set hits accumulated a much larger number of compounds (~6000). This number was much larger than found in the AICAR study because chemical substructure searches, in addition to chemical similarity searches, were used to find compounds from the PP2C initial leads. The substructure searches find compounds that share a particular local chemical substructure, without considering a more global chemical similarity, and these searches resulted in over 90% of the total compounds that were computationally screened. The substructure searches were approximately as successful as the similarity searches in terms of the number of active inhibitors found. Therefore, the substructure search was less efficient than the similarity search when evaluated by "hits per docked compound", since most of the docked compounds were from the substructure search. Approximately 160 compounds were tested from the 6000, and among these compounds the top-ranked from each of the different similarity and substructure searches were represented. Seven additional inhibitors were found, but none of the inhibitors had more potency than those from the initial diversity set screening. It may be that more structure information regarding the interaction of PP2C with the lead compounds is necessary to follow up on these compounds further.

14.6.3 Apo Versus Ligand-bound Docking Models

Protein flexibility is a well-known obstacle to accurate protein–ligand structural predictions.[51,52] Small structural changes within the binding site (as small as

0.5 Å RMSD from a reference structure) that visually seem relatively trivial can have a large effect on ligand binding within an active site. These structural differences affect the docking calculations, as was shown by McGovern and Shoichet[53] who found that enrichment rates for DOCK docking calculations were much better when the calculations were based on protein structures determined with bound ligand, than when the calculations were based on protein structures without bound ligand or based on homology models. For this reason, most successful VLS studies begin with a protein structure with a ligand-bound conformation, as do virtually all of the VLS studies shown in Table 14.4. The PP2C VLS results are a notable exception to this trend, which, as mentioned previously, were based on a protein structure with free phosphate bound in the active site, but no significant organic ligands. The lack of structural information about PP2C ligand-bound conformations was likely to have limited the accuracy of the AutoDock calculations, even though the results from the PP2C VLS report were in line with those of other VLS studies.

In the PP2C study, several structural models for the ligand cofactors were tested, in addition to the two Mg^{2+} ions and six metal-bound crystallographic waters that were used to generate the VLS results. These included models with no metal or waters (cofactorless), no waters (waterless) and all 64 systematic combinations of the six crystallographic waters. Despite the fairly substantial differences in the active-site topology and electrostatic charge of these different docking models, all of the models were able to predict the active inhibitors reasonably accurately, based on enrichment test studies where 10 of the experimentally verified inhibitors were docked alongside 490 docking decoy ligands (unpublished data). It appears that the PP2C inhibitor binding sites described in the VLS study[3] recognize the active inhibitors fairly robustly, despite quite substantial structural variations in the docking models, and that the energetic interactions from these binding sites drive the VLS enrichment. Similarly, docking calculations built on homology models, which also differ in conformation by at least 0.5–2.0 Å RMSD from the target protein structure,[54] have been relatively common and successful,[22,36,42,55,56] despite findings that docking calculations based on homology models are also less accurate than those based on protein structures determined with bound ligand.[53] As with the PP2C VLS, it is likely that the successful homology-model VLS studies have accurately modeled the energetically relevant portions of the protein-binding site.

14.6.4 Binding Site Choices

A factor that can affect AutoDock docking predictions is the selected size and location of the ligand-binding site. Ideally, the grid boundaries that define the binding site for AutoDock are as small as possible and cover chemically relevant and accessible regions of the protein surface, rather than regions within the protein interior or regions far from the surface that would contribute little to the binding energy. A method to define the optimal ligand-binding position was described by Beuscher *et al.*[57] where a three-dimensional region of low binding energy is traced from AutoGrid grid maps. This method starts with a flood-fill algorithm that builds a contiguous volume of low-energy grid points from an initial starting point within the binding pocket by iteratively adding the lowest energy neighboring grid point to the volume until a pre-determined volume size is reached. Next, new binding regions neighboring the flood-fill volume are explored by deleting the worst (highest energy) grid points and randomly adding new neighboring grid points. This process of deleting and adding new grid points is repeated several times and the lowest energy grid volume is retained as the "optimal ligand" volume. By choosing seed points for the optimal ligand approach randomly along the protein surface, the best protein-binding pockets can be identified energetically, using the same energies as those for evaluating protein–ligand interactions. This method can also be used to identify the minimal grid boundaries for a docking calculation.

14.7 Future Work

An accurate treatment of protein flexibility is the current fundamental goal of protein–ligand modeling programs. Most large-scale protein conformational changes are beyond the ability of any current molecular modeling program to model accurately with any certainty. There has been some success in modeling small conformational changes that accompany ligand binding using molecular dynamics programs or trying to capture protein flexibility with a discrete number of protein conformations.[58–60] The difficulties in modeling protein flexibility are further constrained by time for VLS; the method must be relatively quick and robust to process several thousand compounds. Along these lines, work is being completed on AutoDock 4, which will incorporate protein flexibility into the docking process, as well as new search and scoring functions.

As computational power continues to increase and processors multiply, previously infeasible computations become possible. Matrix screens of drug libraries versus large panels of proteins are now starting to be run to explore target specificity and evolution of drug resistance. AutoDock 4 has recently been implemented in an internet distributed project, FightAIDS@Home, running on World Community Grid (http://worldcommunitygrid.org) utilizing over 200 000 processors to run a massive drug-candidate screen against a large panel of HIV protease mutants. The "embarrassingly parallel" nature of docking calculations makes these loosely coupled computational resources ideal for such large-scale virtual screening projects. The problem of protein flexibility can be addressed with the same kind of computational resource, by dividing the dimensionality of the configurational search among the large number of available processors.

References

1. G. M. Morris, D. S. Goodsell, R. S. Halliday, R. Huey, W. E. Hart, R. K. Belew and A. J. Olson, *J. Comp. Chem.*, 1998, **19**, 1639.

2. C. Li, L. Xu, D. W. Wolan, I. A. Wilson and A. J. Olson, *J. Med. Chem.*, 2004, **47**, 6681.
3. J. P. Rogers, A. E. Beuscher IV, M. Flajolet, T. McAvoy, A. C. Nairn, A. J. Olson and P. Greengard, *J. Med. Chem.*, 2006, **49**, 1658.
4. J. R. Somoza, A. G. Skillman, Jr., N. R. Munagala, C. M. Oshiro, R. M. Knegtel, S. Mpoke, R. J. Fletterick, I. D. Kuntz and C. C. Wang, *Biochemistry*, 1998, **37**, 5344.
5. A. Laederach and P. J. Reilly, *J. Comput. Chem.*, 2003, **24**, 1748.
6. R. Huey, D. S. Goodsell, G. M. Morris and A. J. Olson, *Lett. Drug Des. Dev.*, 2004, **1**, 178.
7. W. E. Hart, C. R. Rosin, R. K. Belew and G. M. Morris, Improved evolutionary hybrids for flexible ligand docking in AutoDock, in *Proceedings of the International Conference on Optimization in Computational Chemistry and Molecular Biology*, ed. C. A. Floudas and P. M. Pardalos, Springer, Berlin, 2000, p. 209.
8. S. Bjelic and J. Aqvist, *Biochemistry*, 2004, **43**, 14521.
9. F. Osterberg and J. Aqvist, *FEBS Lett*, 2005, **579**, 2939.
10. K. Ersmark, M. Nervall, E. Hamelink, L. K. Janka, J. C. Clemente, B. M. Dunn, M. J. Blackman, B. Samuelsson, J. Aqvist and A. Hallberg, *J. Med. Chem.*, 2005, **48**, 6090.
11. D. W. Wolan, C. G. Cheong, S. E. Greasley and I. A. Wilson, *Biochemistry*, 2004, **43**, 1171.
12. S. E. Greasley, P. Horton, J. Ramcharan, G. P. Beardsley, S. J. Benkovic and I. A. Wilson, *Nat. Struct. Biol.*, 2001, **8**, 402.
13. D. W. Wolan, S. E. Greasley, G. P. Beardsley and I. A. Wilson, *Biochemistry*, 2002, **41**, 15505.
14. J. H. Voigt, B. Bienfait, S. Wang and M. C. Nicklaus, *J. Chem. Inf. Comput. Sci.*, 2001, **41**, 702.
15. A. McCluskey, A. T. Sim and J. A. Sakoff, *J. Med. Chem.*, 2002, **45**, 1151.
16. A. K. Das, N. R. Helps, P. T. Cohen and D. Barford, *EMBO. J.*, 1996, **15**, 6798.
17. R. X. Wang, Y. Fu and L. H. Lai, *J. Chem. Inf. Comput. Sci.*, 1997, **37**, 615.
18. S. Grüneberg, B. Wendt and G. Klebe, *Angew Chem., Int. Ed.*, 2001, **40**, 389.
19. C. A. Lipinski, F. Lombardo, B. W. Dominy and P. J. Feeney, *Adv. Drug Delivery Rev.*, 2001, **46**, 3.
20. C. Lemmen, T. Lengauer and G. Klebe, *J. Med. Chem.*, 1998, **41**, 4502.
21. M. Rarey, B. Kramer, T. Lengauer and G. Klebe, *J. Mol. Biol.*, 1996, **261**, 470.
22. P. V. Desai, A. Patny, Y. Sabnis, B. Tekwani, J. Gut, P. Rosenthal, A. Srivastava and M. Avery, *J. Med. Chem.*, 2004, **47**, 6609.
23. G. Jones, P. Willett, R. C. Glen, A. R. Leach and R. Taylor, *J. Mol. Biol.*, 1997, **267**, 727.
24. A. V. Filikov, V. Mohan, T. A. Vickers, R. H. Griffey, P. D. Cook, R. A. Abagyan and T. L. James, *J. Comput. Aided Mol. Des.*, 2000, **14**, 593.
25. E. C. Meng, D. A. Gschwend, J. M. Blaney and I. D. Kuntz, *Proteins*, 1993, **17**, 266.

26. B. K. Shoichet and I. D. Kuntz, *Protein Eng.*, 1993, **6**, 723.
27. M. Totrov and R. Abagyan, *Proteins*, 1997, **Suppl. 1**, 215.
28. B. Liu and J. Zhou, *J. Comput. Chem.*, 2005, **26**, 484.
29. J. Wang, X. Kang, I. D. Kuntz and P. A. Kollman, *J. Med. Chem.*, 2005, **48**, 2432.
30. H. Liu, X. Huang, J. Shen, X. Luo, M. Li, B. Xiong, G. Chen, Y. Yang, H. Jiang and K. Chen, *J. Med. Chem.*, 2002, **45**, 4816.
31. M. Cui, X. Huang, X. Luo, J. M. Briggs, R. Ji, K. Chen, J. Shen and H. Jiang, *J. Med. Chem.*, 2002, **45**, 5249.
32. C. E. Atreya, E. F. Johnson, J. J. Irwin, A. Dow, K. M. Massimine, I. Coppens, V. Stempliuk, S. Beverley, K. A. Joiner, B. K. Shoichet and K. S. Anderson, *J. Biol. Chem.*, 2003, **278**, 14092.
33. D. Tondi, U. Slomczynska, M. P. Costi, D. M. Watterson, S. Ghelli and B. K. Shoichet, *Chem. Biol.*, 1999, **6**, 319.
34. P. D. Lyne, P. W. Kenny, D. A. Cosgrove, C. Deng, S. Zabludoff, J. J. Wendoloski and S. Ashwell, *J. Med. Chem.*, 2004, **47**, 1962.
35. M. H. Howard, T. Cenizal, S. Gutteridge, W. S. Hanna, Y. Tao, M. Totrov, V. A. Wittenbach and Y. J. Zheng, *J. Med. Chem.*, 2004, **47**, 6669.
36. M. Schapira, B. M. Raaka, H. H. Samuels and R. Abagyan, *Proc. Natl. Acad. Sci. U. S. A.*, 2000, **97**, 1008.
37. M. Kontoyianni, L. M. McClellan and G. S. Sokol, *J. Med. Chem.*, 2004, **47**, 558.
38. E. Kellenberger, J. Rodrigo, P. Muller and D. Rognan, *Proteins*, 2004, **57**, 225.
39. J. C. Mozziconacci, E. Arnoult, P. Bernard, Q. T. Do, C. Marot and L. Morin-Allory, *J. Med. Chem.*, 2005, **48**, 1055.
40. S. C. Hopkins, R. D. Vale and I. D. Kuntz, *Biochemistry*, 2000, **39**, 2805.
41. T. J. Dickerson, A. E. Beuscher IV, C. J. Rogers, M. S. Hixon, N. Yamamoto, Y. Xu, A. J. Olson and K. D. Janda, *Biochemistry*, 2005, **44**, 14845.
42. T. Hiramoto, Y. Nonaka, K. Inoue, T. Yamamoto, M. Omatsu-Kanbe, H. Matsuura, K. Gohda and N. Fujita, *J. Pharmacol. Sci.*, 2004, **95**, 81.
43. V. J. Gillet, P. Willett, J. Bradshaw and D. V. S. Green, *J. Chem. Inf. Comput. Sci.*, 1999, **39**, 169.
44. M. Glick, A. E. Klon, P. Acklin and J. W. Davies, *Mol. Phys.*, 2003, **101**, 1325.
45. P. M. Andersson, M. Sjostrom, S. Wold and T. Lundstedt, *J. Chemometrics*, 2001, **15**, 353.
46. P. Willett, *Curr. Opin. Biotechnol.*, 2000, **11**, 85.
47. M. S. Lajiness, *Persp. Drug Discovery Des.*, 1997, **7/8**, 65.
48. H. Matter, *J. Med. Chem.*, 1997, **40**, 1219.
49. D. E. Root, S. P. Flaherty, B. P. Kelley and B. R. Stockwell, *Chem. Biol.*, 2003, **10**, 881.
50. P. Willett, J. M. Barnard and G. M. Downs, *J. Chem. Inf. Comput. Sci.*, 1998, **38**, 983.
51. C. W. Murray, C. A. Baxter and A. D. Frenkel, *J. Comput. Aided Mol. Des.*, 1999, **13**, 547.

52. H. A. Carlson, *Curr. Opin. Chem. Biol.*, 2002, **6**, 447.
53. S. L. McGovern and B. K. Shoichet, *J. Med. Chem.*, 2003, **46**, 2895.
54. M. A. Marti-Renom, A. C. Stuart, A. Fiser, R. Sanchez, F. Melo and A. Sali, *Annu. Rev. Biophys. Biomol. Struct.*, 2000, **29**, 291.
55. A. Evers and G. Klebe, *J. Med. Chem.*, 2004, **47**, 5381.
56. A. Evers, G. Hessler, H. Matter and T. Klabunde, *J. Med. Chem.*, 2005, **48**, 5448.
57. A. Beuscher, A. J. Olson and D. S. Goodsell, *Lett. Drug Des. Dev.*, 2005, **2**, 483.
58. F. Osterberg, G. M. Morris, M. F. Sanner, A. J. Olson and D. S. Goodsell, *Proteins*, 2002, **46**, 34.
59. J. Kua, Y. Zhang and J. A. McCammon, *J. Am. Chem. Soc.*, 2002, **124**, 8260.
60. C. F. Wong, J. Kua, Y. Zhang, T. P. Straatsma and J. A. McCammon, *Proteins*, 2005, **61**, 850.
61. J. Shan, D. L. Shi, J. Wang and J. Zheng, *Biochemistry*, 2005, **44**, 15495.
62. M. Forino, D. Jung, J. B. Easton, P. J. Houghton and M. Pellecchia, *J. Med. Chem.*, 2005, **48**, 2278.
63. L. Zhao and R. D. Brinton, *J. Med. Chem.*, 2005, **48**, 3463.
64. L. Toledo-Sherman, E. Deretey, J. J. Slon-Usakiewicz, W. Ng, J. R. Dai, J. E. Foster, P. R. Redden, M. D. Uger, L. C. Liao, A. Pasternak and N. Reid, *J. Med. Chem.*, 2005, **48**, 3221.
65. R. A. Ward, T. D. Perkins and J. Stafford, *J. Med. Chem.*, 2005, **48**, 6991.
66. S. Meiering, O. Inhoff, J. Mies, A. Vincek, G. Garcia, B. Kramer, M. Dormeyer and R. L. Krauth-Siegel, *J. Med. Chem.*, 2005, **48**, 4793.
67. D. Huang, U. Luthi, P. Kolb, K. Edler, M. Cecchini, S. Audetat, A. Barberis and A. Caflisch, *J. Med. Chem.*, 2005, **48**, 5108.
68. A. J. Tervo, S. Kyrylenko, P. Niskanen, A. Salminen, J. Leppanen, T. H. Nyronen, T. Jarvinen and A. Poso, *J. Med. Chem.*, 2004, **47**, 6292.
69. M. Y. Mizutani and A. Itai, *J. Med. Chem.*, 2004, **47**, 4818.
70. O. Kraemer, I. Hazemann, A. D. Podjarny and G. Klebe, *Proteins*, 2004, **55**, 814.
71. Y. Gao, J. B. Dickerson, F. Guo, J. Zheng and Y. Zheng, *Proc. Natl. Acad. Sci. U. S. A.*, 2004, **101**, 7618.
72. S. Y. Wu, I. McNae, G. Kontopidis, S. J. McClue, C. McInnes, K. J. Stewart, S. Wang, D. I. Zheleva, H. Marriage, D. P. Lane, P. Taylor, P. M. Fischer and M. D. Walkinshaw, *Structure*, 2003, **11**, 399.
73. E. Vangrevelinghe, K. Zimmermann, J. Schoepfer, R. Portmann, D. Fabbro and P. Furet, *J. Med. Chem.*, 2003, **46**, 2656.
74. R. Brenk, L. Naerum, U. Gradler, H. D. Gerber, G. A. Garcia, K. Reuter, M. T. Stubbs and G. Klebe, *J. Med. Chem.*, 2003, **46**, 1133.
75. Y. Iwata, M. Arisawa, R. Hamada, Y. Kita, M. Y. Mizutani, N. Tomioka, A. Itai and S. Miyamoto, *J. Med. Chem.*, 2001, **44**, 1718.
76. E. Perola, K. Xu, T. M. Kollmeyer, S. H. Kaufmann, F. G. Prendergast and Y. P. Pang, *J. Med. Chem.*, 2000, **43**, 401.

CHAPTER 15

Challenges and Progresses in Calculations of Binding Free Energies – What Does it Take to Quantify Electrostatic Contributions to Protein–Ligand Interactions?

MITSUNORI KATO, SONJA BRAUN-SAND AND ARIEH WARSHEL

University of Southern California, Department of Chemistry, Los Angeles, CA 90089-1062, USA

15.1 Introduction

The specificity of the interaction between proteins and their ligands plays a major role in most biological processes. Thus, the ability to correlate the structure of protein–ligand complexes with their binding affinity is a challenge of major practical and fundamental importance. A clear example is the need for effective tools in the rational design of drugs that block the activity of proteins involved in specific diseases. The starting point for addressing this challenge is the progress in structural evaluations of protein–ligand complexes by X-ray crystallography and nuclear magnetic resonance (NMR) spectroscopy. The Protein Data Bank (PDB) contains (as of April 2005) about 30 000 protein structures and continues to grow at an enormous rate. The great structural advances call for a corresponding progress in the use of computer modeling in analyzing the energetics of protein–ligand complexes. An obvious step in such studies is the ability to search for the optimal docking of proteins and different drug candidates. Several effective programs have been developed for this purpose (*e.g.* for a review see Muegge and Rarey,[1] Gohlke and Klebe[2] and

Brooijmans and Kuntz[3]) and this issue is out of the scope of the present review. The next step, after finding the binding mode and the ligand orientation in this site, is the evaluation of the binding free energy of this complex. The advances and the problems in addressing this challenge are the main focus of this review. In doing so we emphasize approaches that are based on "first principle" free-energy estimates. Here we focus on the evaluation of the electrostatic contributions to the binding energies, since these contributions are frequently the most important structure-binding correlators.

As much as rational drug design is concerned, the role of first-principle binding calculations is mainly in ranking sets of reasonable candidates, which are obtained from more qualitative approaches. Here we can only consider the present stage as a work in progress since the current situation is far from being satisfactory. However, in our view calculations of binding free energies and, in particular, the corresponding electrostatic contributions should eventually become a major quantitative tool in rational drug design.

15.2 Computational Strategies

The key to energy-based drug design is the ability to obtain reliable results in free-energy calculations of protein–ligand interactions. Accomplishing this task is, however, far from simple due to inherent convergence problems and to inconsistencies of some approaches. In this review we focus on calculations of electrostatic contributions to the overall binding free energy, since we consider this effect to be the most important contribution to the overall binding free energy. However, we also consider below approaches that evaluate all other contributions.

15.2.1 Free-energy Perturbation, Linear Response Approximation and Potential of Mean Force Calculations by All-atom Models

Thermodynamic cycles of the same type as used in chemistry textbooks can describe many biochemical processes. Such cycles allow one to replace the very challenging direct calculations of physical processes by calculations of much simpler either physical or non-physical processes. The use of thermodynamic cycles in microscopic calculations of free energies of biological processes was introduced in the early 1980s,[4] starting with the evaluation of the difference between the pK_a of acids in proteins and in water. This involved considering the non-physical process of "mutating" the protonated acid to its ionized form in the protein and in water, instead of calculating the binding energies of the protonated and ionized forms of the acid.[4] Another example has been provided by studies of enzyme catalysis that compared the physical process of moving the reacting substrate from its ground state (GS) to its transition state (TS) in the protein active site and in solution. This physical process was replaced[4,5] by an alternative cycle that takes the difference between the solvation energy of the

TS in protein and in water and subtracts it from the difference between the solvation energy of the GS in the protein and in water (see discussion in Warshel[6]). Related cycles for studies of ligand binding have been introduced by Wong and McCammon.[7] At any rate, thermodynamic cycles have become a major part of many approaches in free-energy calculations of biological processes (*e.g.* Wong and McCammon,[7] Kollman[8] and Shurki and Warshel[9]).

Of course, the key issue in the evaluation of thermodynamic cycles is the ability to evaluate the relevant free energies. Unfortunately, doing so by statistical mechanical approaches and microscopic simulations is extremely time-consuming, owing to sampling problems. Nevertheless, it is possible in some cases to obtain meaningful results using perturbation approaches. Such calculations are usually done by the so-called free-energy perturbation (FEP) method[10,11] and the related umbrella sampling (US) method.[11] The FEP method evaluates the free energy associated with the change of the potential surface from U_1 to U_2 by gradually changing the potential surface using the relationship

$$U_m(\lambda_m) = U_1(1-\lambda_m) + U_2\lambda_m \tag{15.1}$$

where λ_m is a parameter that changes between $(0 \leq \lambda_m \leq 1)$. The free-energy increment, associated with the change of U_m, can be obtained by[11]

$$\exp\{-\delta G(\lambda_m \rightarrow \lambda_{m'})\beta\} = \langle \exp\{-(U_{m'} - U_m)\beta\}\rangle_m \tag{15.2}$$

where $\langle\ \rangle_m$ indicates that the given average is evaluated by propagating trajectories over U_m. The overall free-energy change is now obtained by changing λ_m in n equal increments and evaluating the sum of the corresponding δG:

$$\Delta G(U_1 \rightarrow U_2) = \sum_{m=0}^{n-1} \delta G(\lambda_m \rightarrow \lambda_{m+1}) \tag{15.3}$$

The FEP approach has been used extensively in studies of free energies of biological systems (*e.g.* Kollman,[8] Shurki and Warshel,[9] Zwanzig,[10] Valleau and Torrie[11] and Warshel and Åqvist[12]) and was introduced quite early[13] in studies of electrostatic free energies in proteins.

In many cases it is very hard to perform converging FEP calculations (*e.g.* binding of large ligands). In such cases, it is extremely useful to estimate the free energy of biological processes by an equation derived by Lee *et al.*[14] and used in studies of ligand binding to proteins. This equation expresses the free energy associated with changing the potential of the system from U_1 to U_2 by

$$\Delta G(U_1 \rightarrow U_2) = \frac{1}{2}(\langle U_2 - U_1\rangle_1 + \langle U_2 - U_1\rangle_2) \tag{15.4}$$

where $\langle\ \rangle_i$ designates molecular dynamics (MD) averages over trajectories obtained with $U = U_i$. The derivation of this equation was based on the assumption that the linear response approximation (LRA)[16] is valid. Namely, the protein and solvent environments respond linearly to the force associated with the given process. This assumption is, in fact, the basis of macroscopic theory where the free energy of charging a positive ion is given by the well known result $\Delta G = U(Q = 1)/2$, where U is the electrostatic potential of the given charge.[17] A geometrical rationale for the LRA approach is given in Figure 15.1 using Marcus-like parabolas. Further discussion and derivations are given in Hummer and Szabo[15] and Sham *et al.*[18] Although it is hard to accept that the LRA can provide a reliable way of describing the energetics of macromolecules or of realistic molecular systems, it was found by simulation studies that it is a reasonable approximation, in particular for processes that depend on

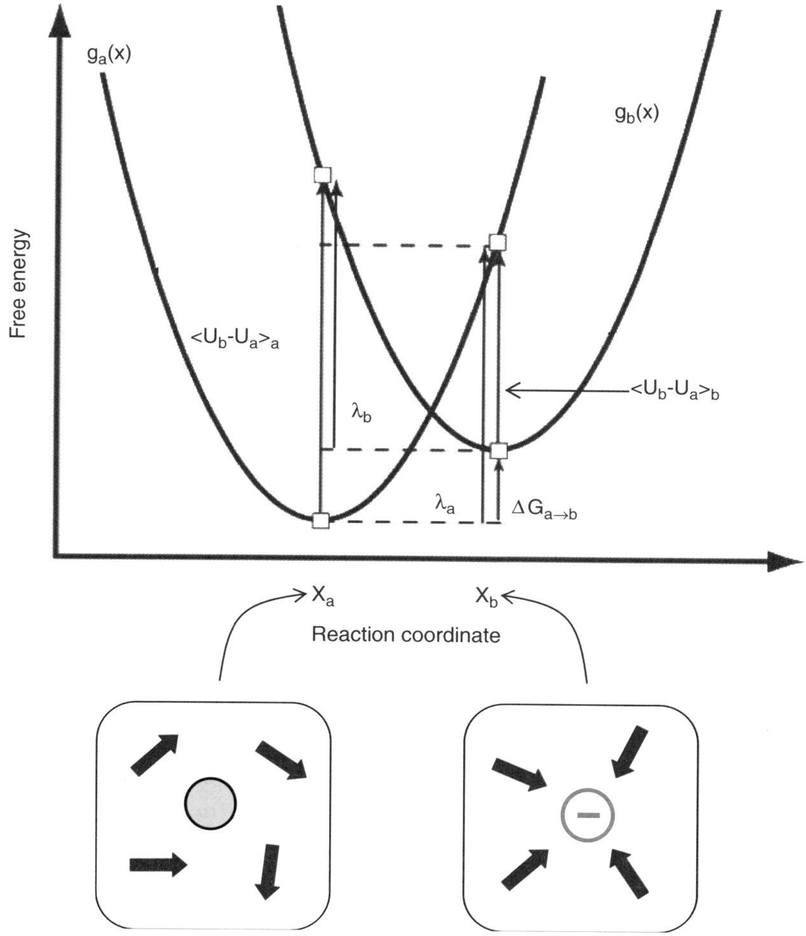

Figure 15.1 A schematic description of the considerations that lead to the LRA. The figure describes the free energy function for the reactant (neutral) and product (charged) states as a function of the solvent coordinates (designated by dipoles). Using either geometrical or analytical considerations leads to Equation (15.4). For more details, see Lee *et al.*[14] and Sham *et al.*[18]

electrostatic effects (Hwang and Warshel,[19] Kuharski *et al.*[20] Åqvist and Hansson,[21] and Morreale *et al.*[22]). It is also useful to note that approaches that only run on one state (*e.g.* Delbuono *et al.*[23]) can also be useful in some cases, although they appear to be less accurate than the LRA approach, which is based on both endpoints.

Many biological processes can be formulated in terms of the probability of being on a given potential surface along a specified reaction coordinate, x. The corresponding free energy $\Delta g(x)$ is obtained by combining the FEP procedure with the US approach.[11] The combined FEP–US approach gives the free-energy function $\Delta g(x)$ as[24–26]

$$\Delta g(x') = \Delta G_m - \beta^{-1} \ln\langle \delta(x - x')\exp\{-\beta(E - U_m)\}\rangle_{U_m} \tag{15.5}$$

where E is the potential surface of interest, β is $1/kT$, and U_m is the mapping potential that keeps the system near x'.

Another quantity that can be used effectively in free-energy calculations is the so-called potential of mean force (PMF). This quantity represents the probability of being at a given x while averaging on all other degrees of freedom, and it can be written as:

$$\Delta G(x_m) = \Delta G(x_{m-1}) + \delta G(x_{m-1} \rightarrow x_m) \tag{15.6}$$

$$\exp\{\delta G(x_{m-1} \rightarrow x_m)\beta\} = \langle \exp\{-[E(x_m) - E(x_{m-1})]\beta\}\rangle_{E(x_{m-1})} \tag{15.7}$$

Interestingly, the rather simple evaluation of the PMF by the FEP–US method using Equation (15.5) has involved some conceptual difficulties. It seems that the idea of "umbrella sampling" was sometimes taken too literally, where it was assumed that one should run different independent runs with different U_m and then try to "overlap" the corresponding results from the second term of Equation (15.5) with an arbitrary constant or with the so-called weighted histogram analysis method (WHAM) approach. Fortunately, the ΔG_m of Equation (15.5) gives this constant and the proper overlap quite reliably (*e.g.* Hwang *et al.*,[24] King and Warshel,[25] and Kato and Warshel[27]).

The above approaches and, in particular, the FEP, US and PMF methods are in principle rigorous. However, it must be emphasized that the convergence of such approaches is extremely slow and that even a perfect convergence does not guarantee appropriate results. For example, FEP calculations with models with inappropriate long-range treatments cannot provide appropriate electrostatic energy. Furthermore, simulation approaches that do not use surface-constraint spherical models require very large simulation systems, and thus involve slow convergence. The best advice in using such methods is to perform validation studies and to see, for example, if the given method can reproduce observed electrostatic energies with different sizes of the explicit simulation system (*e.g.* Kato and Warshel,[27] Sham and Warshel,[28] Alden *et al.*[29]). Interestingly the PMF approach, which has the appearance of the most rigorous approach (here one simply drags the ligand to the active site), may make one "blind" to problems with the simulation model. For example, PMF studies of ion penetration in ion channels do not tell us about the absolute energy of moving the ion from water to the given channel site. Thus, one might be led to believe that the electrostatic model is correct even if the absolute energies are entirely incorrect (see discussion in Burykin *et al.*,[30] and Kato and Warshel[27]).

15.2.2 Proper and Improper Treatments of Long-range Effects in All-atom Models

Explicit all-atom simulations of macromolecules embedded in their actual solvent environment are extremely challenging, even with the current computer power. The key problems are associated with the long-range nature of electrostatic interactions and the slow relaxation times for the protein reorganization following changes in the charges of the protein–substrate system. The number of pair-wise interactions increases with the square of the number of atoms in the system, while the number of atoms in a spherical shell increases like the square of the radius of the system. This leads to a vast number of interactions if all are taken into account in simulations of macromolecules. The customary approach is to include only the interactions within a relatively small cutoff distance. This approach, however, leads to major problems in treatments of energies of charged groups in proteins (for a demonstration see Åqvist[31]).

Early realization of the need for a proper long-range treatment led to developments of a series of models and ideas. The first was probably the realization that solvated systems can be represented by spherical models surrounded by a bulk region.[32] The main conceptual advance of these models was the introduction of buffer (surface) constraints that forced the surface region to behave as if it is part of an infinite system.[32,33] The surface-constraints models and, in particular, the surface-constraint all-atom solvent (SCAAS) model[33] include special polarization constraints that force the surface region to have the angular polarization of the corresponding infinite system. These constraints were found to be crucial for proper treatments of long-range electrostatic effects. The importance of a consistent electrostatic constraint has not been widely appreciated and many popular treatments emphasize the proper treatment of the temperature in the surface region, which can be easily satisfied by any "thermostat" in the so-called "stochastic boundary conditions",[34] rather than considering the polarization of this region and emphasizing proper treatment of electrostatic energies.

Although the use of consistent spherical models offers a convenient way of overcoming the long-range problem, it still requires the expensive evaluation of many interactions when one uses no cutoff with spherical models of significant size. An early attempt to reduce the cutoff problem, and thus to evaluate a smaller number of interactions in spherical models, was introduced by Kuwajima and Warshel[35] by developing a special spherical Ewald-type model. A subsequent development[36] led to the more convenient local reaction field (LRF) model. The stability and reliability of this model has been validated repeatedly.[28,31,36] Alternative long-range treatments for spherical models are also available (*e.g.* Saito[37]), but these models do not yet involve appropriate polarization constraints.

Another way of treating the long-range problem is offered by using periodic boundary conditions and different versions of the Ewald method (Figueirido *et al.*,[38] Bogusz *et al.*,[39] and Hummer *et al.*[40]) The use of the periodic Ewald treatments greatly improves the simulated structural properties, but at present we are not aware of the use and validation of such approaches in the evaluation of electrostatic energies. In fact, it is not clear that the present implementations of the periodic Ewald treatments can give correct electrostatic free energies in proteins (see discussion in Sham and Warshel,[28] and Peter *et al.*[41]). The slow realization that appropriate treatments of long-range effects are crucial for proper modeling of protein functions is still a major problem in the field. Apparently, many workers still emphasize the length of the simulations and the number of atoms in the simulation system instead of validating the calculated electrostatic free energies.

In concluding this section we note that, despite the early emergence of FEP calculations of electrostatic energies in proteins (*e.g.* Warshel *et al.*[13,42]), there are still major convergence problems in both FEP- and LRA-type studies (*e.g.* Sham *et al.*[43] and Buono *et al.*[44]).

15.2.3 Calculations of Electrostatic Energies by Simplified Models

One of the most effective correlations between the structure and function of proteins is provided by the corresponding electrostatic energies (*e.g.* Warshel and Russell[17]). In order to obtain reliable information from such a correlation, it is essential to have a quantitative way for computing electrostatic free energies in macromolecules. The treatments of electrostatic energies in proteins have evolved through different models with different degrees of sophistication and sometimes with major oversimplifications. Early models that treated the protein as a uniform low-dielectric sphere (*e.g.* the influential Tanford–Kirkwood (TK) model[45]) did not consider the crucial contribution of the self-energy of the charged groups and the important role of the protein permanent dipoles (see Warshel *et al.*[46]). The first consistent treatment of electrostatic energies in proteins[47] represented all the relevant electrostatic interactions microscopically, but used a simplified model of Langevin-type dipoles on a grid to represent the solvent. This protein dipoles Langevin dipoles (PDLD) model facilitated consistent studies of the role of electrostatic energies in protein function with the limited computer power of the mid-1970's (see Warshel and Russell[17]). Such studies considered microscopically the shape of the protein and its local polarity as well as the effect of the solvent around the protein and, to a limited extent, the effect of the protein reorganization (which was modeled by energy minimization). The subsequent advance of discretized continuum (DC) models, which are frequently referred to as Poisson–Boltzmann (PB) models, have allowed macroscopic continuum approaches to take into account the shape of the protein.[48,49] Early DC models involved some of the inconsistencies of the TK model. In particular, these models ignored the microscopic nature of the protein permanent dipoles. However, recent PB models have represented these dipoles explicitly, and become increasingly semi-macroscopic.[50]

In many cases one finds that macroscopic and semi-macroscopic models give more stable results than the corresponding microscopic models. This interesting feature is because the microscopic models involve compensation of large numbers while the macroscopic models assume implicitly that the compensation exists by using a protein dielectric constant, ε_p. Thus, PB models give excellent results when dealing with surface groups (where the solvation energy is similar to that for solvation in water), but become more problematic when one deals with groups in the protein interiors. *Here the problem is that the results depend strongly on the assumed* ε_p*, which apparently has little to do with the actual protein dielectric but merely with the contributions that are treated implicitly* (see King *et al.*,[51] Schutz and Warshel[52]).

An effective way to exploit the stability of the semi-macroscopic models, and yet to keep a clear physical picture, is the semi-macroscopic version of the PDLD model (the PDLD/S model[42]). The PDLD/S model takes the PDLD model and scales the corresponding contributions by assuming that the protein has a dielectric constant, ε_p. Now, in order to reduce the unknown factors in ε_p, it is useful to move to the PDLD/S–LRA model, where the PDLD/S energy is evaluated within the LRA approximation of Equation (15.4) with an average on the configurations generated by MD simulation of the charged and the uncharged states (see Sham *et al.*,[43] Schutz and Warshel,[52] Lee *et al.*[53]). In this way one uses MD simulations to automatically generate protein configurations for the charged and uncharged forms of the given "solute" and then average these contributions according to Equation (15.4). Since the protein reorganization is considered explicitly we have fewer uncertainties with ε_p. It is also important to mention that the increasingly popular molecular mechanics PB surface area (MM-PBSA) model,[54] and the related MM generalized Born SA (MM—GBSA) are apparently an adaptation of the PDLD/S–LRA idea of MD generation of configurations for implicit solvent calculations, while only calculating the average over the configurations generated with the charged solute [the first term of Equation (15.4)].

The common tendency to focus on electrostatic benchmarks of surface groups in proteins (whose energy is similar to the corresponding energy in water) has slowed down the realization of the nature of ε_p and the validity of different electrostatic models (see Schutz and Warshel[52]). Apparently, only studies of groups in protein interiors can provide proper validation of macroscopic electrostatic models. With proper benchmarks one finds that the main problem of the PB models is the appropriate selection of ε_p and that different ε_p's should be used for different properties and different interactions (see Schutz and Warshel[52]).

Electrostatic energies in proteins can be formulated in terms of a two-step thermodynamic cycle. The first step involves the transfer of each ionized group from water to its specific protein site where all other groups are neutral, and a second step where the interaction between the ionized groups is turned on.[52] Thus, we can write the energy of moving ionized groups from water to their protein site as

$$\Delta G = \sum_{i=1}^{N} \Delta G_i + \sum_{i>j} \Delta G_{ij} \tag{15.8}$$

where the first and second terms correspond to the first and second steps, respectively. ΔG_i, which represents the change in self-energies of the ith ionized group upon transfer from water to the protein site, can be evaluated by the PDLD/S–LRA approach with a relatively small ε_p. However, the charge–charge interaction terms (the ΔG_{ij}) are best reproduced by using a simple Coulomb's-type law

$$\Delta G_{ij} = 332 Q_i Q_j / r_{ij} \varepsilon_{ij} \tag{15.9}$$

where ΔG is given in kcal mol^{-1}, r_{ij} in Å and ε_{ij} is a relatively high distance-dependent dielectric constant.[46] Although the fact that ε_{ij} is large has been supported by mutation experiments (*e.g.* Alden *et al.*[29] and Johnson and Parson[55]) and conceptual considerations,[17,56] the use of Equation (15.9) is still considered by many as a poor approximation relative to PB treatments (without realizing that the PB approach depends entirely on an unknown ε_p). Interestingly, the so-called generalized Born (GB) model, whose usefulness is now widely appreciated,[57] is basically a combination of Equation (15.9) and the Born energy of the individual charges. More specifically, as pointed out originally in Warshel and Russell[17] and Luzhkov and Warshel,[58] the energy of an ion pair in a uniform dielectric medium can be written as[17]:

$$\Delta G^{+-} = -332/r + \Delta G_{sol} = \Delta G_{sol}^{\infty} - 332/r\varepsilon \tag{15.10}$$

where the free energies are given in kcal mol^{-1} and the distances in Å. ΔG_{sol} is the solvation energy of the ions and ΔG_{sol}^{∞} is the solvation of the ions at infinite separation. The ΔG_{sol} of equation (15.10) gives for a collection of charges (see also Warshel and Russell[17]):

$$\begin{aligned} \Delta G_{sol} &= \Delta G_{sol}^{\infty} - \left(332 Q_i Q_j / r_{ij}\right)(1 - 1/\varepsilon) \\ &= -166\left[\left(Q_i^2/a_i + Q_j^2/a_j\right) + \left(2 Q_i Q_j / r_{ij}\right)(1 - 1/\varepsilon)\right] \end{aligned} \tag{15.11}$$

where a_i and a_j are the Born's radii of the positive and negative ions, respectively. Equation (15.11), with some empirical modifications, leads to the GB treatment.[59] Thus, the widely accepted GB treatment of proteins is merely a version of an earlier treatment.[17] Here, the main issue is the validity of Equation (15.9) and not its legitimization by the seemingly "rigorous" GB formulation. Nevertheless, it is instructive to note that the GB approximation can also be obtained by assuming a *local* model where the vacuum electric field, $\boldsymbol{E}_0$, and the displacement vector, $\boldsymbol{D}$, are identical.[60] This model, which can be considered as a "local Langevin Dipole model"[61] or as a version of the non-iterative LD model, allows one to approximate the energy of a collection of charges and to obtain (with some assumptions about the position dependence of the dielectric constant) the GB approximation.

15.3 Calculating Binding Free Energies

Armed with the computational approaches described in Section 15.2 we may now examine different strategies for calculations of binding free energies. This is done below placing special emphasis on electrostatic calculations.

15.3.1 Studies of Drug Mutations by FEP Approaches

In principle we can evaluate binding energies by the two paths depicted in Figure 15.2. The first path involves a direct PMF calculation of Equations (15.6) and (15.7). This approach involves major convergence problems associated with the transfer of the ligand from the bulk water to the protein-binding site. The second path involves a cycle that can be evaluated by the FEP approach of Equations (15.1)–(15.3). This approach gives reasonable results in studies of differences in binding due to small "mutations" of the given ligand[7,8,62] (*e.g.* replacement of NH_2 by OH), but it involves major convergence problems when the absolute binding free energy of medium-size ligands is required. In such cases, one has to "mutate" the ligand to "nothing", and at present the reported results are quite disappointing except in cases of very small ligands. Nevertheless, this approach can be used in some well-defined test cases of polar or charged ligands (*e.g.* Lee *et al.*,[14] Florian *et al.*,[63] and Dolenc *et al.*[64]) and it can be considered, at least in principle, as the "cleanest" approach for calculations of absolute binding energies. Thus, one should keep in mind the FEP approach in performing validation studies on small, rigid ligands where the converging FEP results can be considered as the "gold standard" for calibrating and validating other approaches.

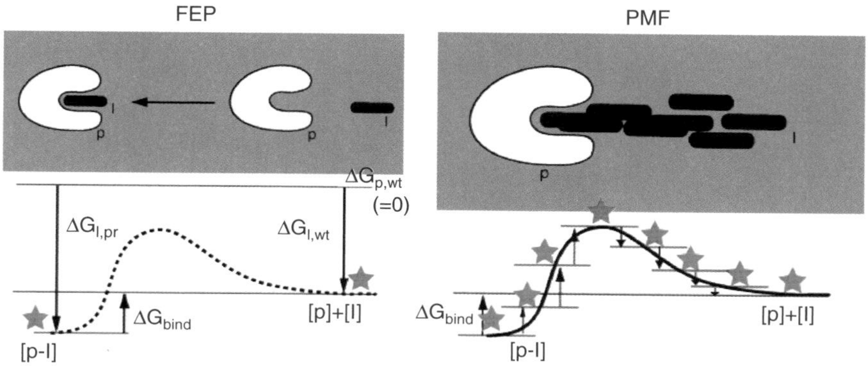

Figure 15.2 Showing two alternative ways (FEP and PMF) of evaluating protein–ligand binding free energy.

15.3.2 Evaluation of Absolute Binding Energies by the LRA and LIE Approaches

In the general case of large ligands one cannot use the FEP approach and the best alternative is probably provided by the LRA method of Equation (15.4). This approach is particularly effective in calculating the electrostatic contribution to the binding energy.[14,18] In considering the LRA approach we start with the cycle of Figure 15.3.

This cycle is composed of electrostatic and non-electrostatic components and expresses the binding free energy as

$$\Delta G_{bind} = \frac{1}{2}\left[\left\langle U^{p}_{elec,\ell}\right\rangle_{\ell} + \left\langle U^{p}_{elec,\ell}\right\rangle_{\ell'} - \left\langle U^{w}_{elec,\ell}\right\rangle_{\ell} - \left\langle U^{w}_{elec,\ell}\right\rangle_{\ell'}\right] + \Delta G^{nonelec}_{bind} \quad (15.12)$$

where $U^{p}_{elec,\ell}$ is the electrostatic contribution for the interaction between the ligand and its surroundings, p and w designate protein and water, respectively,

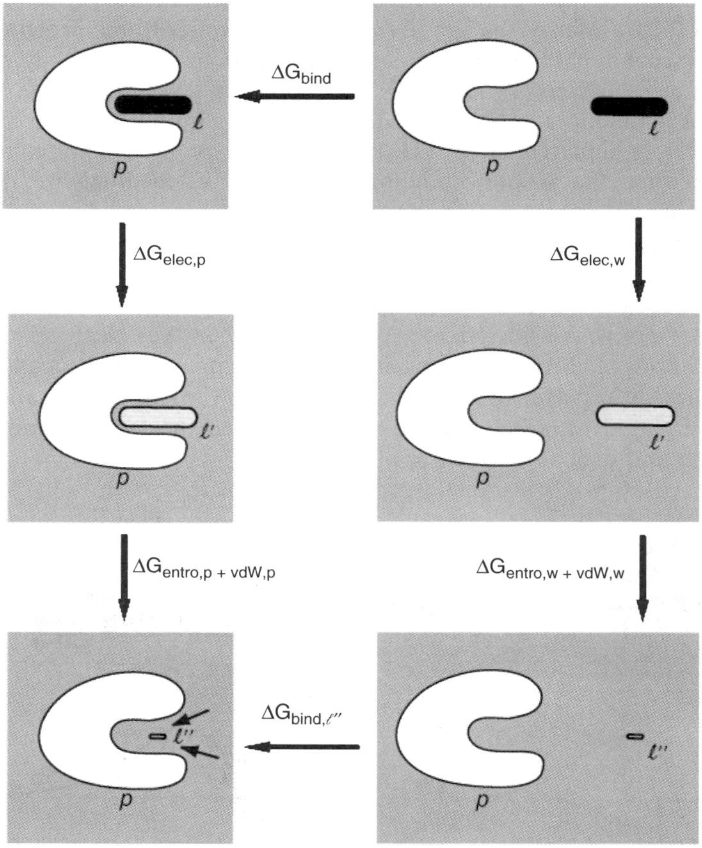

Figure 15.3 Thermodynamic cycle for evaluating ligand binding energies. The cycle is divided into electrostatic and non-electrostatic parts. The free energy for the lowest part of the cycle is zero (see Ref. 18).

and ℓ and ℓ' designate the ligand in its actual charged form and the "non-polar" ligand where all the residual charges are set to zero. In this expression the terms $\langle U_{elec,\ell} - U_{elec,\ell'} \rangle$, which are required by Equation (15.4), are replaced by $\langle U_{elec,\ell} \rangle$, since $U_{elec,\ell'} = 0$. Now, the evaluation of the nonelectrostatic contribution $\Delta G_{bind}^{nonelec}$ is still very challenging, since these contributions might not follow the LRA. A useful option, which was used in references Lee *et al.*[14] and Sham *et al.*[18] is to evaluate the contribution to the binding free energy from hydrophobic effects, van der Waals and water penetration. Another powerful option is the so-called linear interaction energy (LIE) approach.[21,65] This approach adopts the LRA approximation for the electrostatic contribution, but neglects the $\langle U_{elec,\ell} \rangle_{\ell'}$ terms. The binding energy is then expressed as

$$\Delta G_{bind} \approx \alpha \left[\left\langle U^{p}_{elec,\ell} \right\rangle_{\ell} - \left\langle U^{w}_{elec,\ell} \right\rangle_{\ell} \right] + \beta \left[\left\langle U^{p}_{vdW,\ell} \right\rangle_{\ell} - \left\langle U^{w}_{vdW,\ell} \right\rangle_{\ell} \right] \tag{15.13}$$

where α is a constant that is around 1/2 in many cases and β is an empirical parameter that scales the van der Waals (vdW) component of the protein–ligand interaction. A careful analysis of the relationship between the LRA and LIE approaches and the origin of the α and β parameters is given in Sham *et al.*[18] This analysis shows that β can be evaluated in a deterministic way provided one can determine the entropic contribution and preferably the water-penetration effect in a microscopic way (see discussion in Section 15.4).

The idea that conformational averaging is useful for free energy calculations seems quite obvious, generally understood and widely used.[54,66] What is much less understood and somehow still ignored by a large part of the computational community is the realization that a proper LRA treatment requires averaging on both the charged (or polar) and non-polar states of the ligand. The frequently neglected average on the non-polar state [the second term of Equation (15.4)] plays a crucial role in proteins as it reflects the effect of the protein preorganization (see also Section 15.5). The role of this term in binding calculations has been discussed in Sham *et al.*[18] and its importance has been illustrated in an impressive way in studies of the fidelity of DNA polymerase.[67] In this case (*e.g.* see Figure 15.4) it was found that the average of the charge distribution of the non-polar state (the $\langle U_{\ell} \rangle_{\ell'}$) contributes in a major way to the binding of the incoming base. This means that the crucial process of discrimination between correct and incorrect bases is determined to a significant extent by the pre-organization of the binding site plus the template. It also provides an excellent example of the possible importance of including the pre-organization term in calculations of binding energies.

15.3.3 Using Semi-macroscopic and Macroscopic Approaches in Studies of Ligand Binding

Although microscopic models are rigorous in principle they may encounter major convergence problems and thus it is frequently very useful to explore binding energies by semi-macroscopic and macroscopic approaches. This

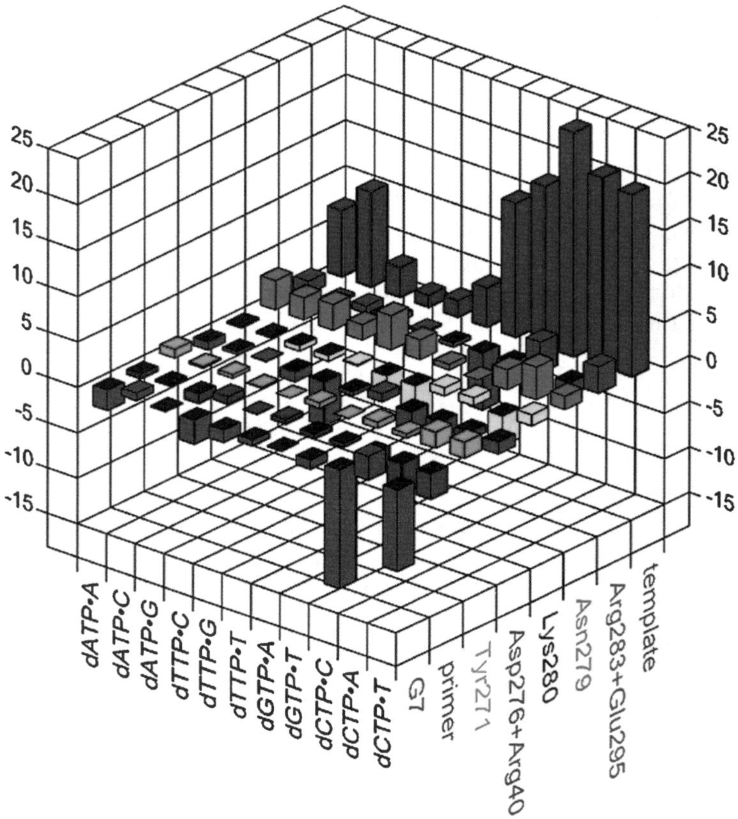

Figure 15.4 Calculated group contributions to the binding energy of dNTP to Pol β. The largest contributions (back panel) are due to the interaction with the template. These contributions include major components from the pre-organization term of Equation (15.12) ($\langle\langle U_\ell\rangle_{\ell'}\rangle$). This term is ignored in the LIE treatment (see Florian *et al.*[67] for discussion).

includes the use of the PDLD/S–LRA model for both the electrostatic and non-electrostatic components (see Lee *et al.*[14] and Sham *et al.*[18]) and the PB approach augmented by estimates of the hydrophobic contributions using the calculated surface area of the ligand.[68]

Here the most rigorous approach is probably the PDLD/S–LRA method mentioned in Section 15.2.3. This approach considers explicitly the protein reorganization and thus allows one to use a less arbitrary ε_p than the one used by more macroscopic approaches. An example of the performance of the PDLD/S–LRA approach in binding calculations is given in Figure 15.5.

The results reported in Figure 15.5, which were obtained in the systematic study of binding of human immunodeficiency virus (HIV) protease inhibitors,[18] may look encouraging, and they are as good as the results obtained by the best current state-of-the-art methods. Unfortunately, one can see that the agreement

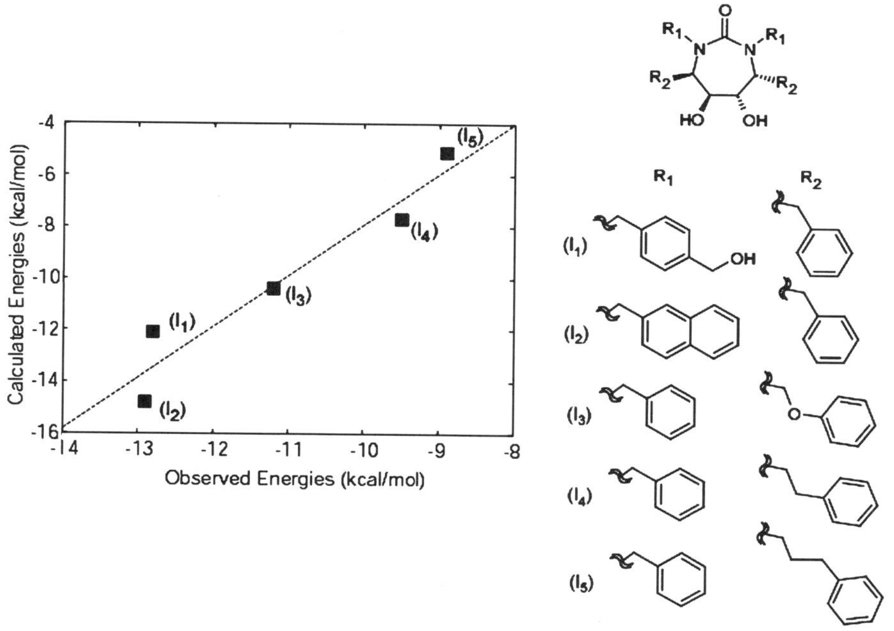

Figure 15.5 The correlation between the PDLD/S–LRA calculations and the observed binding energies of cyclic urea based HIV protease inhibitors (see Sham *et al.*[18] for more details).

is not perfect and that in some cases (*e.g.* ℓ_1 and ℓ_2) we have a significant disagreement. This is quite a general situation in binding calculations, where the trend in cases of ligands with very different binding affinity is easily reproduced, but where the binding affinity is similar it was frequently found that the calculations could not produce the observed differences.

15.3.4 Protein–protein Interactions

One of the most exciting aspects of rational drug design involves the search for drugs that would disrupt the interaction between proteins that transmit different signals in the cell cycle. Here the hope is that understanding the nature of the interaction between two proteins will allow one to design drugs that would exploit the same type of interactions, and thus will bind to one of the proteins and compete with the binding between these proteins. Attempts to analyze protein–protein interactions were reported by several groups (*e.g.* Gohlke and Case[69]). Perhaps the most systematic studies were reported by Muegge *et al.*[70] who explored carefully the effect of the protein reorganization in the Ras–Raf complex. Several approaches and several levels of the PDLD/S–LRA model were explored. However, disappointing results were obtained from the formally most rigorous treatments that consider explicitly the reorganization of the

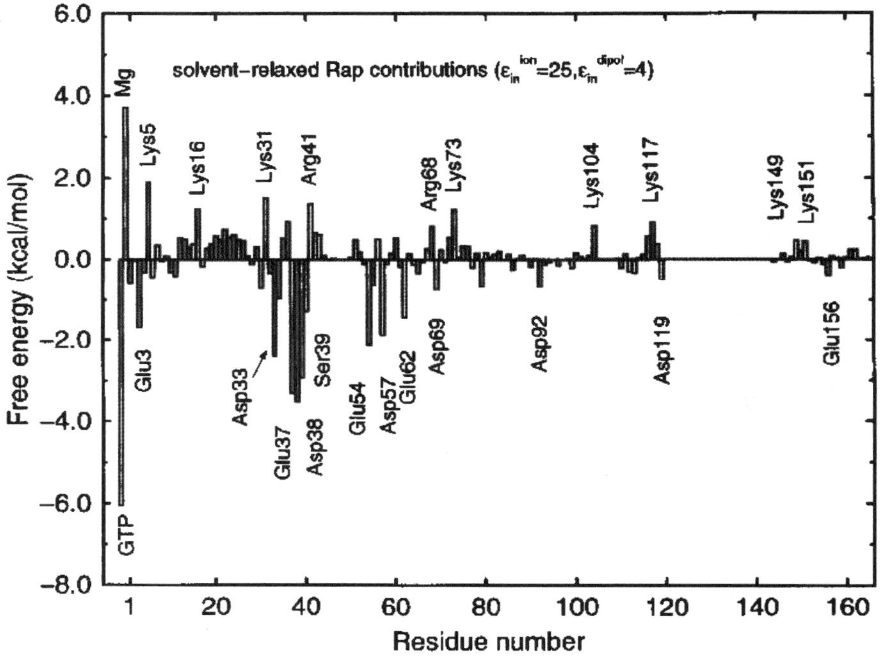

Figure 15.6 Electrostatic contributions of individual residues of Rap to the binding energy of the Raf–Rap complex (see Muegge *et al.*[70] for more details).

proteins during the binding process while using small ε_p. The most promising results were obtained by using the PDLD/S with $\varepsilon_p = 4$ for dipole interaction and Equation (15.9) with $\varepsilon_{eff} = 23$ for charge–charge interaction. This approach was found to be effective in both predicting the effects of mutations on the protein–protein binding energy and in estimating the overall absolute binding energies. Thus, the use of small ε_p for dipole and large ε_{eff} for charge–charge interaction can be quite effective in providing estimates of the contributions of different residues to protein–protein interactions (Figure 15.6). These types of calculations can be quite useful for designing drugs that would disrupt the binding of Ras and Raf and help fight cancer.

15.4 Challenges and New Advances

Despite the qualitative and semi-quantitative advances in the calculation of binding energies, the progress in getting actual quantitative predictions has been quite slow. Here, as stated above, the discrimination between the binding affinities of different ligands in the 2 kcal mol^{-1} limit has been disappointing. As an instructive example it is useful to consider the benchmark of chorismate mutase[71] (CM) where we consider a highly charged active site and ligands in three different charged states (see Figure 15.7). Our attempts to actually evaluate the binding free energies by the fully microscopic FEP approach were very disappointing even with the use of a polarizable force field and the use of proper long-range treatments.

Attempts to use the PDLD/S–LRA semi-macroscopic approach were more successful (see Figure 15.7), but still far from being satisfactory. In the fully microscopic cycle we must have significant convergence problems. In fact, we were forced to use $\varepsilon_p = 6$ for the doubly charged ligand. What comes to mind is the incomplete protein reorganization and water penetration that seems to lead to a major overestimate of charge–charge interactions.[55,70] In dealing with the semi-macroscopic estimate we can handle the electrostatic problem by using an appropriate effective dielectric constant for charge–charge interactions, but the non-electrostatic term still presents a problem.

The above problem presents, in fact, an opportunity since it provides a clear benchmark that must be resolved unless one believes in metaphysical missing effects. That is, one can and should require that the PDLD/S–LRA and the microscopic simulations both converge to the experimental results. Note that, at present, only the PDLD/S–LRA formulation can provide a clear bridge between the semi-macroscopic and microscopic models.[52] At any rate, requiring that the microscopic and macroscopic simulations give the same results for each part of the overall thermodynamic cycle should eventually provide clear clues about the origin of the problems with these approaches. As stated in Section 15.3.2, the connection between these two approaches is provided by evaluating the parameter β of Equation (15.13) using the approach of Sham *et al.*[18] Once we insist on microscopic evaluation of the key parameters in the binding of the uncharged form of the ligand, probably the two biggest

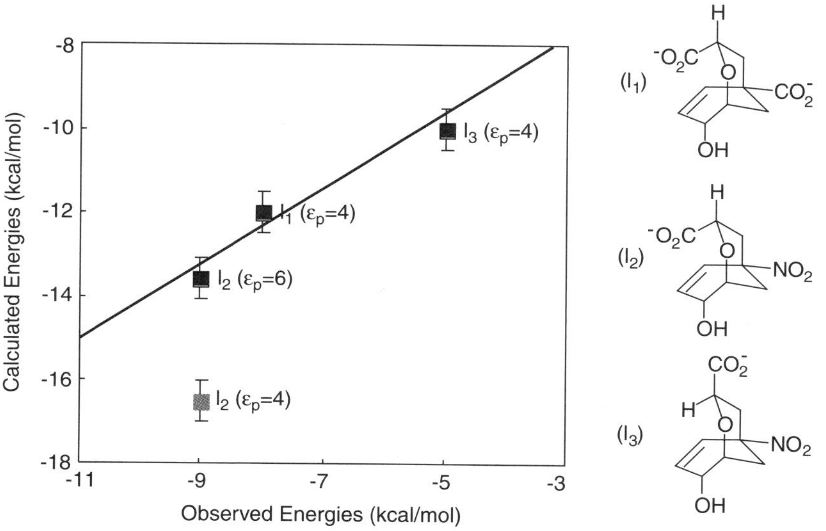

Figure 15.7 The calculated and observed binding energies of different ligands of CM.

challenges involve the evaluation of the relevant binding entropy and the effect of the water penetration. The challenge of evaluating these two contributions is considered briefly below.

One of the most serious problems in obtaining a converging result in calculations of binding free energies is that the binding process involves a change in the number of internal water molecules and that this process is probably associated with a slow relaxation time. In order to overcome this challenge we introduced recently the cycle shown in Figure 15.8 where we actually mutated the ligand to water molecules in both the protein site and in solution. This type of treatment is expected to provide a clearer insight on the nature of β and significantly improve the microscopic results.

Another key problem is the evaluation of the binding entropy. Here it was realized that the frequently used harmonic or quasiharmonic approximations[72,73] are unlikely to give quantitative results, and thus drawing from our early idea[74] and from a related idea of Hermans and Wang,[75] we developed a variational restraint release (RR) approach. This approach, which is described in detail elsewhere,[76] is more general than the Hermans approach since it is

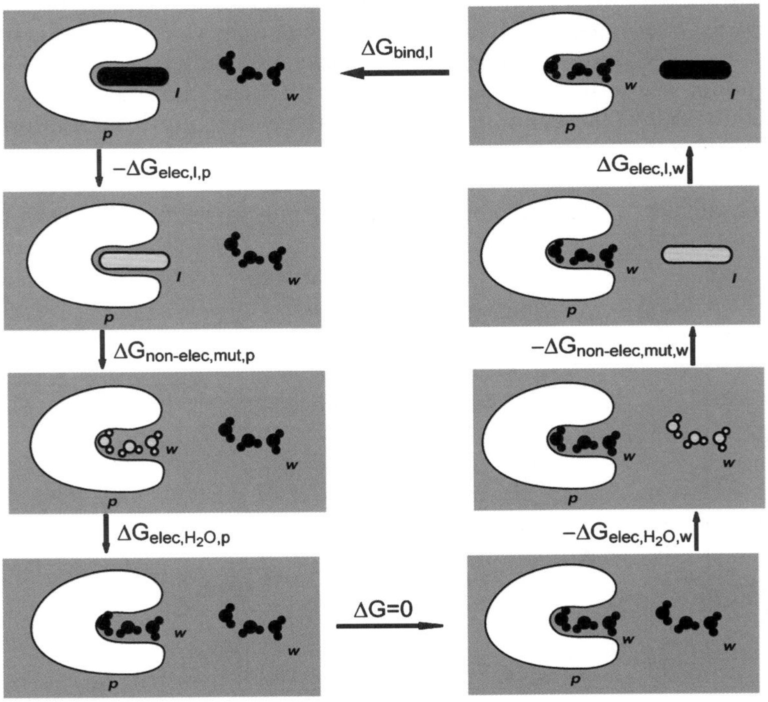

Figure 15.8 A thermodynamic cycle that deals specifically with the water-penetration effect. The cycle involves mutation of the non-polar form of the substrate to non-polar water molecules and then to polar water molecules. The polar and non-polar forms are designated by black and light colors, respectively.

formulated in general Cartesian coordinates. More importantly, it recognizes the fact that the RR free energy depends on the constraint coordinates and thus minimizes the enthalpic contributions to the RR by searching for the lowest limit of the RR results. This approach has already been used in binding calculations[18] and more systematic studies are underway.

15.5 Perspectives

Despite the above problems there is no doubt that computer-aided free-energy calculations will eventually become a key quantitative factor in rational drug design. It is also clear that reliable calculations of electrostatic free energies will be one of the most important factors in such a progress. Once a more quantitative level is achieved we can see several promising directions. For example, it will be possible to augment automated screening procedures and to refine the final drug candidates. Here the ability to understand what interactions are responsible for the optimal binding may help in coming up with better ideas about new drugs that were not included in the original refinement process.

Another promising direction where computational approaches can, in fact, compete with combinatorial drug screening is the fight against drug resistance. Here the challenge is to anticipate the mutations that will be developed by the resistant strain (*e.g.* the HIV virus) and create a drug that will still be bound strongly despite the new mutations. A preliminary advance in this direction has been made in our laboratory by calculating the so-called vitality value [$\gamma = (k_{cat}/K_M) \times K_i$], which expresses the competition between the attempt of the mutated virus to increase K_i and the requirement that the specific protein (*e.g.* an HIV protease) will still perform its usual catalytic function. Evaluating simultaneously the PDLD/S contribution of each residue to k_{cat}/K_M (the binding energy of the TS) for the neutral substrate and the K_i for the given drug produces the corresponding contribution to the vitality factor. With these predicted values one can look for mutations that will maximize γ, and thus predict the optimal drug resistance process. Once the next move of the virus is anticipated, the problem is reduced to the design of drugs that will bind strongly to the new, predicted mutants.

Other major advances in computer-aided drug design will clearly emerge in the future and the only question is how long it would take to make such approaches fully quantitative.

Acknowledgement

This work was supported by NIH Grants GM 24492 and GM 40283.

References

1. I. Muegge and M. Rarey, Small molecule docking and scoring, in *Reviews in Computational Chemistry*, ed. K. B. Lipkowitz and D. B. Boyd, 2001, Wiley-VCH, John Wiley and Sons, New York, pp. 1–60.

2. H. Gohlke and G. Klebe, Approaches to the description and prediction of the binding affinity of small-molecule ligands to macromolecular receptors, *Angew. Chem., Int. Ed.*, 2002, **41**(15), 2645–2676.
3. N. Brooijmans and I. D. Kuntz, Molecular recognition and docking algorithms, *Annu. Rev. Biophys. Biomol. Struct.*, 2003, **32**, 335–373.
4. A. Warshel, Calculations of enzymic reactions: calculations of pK_a, proton transfer reactions, and general acid catalysis reactions in enzymes, *Biochemistry*, 1981, **20**, 3167–3177.
5. A. Warshel, Simulating the energetics and dynamics of enzymatic reactions, In *Specificity in Biological Interactions*, ed. C. Chagas and B. Pullman, Pontificiae Academiae Scientiarum Scripta Varia, Citta del Vaticano, 1984, vol. 55, pp. 59–81.
6. A. Warshel, Computer simulations of enzymatic reactions, *Curr. Opin. Struct. Biol.*, 1992, **2**, 230–236.
7. C. F. Wong and J. A. McCammon, Dynamics and design of enzymes and inhibitors, *J. Am. Chem. Soc.*, 1986, **108**, 3830–3832.
8. P. Kollman, Free energy calculations:applications to chemical and biochemical phenomena, *Chem. Rev.*, 1993, **93**, 2395–2417.
9. A. Shurki and A. Warshel, Structure/function correlations of proteins using MM, QM/MM and related approaches; methods, concepts, pitfalls and current progress, *Adv. Protein Chem.*, 2003, **66**, 249–312.
10. R. W. Zwanzig, High-temperature equation of state by a perturbation method. I. Nonpolar gases, *J. Chem. Phys.*, 1954, **22**, 1420.
11. J. P. Valleau and Torrie. A guide to Monte Carlo for statistical mechanics. 2. Byways, in *Modern Theoretical Chemistry*, ed. B. J. Berne, Plenum Press, New York, 1977, vol. 5, pp. 169–194.
12. A. Warshel and J. Åqvist, Electrostatic energy and macromolecular function, *Annu. Rev. Biophys. Chem.*, 1991, **20**, 267–298.
13. A. Warshel, F. Sussman and G. King, Free energy of charges in solvated proteins:microscopic calculations using a reversible charging process, *Biochemistry*, 1986, **25**, 8368–8372.
14. F. S. Lee, Z. T. Chu, M. B. Bolger and A. Warshel, Calculations of antibody–antigen interactions: microscopic and semi-microscopic evaluation of the free energies of binding of phosphorylcholine analogs to McPC603, *Protein Eng.*, 1992, **5**, 215–228.
15. G. Hummer and A. Szabo, Calculation of free energy differences from computer simulations of initial and final states, *J. Chem. Phys.*, 1996, **105**(5), 2004–2010.
16. R. Kubo, M. Toda and N. Hashitsume, *Statistical Physics II: Nonequilibrium Statistical Mechanics*, Springer-Verlag, Berlin, 1985.
17. A. Warshel and S. T. Russell, Calculations of electrostatic interactions in biological systems and in solutions, *Q. Rev. Biophys.*, 1984, **17**, 283–421.
18. Y. Y. Sham, Z. T. Chu, H. Tao and A. Warshel, Examining methods for calculations of binding free energies:LRA, LIE PDLD–LRA, and PDLD/S–LRA calculations of ligands binding to an HIV protease, *Proteins: Struct., Funct., Genet.*, 2000, **39**, 393–407.

19. J.-K. Hwang and A. Warshel, Microscopic examination of free energy relationships for electron transfer in polar solvents, *J. Am. Chem. Soc.*, 1987, **109**, 715–720.
20. R. A. Kuharski, J. S. Bader, D. Chandler, M. Sprik, M. L. Klein and R. W. Impey, Molecular model for aqueous ferrous ferric electron transfer, *J. Chem. Phys.*, 1988, **89**, 3248–3257.
21. J. Åqvist and T. Hansson, On the validity of electrostatic linear response in polar solvents, *J. Phys. Chem.*, 1996, **100**, 9512–9521.
22. A. Morreale, X. de la Cruz, T. Meyer, J. L. Gelpi, F. J. Luque and M. Orozco, Partition of protein solvation into group contributions from molecular dynamics simulations, *Proteins: Struct., Funct., Genet.*, 2005, **58**(1), 101–109.
23. G. S. Del Buono, F. E. Figueirido and R. M. Levy, Intrinsic pK_as of ionizable residues in proteins – an explicit solvent calculation for lysozyme, *Proteins: Struct., Funct., Genet.*, 1994, **20**(1), 85–97.
24. J.-K. Hwang, G. King, S. Creighton and A. Warshel, Simulation of free energy relationships and dynamics of S_N2 reactions in aqueous solution, *J. Am. Chem. Soc.*, 1988, **110**(16), 5297–5311.
25. G. King and A. Warshel, Investigation of the free energy functions for electron transfer reactions, *J. Chem. Phys.*, 1990, **93**, 8682–8692.
26. A. Warshel, Dynamics of reactions in polar solvents. Semiclassical trajectory studies of electron-transfer and proton-transfer reactions, *J. Phys. Chem.*, 1982, **86**, 2218–2224.
27. M. Kato and A. Warshel, Through the channel and around the channel: Validating and Comparing Microscopic Approaches for Evaluation of Free Energy Profiles for Ion Penetration through Ion Channels, *J. Phys. Chem. B*, 2005, **109**, 19516–19522.
28. Y. Y. Sham and A. Warshel, The surface constrained all-atom model provides size independent results in calculations of hydration free energies, *J. Chem. Phys.*, 1998, **109**, 7940–7944.
29. R. G. Alden, W. W. Parson, Z. T. Chu and A. Warshel, Calculations of electrostatic energies in photosynthetic reaction centers, *J. Am. Chem. Soc.*, 1995, **117**, 12284–12298.
30. A. Burykin, C. N. Schutz, J. Villa and A. Warshel, Simulations of ion current in realistic models of ion channels: The KcsA potassium channel, *Proteins: Struct., Funct., Genet.*, 2002, **47**, 265–280.
31. J. Åqvist, Calculation of absolute binding free energies for charged ligands and effects of long-range electrostatic interactions, *J. Comput. Chem.*, 1996, **17**(14), 1587–1597.
32. A. Warshel, Calculations of chemical processes in solutions, *J. Phys. Chem.*, 1979, **83**, 1640–1650.
33. G. King and A. Warshel, A surface constrained all-atom solvent model for effective simulations of polar solutions, *J. Chem. Phys.*, 1989, **91**(6), 3647–3661.
34. C. L. Brooks III and M. Karplus, Deformable stochastic boundaries in molecular dynamics, *J. Chem. Phys.*, 1983, **79**, 6312–6325.

35. S. Kuwajima and A. Warshel, The extended Ewald method: a general treatment of long-range electrostatic interactions in microscopic simulations, *J. Chem. Phys.*, 1988, **89**, 3751–3759.
36. F. S. Lee and A. Warshel, A local reaction field method for fast evaluation of long-range electrostatic interactions in molecular simulations, *J. Chem. Phys.*, 1992, **97**, 3100–3107.
37. M. Saito, Molecular dynamics simulations of proteins in solution: artifacts caused by the cutoff approximation, *J. Chem. Phys.*, 1994, **101**, 4055–4061.
38. F. Figueirido, G. S. Del Buono and R. M. Levy, On the finite size corrections to the free energy of ionic hydration, *J. Phys. Chem. B*, 1997, **101**(29), 5622–5623.
39. S. Bogusz, T. E. Cheatham III and B. R. Brooks, Removal of pressure and free energy artifacts in charged periodic systems via net charge corrections to the Ewald potential, *J. Chem. Phys.*, 1998, **108**, 7070–7084.
40. G. Hummer, L. R. Pratt and A. E. Garcia, Free energy of ionic hydration, *J. Phys. Chem.*, 1996, **100**, 1206–1215.
41. C. Peter, W. F. van Gunsteren and P. H. Hunenberger, A fast-Fourier transform method to solve continuum-electrostatics problems with truncated electrostatic interactions: algorithm and application to ionic solvation and ion–ion interaction, *J. Chem. Phys.*, 2003, **119**(23), 12205–12223.
42. A. Warshel, G. Naray-Szabo, F. Sussman and J. -K. Hwang, How do serine proteases really work?, *Biochemistry*, 1989, **28**, 3629–3673.
43. Y. Y. Sham, Z. T. Chu and A. Warshel, Consistent calculations of pK_a's of ionizable residues in proteins: semi-microscopic and macroscopic approaches, *J. Phys. Chem. B*, 1997, **101**, 4458–4472.
44. G. S. Buono, F. E. Figueirido and R. M. Levy, Intrinsic pK_a's of ionizable residues in proteins: an explicit solvent calculation for lysozyme, *Proteins: Struct., Funct., Genet.*, 1994, **20**, 85–97.
45. C. Tanford and J. G. Kirkwood, Theory of protein titration curves. I. General equations for impenetrable spheres, *J. Am. Chem. Soc.*, 1957, **79**, 5333.
46. A. Warshel, S. T. Russell and A. K. Churg, Macroscopic models for studies of electrostatic interactions in proteins: limitations and applicability, *Proc. Natl. Acad. Sci. U. S. A.*, 1984, **81**, 4785–4789.
47. A. Warshel and M. Levitt, Theoretical studies of enzymic reactions: dielectric, electrostatic and steric stabilization of the carbonium ion in the reaction of lysozyme, *J. Mol. Biol.*, 1976, **103**, 227–249.
48. J. Warwicker and H. C. Watson, Calculation of the electric potential in the active site cleft due to alpha-helix dipoles, *J. Mol. Biol.*, 1982, **157**, 671–679.
49. K. A. Sharp and B. Honig, Electrostatic interactions in macromolecules: theory and applications, *Annu. Rev. Biophys. Biophys. Chem.*, 1990, **19**, 301–332.
50. A. Warshel and A. Papazyan, Electrostatic effects in macromolecules: fundamental concepts and practical modeling, *Curr. Opin. Struct. Biol.*, 1998, **8**, 211–217.

51. G. King, F. S. Lee and A. Warshel, Microscopic simulations of macroscopic dielectric constants of solvated proteins, *J. Chem. Phys.*, 1991, **95**, 4366–4377.
52. C. N. Schutz and A. Warshel, What are the dielectric "constants" of proteins and how to validate electrostatic models, *Protein: Struct., Funct., Genet.*, 2001, **44**, 400–417.
53. F. S. Lee, Z. T. Chu and A. Warshel, Microscopic and semimicroscopic calculations of electrostatic energies in proteins by the POLARIS and ENZYMIX programs, *J. Comput. Chem.*, 1993, **14**, 161–185.
54. P. A. Kollman, I. Massova, C. Reyes, B. Kuhn, S. Huo, L. Chong, M. Lee, T. Lee, Y. Duan, W. Wang, O. Donini, P. Cieplak, J. Srinivasan, D. A. Case and T. E. Cheatham III, Calculating structures and free energies of complex molecules: combining molecular mechanics and continuum models, *Acc. Chem. Res.*, 2000, **33**, 889–897.
55. E. T. Johnson and W. W. Parson, Electrostatic interactions in an integral membrane protein, *Biochemistry*, 2002, **41**, 6483–6494.
56. Y. Y. Sham, I. Muegge and A. Warshel, The effect of protein relaxation on charge–charge interactions and dielectric constants of proteins, *Biophys. J.*, 1998, **74**, 1744–1753.
57. D. Bashford and D. A. Case, Generalized Born models of macromolecular solvation effects, *Annu. Rev. Phys. Chem.*, 2000, **51**, 129–152.
58. V. Luzhkov and A. Warshel, Microscopic models for quantum mechanical calculations of chemical processes in solutions: LD/AMPAC and SCAAS/AMPAC calculations of solvation energies, *J. Comput. Chem.*, 1992, **13**, 199–213.
59. W. C. Still, A. Tempczyk, R. C. Hawley and T. Hendrickson, Semianalytical treatment of solvation for molecular mechanics and dynamics, *J. Am. Chem. Soc.*, 1990, **112**, 6127–6129.
60. D. Borgis, N. Levy and M. Marchi, Computing the electrostatic free-energy of complex molecules: The variational Coulomb field approximation, *J. Chem. Phys.*, 2003, **119**(6), 3516–3528.
61. T. HaDuong, S. Phan, M. Marchi and D. Borgis, Electrostatics on particles: Phenomenological and orientational density functional theory approach, *J. Chem. Phys.*, 2002, **117**(2), 541–556.
62. M. Graffner-Nordberg, J. Marelius, S. Ohlsson, A. Persson, G. Swedberg, P. Andersson, S. E. Andersson, J. Aqvist and A. Hallberg, Computational predictions of binding affinities to dihydrofolate reductase: synthesis and biological evaluation of methotrexate analogues, *J. Med. Chem.*, 2000, **43**(21), 3852–3861.
63. J. Florian, M. F. Goodman and A. Warshel, Free-energy perturbation calculations of DNA destabilization by base substitutions: The effect of neutral guanine thymine, adenine cytosine and adenine difluorotoluene mismatches, *J. Phys. Chem. B*, 2000, **104**(43), 10092–10099.
64. J. Dolenc, C. Oostenbrink, J. Koller and W. F. van Gunsteren, Molecular dynamics simulations and free energy calculations of netropsin and

distamycin binding to an AAAAA DNA binding site, *Nucleic Acids Res.*, 2005, **33**(2), 725–733.
65. J. Åqvist, V. B. Luzhkov and B. O. Brandsdal, Ligand binding affinities from MD simulations, *Acc. Chem. Res.*, 2002, **35**(6), 358–365.
66. T. J. You and D. Bashford, Conformation and hydrogen ion titration of proteins: a continuum electrostatic model with conformational flexibility, *Biophys. J.*, 1995, **69**, 1721–1733.
67. J. Florian, M. F. Goodman and A. Warshel, Theoretical investigation of the binding free energies and key substrate-recognition components of the replication fidelity of human DNA polymerase β, *J. Phys. Chem. B*, 2002, **106**, 5739–5753.
68. N. Froloff, A. Windemuth and B. Honig, On the calculation of binding free energies using continuum methods: Application to MHC class I protein–peptide interactions, *Protein Sci.*, 1997, **6**(6), 1293–1301.
69. H. Gohlke and D. A. Case, Converging free energy estimates: MM-PB(GB)SA studies on the protein–protein complex Ras–Raf, *J. Comput. Chem.*, 2004, **25**(2), 238–250.
70. I. Muegge, T. Schweins and A. Warshel, Electrostatic contributions to protein–protein binding affinities: application to Rap/Raf interaction, *Proteins: Struct., Funct., Genet.*, 1998, **30**, 407–423.
71. A. Mandal and D. Hilvert, Charge optimization increases the potency and selectivity of a chorismate mutase inhibitor, *J. Am. Chem. Soc.*, 2003, **125**(19), 5598–5599.
72. D. Bakowies and W.F. van Gunsteren, Simulations of apo and holo-fatty acid binding protein: Structure and dynamics of protein, ligand and internal water, *J. Mol. Biol.*, 2002, **315**(4), 713–736.
73. M. Karplus and J. N. Kushick, Method for estimating the configurational entropy of macromolecules, *Macromolecules*, 1981, **14**(2), 325–332.
74. A. Warshel, Computer Simulation of Chemical Reactions in Synthetic Model Compounds and Genetically Engineered Active Sites. Proposal to the Office of Naval Research, Award Number N00014–91–J-1318, 1991.
75. J. Hermans and L. Wang, Inclusion of loss of translational and rotational freedom in theoretical estimates of free energies of binding. Application to a complex of benzene and mutant T4 lysozyme, *J. Am. Chem. Soc.*, 1997, **119**, 2707–2714.
76. M. Strajbl, Y. Y. Sham, J. Villa, Z. T. Chu and A. Warshel, Calculations of activation entropies of chemical reactions in solution, *J. Phys. Chem.*, 2000, **104**, 4578–4584.

Section 5
Fragment-Based Design

CHAPTER 16

Discovery and Extrapolation of Fragment Structures towards Drug Design

ALESSIO CIULLI[1], TOM L. BLUNDELL[2] AND CHRIS ABELL[1]

[1]University Chemical Laboratory, Lensfield Road, Cambridge, CB2 1EW, UK
[2]Department of Biochemistry, 80 Tennis Court Road, Cambridge, CB2 1GA, UK

16.1 Structure-based Approaches to Drug Discovery

Structure-guided design of protein ligands has its origins 30 years ago in some visionary work using the first three-dimensional structures of globins.[1,2] However, the area began to transform in the 1980s as the first structures of real target proteins or their close homologues became available.[3] Three-dimensional structures proved useful in defining topographies of the complementary surfaces of ligands and their protein targets, and could be exploited to optimize potency and selectivity.[4] Later the crystal structures of real drug targets became available; acquired immunodeficiency syndrome (AIDS) drugs, such as Agenerase and Viracept, were developed using the crystal structure of human immunodeficiency virus (HIV) protease,[5,6] and the flu drug Relenza was designed using the crystal structure of neuraminidase.[7] Hardy and Malikayil list more than 40 compounds that have been discovered with the aid of structure-guided methods and that have entered clinical trials;[8] around one-fifth have become approved and marketed drugs.

For more than a decade, lead discovery has been dominated by random high-throughput screening (HTS).[9] In an attempt to increase the number and quality of the compounds available for screening, companies established large libraries using both solid phase and solution phase chemistry approaches, coupled with high-throughput purification platforms.[10–12] However, the quality of libraries has been difficult to improve and some targets have not been very tractable.[9] More focused high-throughput screens have often proved useful, especially where there is information about ligands and protein structure.[13] Quite often it is possible to use virtual screening and *de novo* design to select a sub-set of samples from the larger compound file,[14–17] or to design a new lead using knowledge of previously identified ligands. Chemists have begun to produce target-class specific compound libraries, directed by known ligands or by knowledge of the active site of at least one member of the protein family.[12]

Partly in response to the limited success of random screening, a range of different approaches to lead generation have recently emerged, based on identifying small molecules that are much smaller and functionally simpler than drugs themselves, referred to as 'fragment-based drug discovery'.[18–21] Fragments have low molecular weight (MW typically <250 Da) and as a consequence will usually bind weakly to proteins ($K_d > 100\ \mu M$). Nevertheless, fragments are suitable starting points for the assembly of more potent compounds. Knowledge of how the fragment binds in the active site of the target provides the basis for deriving good-quality leads. In this context, fragment-based drug discovery may be seen as the convergence of screening and structure-based design.[22,23]

16.2 Properties of Molecular Fragments

Fragment-based approaches to drug discovery begin with the identification of small molecules (referred to primarily as 'fragments',[18–21] but also as 'scaffolds',[24] 'needles'[25] or 'shapes'[26]) that bind to proteins, and their subsequent rational elaboration into more potent drug leads. The chemical and physical properties of fragments have received attention with focus to applications in library design.[27] On a similar framework to Lipinski's 'rule of five'[28] that described optimal drug-like properties with respect to solubility, absorption and permeability, Jhoti *et al.* have proposed a 'rule of three' for fragment libraries[29] (Table 16.1). According to this rule, the MW of fragments should be <300, the number of hydrogen bond donors and of hydrogen bond acceptors ≤3, and the clogP ≤3.

Fragments (Figure 16.1) typically possess fewer functional groups than hits from HTS. Moreover, fragments need to be more soluble, allowing screening at higher concentrations to identify inherently weaker affinities. Fragments are less prone to the kind of aggregation that can lead to false positives in HTS due to non-specific enzyme inhibition.[30,31]

Table 16.1 Rules for drug-like and fragment-like compound libraries.

	Rule of five[28]	*Rule of three*[29]
MW (Da)	≤ 500	≤ 300
Hydrogen bond donors	≤ 5	≤ 3
Hydrogen bond acceptors	≤ 10	≤ 3
clogP	≤5	≤ 3

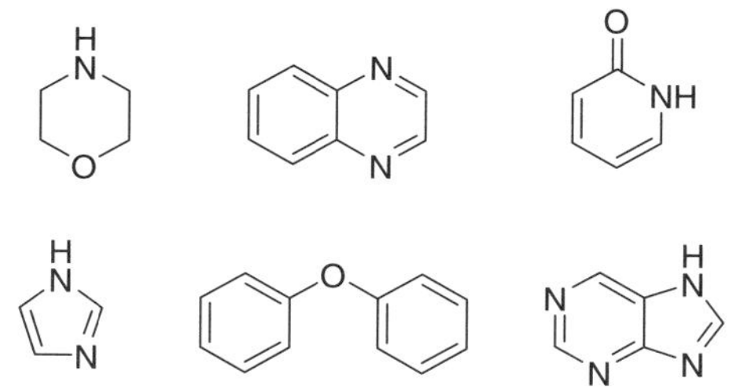

Figure 16.1 Typical fragments.

Libraries of typically less than 10^3 fragments are generally employed, enabling relatively fast screening using low-throughput information-rich biophysical methods (see later). The key to fragment screening is that a significant amount of chemical space can be sampled with relatively few compounds because fragments are free to adopt different orientations and conformations within the enzyme-binding site. The discrimination between a binding event and a non-binding event is solely dependent on the molecular complementarity between the fragment and the pocket surface and on the binding affinity. This would not be necessarily true if the fragment was constrained as part of a larger molecule. In this case its binding interactions may not be optimal and the fragment contribution to the total affinity of the ligand would be reduced.

Fragments typically bind weakly to proteins, often with dissociation constants of 100 μM–10 mM. Although affinities are low, the majority of the ligand is likely to make favorable contacts with the protein, and key binding interactions are formed. The aim is to retain these when the fragment is subsequently modified into a larger compound. When a ligand binds to a protein, it loses a significant amount of translational and rotational entropy. Murray and Verdonk[32] have estimated this rigid-body entropic barrier to be 15–20 kJ mol^{-1}, corresponding to about three orders of magnitude in the binding affinity. A fragment binding with a K_d of 1 mM would hence possess a much higher intrinsic binding energy, corresponding to a K_d of about 1 μM. This unfavorable contribution to the free energy of binding is already paid for by the fragment and will not be expected to increase dramatically for larger molecules derived from the initial fragment.

The favorable binding interactions, combined with their low MW, make fragments highly efficient ligands.[33] Ligand efficiency Δg is defined as affinity relative to MW.[34] The actual parameter of binding energy per heavy atom ($\Delta g = \Delta G/N_{\text{non-hydrogen atoms}}$) was first introduced by Kuntz *et al.* in their account on the maximal affinity of ligands.[35] Based on experimental data available at the time for a large number of strong binders clustered by their number of non-hydrogen atoms (NHAs), they noted that the affinity of small molecules for protein targets increased linearly with respect to NHAs for very small compounds (NHAs < 10–15). However, this linear tendency is broken for larger compounds, as adding more atoms no longer leads to marked increases in binding energy.[35] Therefore, the resulting ligand efficiency Δg tends to be maximal for small molecules (*e.g.* fragments) and then steadily decreases as more heavy atoms are added.

The concept of ligand efficiency and its importance as a useful parameter to assess and select leads for drug discovery has been discussed.[34] Simple calculations based on average molecular properties of drug-like compounds obeying the Lipinski 'rule of five'[28] defines the lowest acceptable limit for ligand efficiency of hits and drug leads around 1.2 kJ mol^{-1} per NHA.[34] Analysis of several examples, in which lead compounds have been derived from the identification of starting fragments, has shown that in all cases both initial fragments and the leads derived from them had ligand efficiencies between 1.2 and 2.1.[20]

16.3 From Molecular Fragments to Drug Leads

The elaboration of fragment hits into lead compounds must improve the binding affinity from the mM range to nM. Throughout the process, the chemistry of fragment elaboration is guided by structural information on the binding mode of the fragment(s) and by inspection of additional potential interactions with the protein. Knowledge of binding thermodynamics can also help understanding the effect of particular structural modifications. Different strategies can be employed to elaborate small fragments into larger lead compounds: fragment growing, fragment linking, and fragment self-assembly.[18]

16.3.1 Fragment Growing

Fragments are used as starting points for extending or 'growing' an inhibitor in the binding pocket (illustrated schematically in Figure 16.2A). An initial fragment is steadily built up to explore favorable interactions with adjacent regions of the binding site. It has been found that the binding mode of a fragment is generally maintained during the optimization process.[33,36] Examples of this

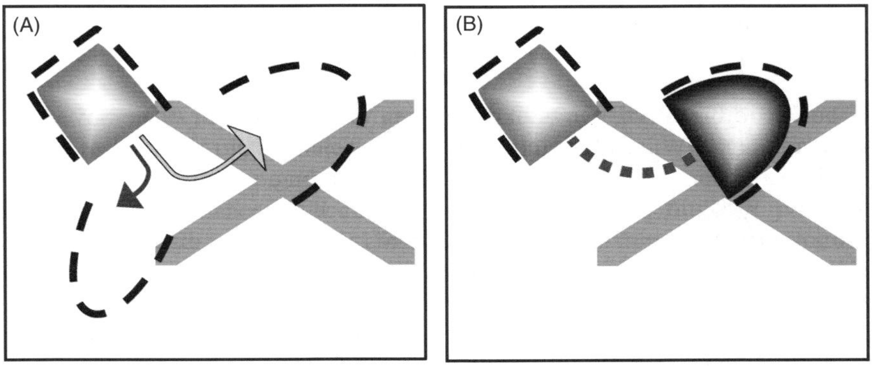

Figure 16.2 Schematic of the fragment growing (A) and linking (B) approaches.

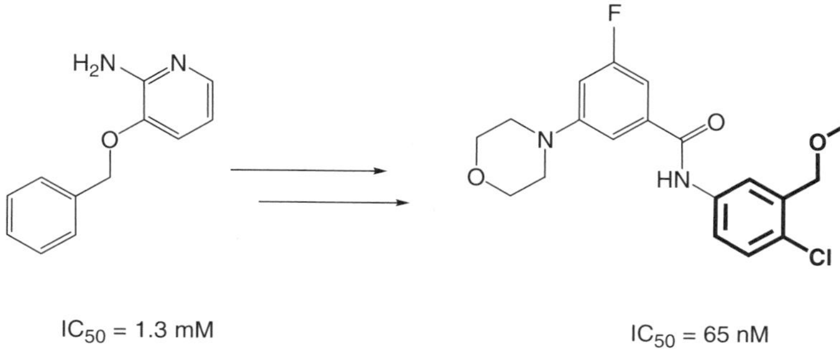

Scheme 16.1 A potent inhibitor of p38α MAP kinase was developed using the fragment-growing approach.[39]

approach include the development of inhibitors against deoxyribonucleic acid (DNA) gyrase,[25] Erm methyltransferase,[37] and urokinase.[38] More recently, researchers at Astex Therapeutics elaborated an initial fragment (IC_{50} of 1.3 mM) in successive steps to yield a novel p38α mitogen-activated protein (MAP) kinase inhibitor with IC_{50} of 65 nM[39,40] (see Scheme 16.1).

16.3.2 Fragment Linking

A high-affinity compound can be designed through conjugation of low-affinity fragments that bind to adjacent sites (illustrated schematically for two fragments in Figure 16.2B). Page and Jencks provided the theoretical framework for the fragment-linking approach by pointing out that the expected binding energy of the joined molecule could be greater than the sum of the components.[41,42] Murray and Verdonk developed further this concept with particular reference to fragment-based drug discovery.[32] The energetic advantage of joining two fragments A and B into a larger molecule A–B can be readily appreciated by describing the Gibbs free energy of each molecule in terms of a favorable 'intrinsic binding energy', ΔG_{i}, and an unfavorable term, ΔG_{r}, representing the loss of translational and rotational entropy (Figure 16.3). Although ΔG_{i} is more-or-less additive, the respective ΔG_{r} are not expected to increase much for the larger compound. Therefore, the unfavorable ΔG_{r} term is paid for only once in molecule A–B, accounting for the energetic advantage of fragment linking. Murray and Verdonk have estimated ΔG_{r} to be a significant energetic barrier of 15–20 kJ mol^{-1}, thus providing up to three orders of magnitudes in binding affinity.[32]

Crucial to the fragment-linking approach is the design of a suitable linker capable of maintaining the preferred orientations and thus optimal binding interactions of the single fragments. Linker design is most effective when driven by structural knowledge of the binding modes and relative orientations of the fragments.

Various examples are available which support theoretical predictions of significant improvements in binding energy upon joining fragments. Hajduk *et al.*

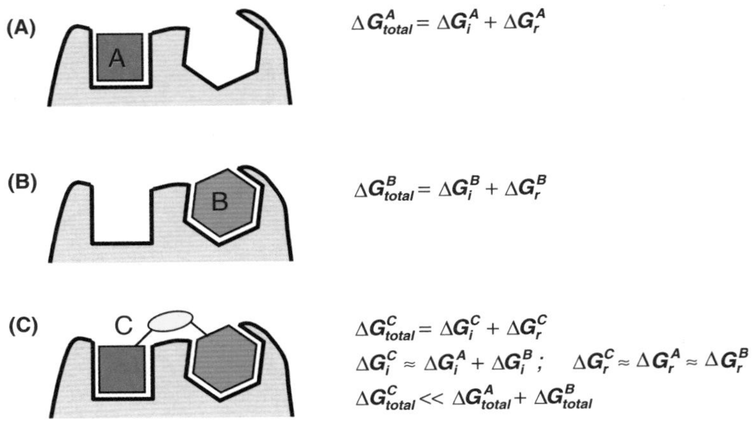

Figure 16.3 Schematic illustration of the energetic gain expected during fragment linking.

$$\Delta G_{total}^A = \Delta G_i^A + \Delta G_r^A$$

$$\Delta G_{total}^B = \Delta G_i^B + \Delta G_r^B$$

$$\Delta G_{total}^C = \Delta G_i^C + \Delta G_r^C$$
$$\Delta G_i^C \approx \Delta G_i^A + \Delta G_i^B \; ; \quad \Delta G_r^C \approx \Delta G_r^A \approx \Delta G_r^B$$
$$\Delta G_{total}^C \ll \Delta G_{total}^A + \Delta G_{total}^B$$

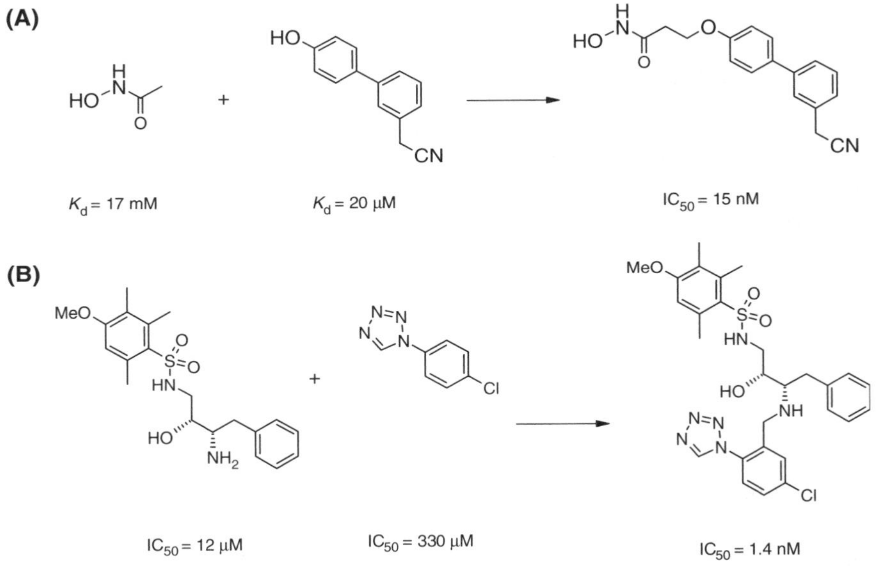

Scheme 16.2 Examples of the fragment-linking approach to discover nM inhibitors of (A) stromelysin[43] and (B) thrombin.[44]

at Abbott Laboratories identified two weak-affinity fragments by nuclear magnetic resonance (NMR) spectroscopy with individual K_d of 17 mM and 20 μM against stromelysin (Scheme 16.2A). The resulting linked molecule was a potent inhibitor with IC_{50} of 15 nM.[43] More recently a potent novel inhibitor of thrombin ($IC_{50} = 1.4$ nM) was discovered as a result of joining two individual

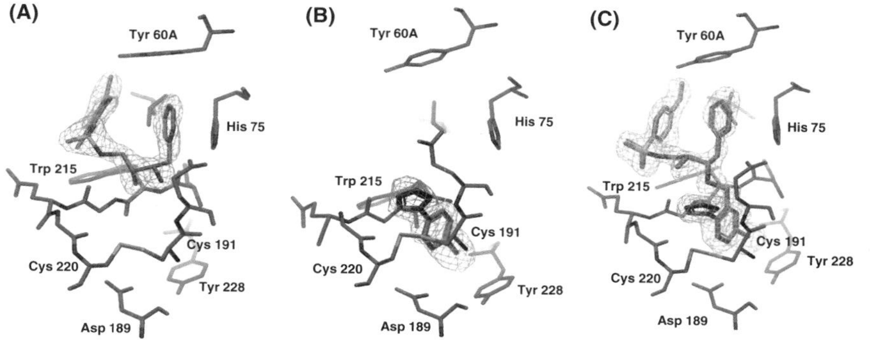

Figure 16.4 Fragment linking to generate a nM inhibitor of thrombin. The crystal structures in (A)–(C) are those of the corresponding compounds shown in Scheme 16.2B, the F_o–F_c electron density identifying the presence of the ligand (light blue) and contoured at 3σ. Key protein residues are shown with green carbon atoms. Nitrogen atoms are shown in blue, oxygen atoms in red, and sulfur in yellow.

compounds which showed IC_{50} of 330 and 12 μM, respectively (Scheme 16.2B; see also Figure 16.4).[44]

16.3.3 Fragment Assembly

An alternative to building up fragments is to use fragments containing complementary functional groups which are allowed to react reversibly in the presence of the protein target. The idea is that the most potent conjugated ligand will bind to the protein and ultimately be identified (Figure 16.5). Self-assembly of dynamic combinatorial libraries (DCLs) in the presence of a template is a large field of its own[45–47] with applications to drug discovery.[48,49] Pioneering work in this area using enzymes to template ligand assembly has been described by Lehn[46] and Sharpless *et al.*[50] to make potent enzyme inhibitors. In a DCL all compounds are in thermodynamic equilibrium, allowing the library composition to adapt to external influences. However, non-covalent interactions between the library components and the protein receptor drive the selection, stabilization and, thereby, amplification of the most potent compound. The protein may catalyze the formation of its own best ligand.[47,49] DCL approaches are particularly sensitive to linker design. Two optimal binding fragments may not form a viable conjugated ligand due to inappropriate linkage during the assembly.

Several reactions are suitable for reversible library self-assembly. These include the reaction of amines with carbonyls to form imines, where the product distribution can be fixed by addition of cyanoborohydride (reductive amination); likewise, reaction of hydroxylamines and hydrazines with carbonyls to form hydrazones and oximes. Hydrazones and oximes have the advantage of being more hydrolytically stable so no reductive step is required. Imine

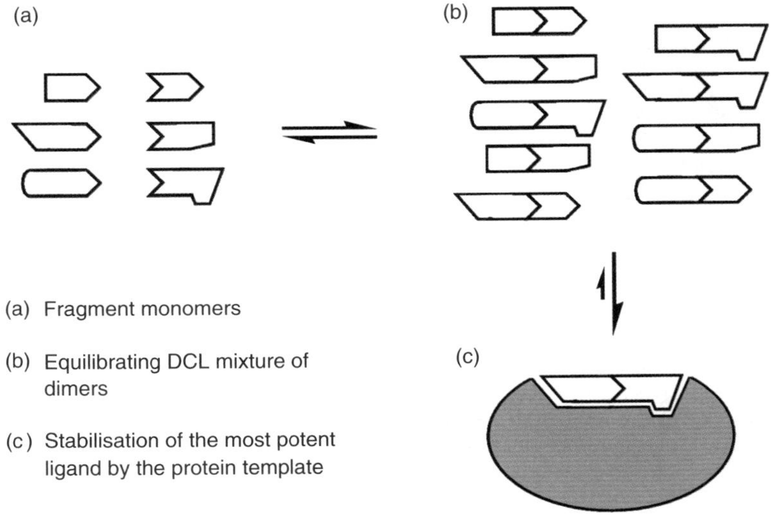

Figure 16.5 Schematic of the fragment assembly approach.

chemistry has been used for dynamic combinatorial assembly to discover inhibitors of neuraminidase[51] and cyclin-dependent kinase 2.[52]

Another popular reaction for DCL assembly is the reversible formation of disulfides.[53] Disulfide exchange with thiols takes place under mildly basic conditions (*e.g.* pH 7.5) in the presence of a catalytic amount of thiolate anion. The exchange can be easily quenched by shifting to slightly acidic conditions (*e.g.* pH 4) and then the mixture analyzed. Erlanson *et al.* at Sunesis Pharmaceuticals used disulfide exchange to discover nM inhibitors of caspase-3.[54]

A general limitation of dynamic covalent chemistry (DCC) is that relatively large amounts of protein are required to sufficiently perturb the product distribution in the DCL, and so identify the key binder. An alternative approach has been developed by Astex Therapeutics, whereby the DCC experiment is performed in the presence of a single protein crystal, an approach referred to as dynamic combinatorial X-ray crystallography (DCX).[52] There is very little enzyme in a crystal compared to the ligands, and therefore the distribution of products in solution is not significantly perturbed. However, this is not important, because the most potent ligand can be identified directly by inspection of the X-ray crystallographic density map.

16.4 Screening and Identification of Fragments

The starting point to any fragment-based approach is the identification of fragments that bind to the target protein and the characterization of their binding interactions. The molecular properties of fragments present specific experimental challenges. Their weak affinities are often outside the detection range of bioassays, in which compounds are typically tested at micromolar concentrations. Furthermore, high ligand concentrations can generate artifacts in the complex assays, leading to false positives. Rather than trying to adapt bioassays to detect fragment binding, biophysical methods are used to detect fragment binding.[55] Although such approaches can be material-intensive and time-demanding – and therefore not suited to high-throughput application, they have advantages over HTS assays for fragment screening:

(i) The hit rate from screening fragments is typically higher than that from screening large HTS libraries. This is due to the low molecular complexity of fragments, which increases the probability of the ligand finding a 'good fit' at the protein binding pocket.[56] A direct consequence of the high hit rate for fragments is that only a relatively small number of compounds (typically <1000) need to be tested.
(ii) Formation of the protein–ligand complex and/or the detection of an association event are monitored in a direct way, which enables control over the system under investigation and minimizes any potential interference from non-specific binding.
(iii) The structure of the protein–ligand complex or, at least, some details on the orientation of fragments and location of the binding site can be obtained. This information is critical for structure-guided chemical optimization of fragments and lead design.
(iv) The thermodynamics of the interaction can be determined.
(v) Quality control on the screening assay is provided and any problems arising from compound aggregation or degradation can be quickly detected.

A growing repertoire of techniques can be used to characterize fragment binding. The most important are X-ray crystallography, NMR spectroscopy, and mass spectrometry. However, other techniques, such as isothermal titration calorimetry (ITC), thermal shift, and surface plasmon resonance (SPR) are gaining in importance.

16.5 X-ray Crystallography for Fragment-based Lead Identification

X-ray crystallography is one of the most useful techniques to identify fragments bound to a protein target.[33] In one respect, this is surprising because protein crystallography was considered a relatively slow, resource-intensive technique that has traditionally been used for lead optimization.[3] The ability to reliably and reproducibly form crystals of protein–ligand complexes was often a limiting factor. However, recent years have seen major improvements in protein expression, crystallization, and methods for structure determination that the concept of high-throughput crystallography has emerged.[57–59] This transformation has allowed X-ray crystallography to impact more broadly in drug discovery and to be implemented as a biophysical screening tool for fragment-based lead identification.[57,60,61] Industrial examples of pioneering use of X-ray crystallography in fragment-based drug design include early work at Abbott Laboratories in the USA[62] and the Pyramid™ crystallographic approach developed at Astex Therapeutics.[33,40,57] More recently, related scaffold-based approaches using X-ray crystallography have been developed by Plexxicon in the USA.[24]

X-ray crystallography has been developed to screen molecular fragments and identify their precise binding interaction with the protein.[36,57,61,62] This approach has the advantage of defining the ligand-binding site and directly elucidating the binding mode of the ligand, which can guide chemistry towards the elaboration and modification of the bound fragments. Several examples are available showing that the binding mode of fragments is generally retained in the larger compound.[33,36] In crystallographic screenings, fragments are typically dissolved in dimethylsulfoxide (DMSO), diluted in relevant buffer, and then soaked, individually or as mixtures, into the target protein crystal.[62] The concentration of the molecular fragments is very high, usually over 20 mM and even up to 100 mM, to allow for the weakness of the interaction being investigated as well as for the intrinsically high concentration of protein in the crystal (~10 mM).[57] However, several drawbacks are associated with such extreme and unphysiological experimental conditions.[22,63] Protein crystals may not tolerate the high concentrations of compounds and DMSO used and can be damaged. Sometimes the crystal form of the unliganded protein is not suitable for the soaking of fragments, requiring laborious co-crystallization trials for each ligand or mixture.

A related crystallographic approach to probe binding sites and 'hot spots' on protein surfaces is solvent mapping, in which the location and affinity of small organic molecules (*e.g.* acetone, methanol, or DMSO) on a protein surface are identified.[64]

16.6 NMR Spectroscopy

NMR spectroscopy is particularly suitable to study protein–ligand and protein–protein interactions in solution,[65] and is being used extensively in the pharmaceutical industry for hit identification.[66–68] Different experimental formats have been used which are either ligand-based or protein-based, depending on whether the resonance signals of the small molecule or those of the macromolecule are monitored. Improvements in instrumentation and advances in automation facilitate screening of increasingly large compound libraries.[69–71]

16.6.1 Protein-based Methods: Structure–activity Relationship by NMR

NMR spectroscopic methods were amongst the first to be applied for fragment screening due to their flexibility and ability to detect weak interactions.

Fesik *et al.* pioneered the 'structure–activity relationship (SAR) by NMR' method in which perturbations of the two-dimensional ^{1}H–^{15}N heteronuclear single quantum correlation (HSQC) spectrum of a protein are used to detect ligand binding.[72] Although the ^{1}H–^{15}N HSQC experiment was not new for studying proteins, 'SAR by NMR' first demonstrated that it could be used in a medium-throughput format to detect hits and suggest ways to link them to form high-affinity compounds. This approach was subsequently implemented for screening libraries of compounds.[69] Although information on the ligand-binding site can be obtained, various drawbacks are associated with 'SAR by NMR' as well as with other protein-based NMR methods. Primarily, many pharmaceutical targets are not amenable to NMR spectroscopic observation because of their size and physicochemical properties. The protein is typically present at around 0.1 mM concentration and must be isotopically labeled with ^{15}N, and in some cases also with ^{13}C. Although pulse sequences have been developed to extend the MW range accessible to NMR spectroscopy,[73] only proteins of MW $<$30–40 kDa are routinely accessible. Moreover, experiments are time-consuming and the peaks from the two-dimensional protein NMR spectra must be assigned to obtain structural information. One of the first examples of 'SAR by NMR' was the discovery of potent non-peptidic inhibitors of stromelysin[43] (Scheme 16.2A). The 'SAR by NMR' method was used to validate low MW hits identified by virtual screening and biochemical assay against DNA gyrase. Approximately 50% of the classes of compounds identified were validated as DNA gyrase inhibitors and further structure-guided optimization led to the discovery of a novel, potent inhibitor.[25]

16.6.2 Ligand-based Methods

Ligand-based NMR techniques are generally faster, require less protein, and enable direct identification of binders, even in mixtures, using one-dimensional (1D) NMR methods. These approaches render the MW of the target receptor irrelevant, although small proteins (MW $<$ 10 kDa) can be problematic. Usually, an NMR spectrum of the ligand or the ligand mixture is recorded in the presence of the protein and compared to that recorded in the absence of the protein under identical experimental conditions. The ligands are present at total mM concentrations, in 10–1000 fold excess with respect to the protein. If a ligand binds to the protein, chemical exchange between bound and free ligand occurs. This exchange is generally fast on the NMR time scale and hence the properties of the bound ligand are transferred to the free ligand for detection. In all cases, the average NMR properties of the ligand (*i.e.* free + bound) are observed. These properties may include chemical shifts, relaxation times, and cross-relaxation rates such as the transferred Nuclear Overhauser Effect (tr-NOE) which can monitor the bound ligand conformation, and therefore provide some information on the binding mode of the molecule.[74,75]

The fast-exchange approximation holds true for moderate-to-weak affinity ligands, *i.e.* $K_d > 1\ \mu M$. Owing to the intrinsic weak affinities of small molecular fragments, ligand-based NMR methods are well suited for fragment-based drug design. When chemical exchange is too slow relative to the NMR time scale, as in the case of tighter binding ligands with slower off rates, the bound-state information is not transferred to the free state during the acquisition. In this case, only the free-state properties of the ligand would be observed, resulting in false negatives. To overcome these potential pitfalls, several approaches using competition experiments have been developed which extend the applicability of ligand-based NMR methods towards detecting high-affinity ligands.[76–79]

Most ligand-based methods employ ^{1}H-NMR spectroscopic experiments. In addition to ^{1}H nuclei other NMR-active nuclei can be employed to monitor protein–ligand interactions. As an example, the application of ^{19}F-NMR spectroscopy to ligand screening has been described by Dalvit *et al.*[80,81] Fluorine-NMR spectroscopy is attractive since ^{19}F has a natural abundance of 100% and has a sensitivity to NMR spectroscopic detection that is comparable (83%) to that of ^{1}H.

Generally, fragments in NMR screening libraries exhibit slow relaxation, *i.e.* long longitudinal (T_1) and transverse (T_2) relaxation times; small and positive NOEs; and fast diffusion, *i.e.* large translational diffusion coefficients (D_t). In contrast, bound ligands share the NMR properties of the protein which are distinctly different: fast relaxation, *i.e.* short T_1 and T_2 relaxation times; large and negative NOE; and slow diffusion, *i.e.* small D_t coefficients. These differences give rise to changes in the average NMR parameters of the ligands which can be monitored to assess binding to the protein target. Ligand-based NMR experiments have been described which take advantage of all these properties. Relaxation- and diffusion-edited NMR methods exploit the differential rotational and translational mobility of the ligands as they associate with their receptors.[82] Both relaxation-edited and tr-NOE NMR methods have been used by Moore *et al.* at Vertex Pharmaceuticals in their SHAPES strategy.[26,83]

Radical compounds have been used as 'spin labels' for paramagnetic relaxation enhancement by Jahnke *et al.* to increase the sensitivity of these approaches for hit identification.[84,85] A common approach is to covalently modify the target protein with a spin label such as 2,2,6,6-tetramethyl-piperidine-1-oxyl (TEMPO).[85,86] This approach has been subsequently implemented in 'second-site screening', in which a spin-labeled binding ligand aids detection of fragments which bind to the target protein at an adjacent binding site.[86,87] This strategy has been used to discover and optimize a protein kinase inhibitor targeting an allosteric site next to the ATP-binding site.[88]

Other methods, such as saturation transfer difference (STD)[65,89] and water–ligand observed via gradient spectroscopy (WaterLOGSY)[67,90] exploit a magnetization transfer process from the protein to detect binders. Observation of negative NOEs for small organic molecules in the presence of the protein is used as indication of binding.[66] These techniques rely on excitation of the protein–ligand complex either directly (STD) or indirectly *via* water molecules.

WaterLOGSY is a particularly rapid and powerful 1D ^{1}H-NMR spectroscopic method to discriminate binders from non-binders in solution.[67,90] Like in STD, WaterLOGSY relies on excitation of the protein–ligand complex through a selective pulse scheme. However, WaterLOGSY achieves this indirectly by irradiating the bulk water rather than selectively saturating a subset of receptor protons as in STD.[68,89,90] Therefore, the transfer of magnetization for binding ligands is bulk water → protein–ligand complex → free ligand for detection.[68] Magnetization transfer from water occurs primarily via direct ^1H–^1H cross-relaxation between the bound ligand and long-lived waters 'bound' at the active site. Owing to the very different tumbling times of free ligands and protein–ligand complex, the intermolecular NOEs from water have opposite sign (positive for the former, negative for the latter), and tend to be stronger for the interacting ligand.[91] As a consequence, binders and non-binders display WaterLOGSY signal intensities of opposite sign, which provides a direct means to discriminate between them (Figure 16.6).[92]

Like all ligand-based screening experiments, WaterLOGSY is biased towards the detection of weak-affinity ligands, and so is particularly suited for fragment-based approaches. Ligands binding more tightly are likely to show slower exchange and will correspondingly spend longer times bound to the protein.

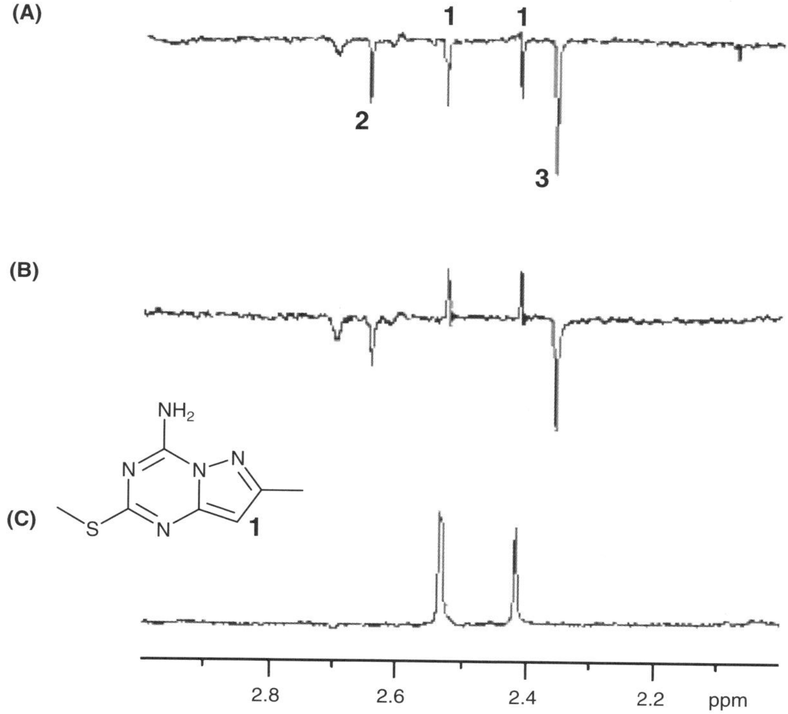

Figure 16.6 WaterLOGSY ^{1}H-NMR spectra are shown. (A) Cocktail of three compounds each at 0.5 mM concentration. (B) Same cocktail in the presence of 10 μM ketopantoate reductase.[92] Compound **1** binds to the protein whereas **2** and **3** do not. (C) **1** in the presence of the protein is confirmed as a hit compound.

If the residence times become too long, the transferred spin inversion will vanish due to longitudinal relaxation (short T_1 of the protein–ligand complex) before the ligand can dissociate back into free solution. Under these conditions, the negative LOGSY contribution to signal intensity from the free ligand state can overwhelm the positive one of bound state, resulting in a false negative.[68] The estimated lower limit on K_d for WaterLOGSY is $\approx 0.1\ \mu M$.[91]

The development of novel NMR spectroscopic methods for screening protein–ligand interactions is a large and rapidly growing area of research.[93] Several NMR methods have been recently described, including target immobilized NMR screening (TINS) and rapid analysis and multiplexing of experimentally discriminated uniquely labeled proteins using NMR (RAMPED-UP NMR). TINS[94] is a relaxation-edited NMR spectroscopic method for screening ligands against target receptors which have been immobilized onto a solid support. Advantages of TINS include the ability to re-use the receptor sample several times, the considerably reduced amount of target required, and the possibility of screening targets that are insoluble, *e.g.* membrane proteins. In RAMPED-UP NMR[95] different proteins are uniquely labeled with one amino acid type, which allows screening of multiple proteins simultaneously. This approach is particularly powerful at addressing the question of ligand selectivity.

16.7 Mass Spectrometry

Mass spectrometry is sensitive, fast, and can be used to analyze mixtures of small molecules. Consequently, small-molecule mass spectrometry has played a key role in many phases of the drug-discovery process.[96,97] In contrast, the use of protein mass spectrometry within the pharmaceutical industry has been much more limited.

Different mass spectrometric methods have been reported to monitor ligand binding, depending on whether the ligand forms a covalent bond to the target protein or not.[96,98] The increase in mass due to formation of the covalent protein–ligand complex is directly detected by mass spectrometry. Mass spectrometric detection is usually more challenging for non-covalent systems and different approaches have been described in which either the protein, the ligand, or the non-covalent protein–ligand complex is monitored by mass spectrometry.

16.7.1 Covalent Mass Spectrometric Methods

The approach of site-directed ligand discovery or tethering, developed at Sunesis Pharmaceuticals, employs a covalent bond to stabilize the low-affinity interaction between a small molecular fragment and the target protein.[99] The strategy consists in exposing the protein to a library of disulfide-containing fragments to probe a binding site of interest containing a (required) cysteine residue. Binders that form a disulfide bond (tether) at the active site are then detected by mass spectrometry. Small-molecule inhibitors of thymidylate synthase[100] and modulators of protein–protein interactions, as in the case of the complex between cytokine interleukin-2 (IL-2) and the IL-2R receptor,[101–102] have been successfully discovered using *in situ* fragment assembly by tethering. An elaboration of the tethering approach employs a known specific-site binder referred to as an 'extender' to probe for fragments at adjacent binding sites. This was used to discover potent small molecules inhibitors of an important class of cysteine proteases, the caspases.[54]

In a different approach to covalent trapping of binders on an enzyme, reductive amination was used to trap imines formed at the active-site pyruvoyl group of aspartate decarboxylase.[103] A library of amines was screened against the enzyme and binders were detected using matrix-assisted laser desorption/ionization–time of flight (MALDI-TOF) mass spectrometry after treatment with sodium cyanoborohydride.

16.7.2 Non-covalent Mass Spectrometric Methods

Nano electrospray ionization mass spectrometry (nano ESI-MS) is presently the state-of-the-art technique to look at non-covalent complexes by mass spectrometry.[104–107] The sample is introduced through a capillary of internal tip diameter of 2–5 μm, compared to a normal ESI capillary of 100 μm diameter. The thinner nanoflow capillary produces considerably smaller droplets, consequently the desolvation process is more efficient and hence gentler conditions can be used.[104] Nano ESI-MS is particularly effective at maintaining intact non-covalent complexes in the gas phase.[108,109] It has been used to study protein–ligand complexes,[110] multi-protein complexes,[111] and even to observe intact ribosomes[112] and viruses[113] in the gas phase.

The use of nano ESI-MS to monitor protein complexes and to determine their stoichiometry has become well established.[107,114] However, the ability to monitor binding of low-affinity small molecules, *e.g.* fragments to proteins or even their effect within large macromolecular assemblies, is more challenging. This challenge arises since it is difficult to resolve binding of small molecules within a background of water molecules and buffer salts that adhere to the protein target after its ionization from solution under physiological conditions. Using nano ESI-MS, McCammon *et al.* observed cooperative binding of tryptophan to a protein assembly with 12 subunits.[115] Nano ESI-MS has also been used to screen small ligands against tetrameric transthyretin.[116] The different stabilities of the complexes in the gas phase were used to predict a rank order of affinities.[116] De Vriendt *et al.* reported a good agreement between dissociation constants of protein–ligand complexes measured by mass spectrometry and those determined in solution.[117] Therefore, these approaches are not only useful to generate qualitative information, *i.e.* identify hit compounds, but it is also possible to obtain quantitative estimates of affinity.

Of particular relevance to drug-discovery applications are examples of using nano ESI-MS for automated protein–ligand screening using a microchip device.[118,119] This method has proved to be successful for individual compounds or very small cocktails. However, using larger cocktails of small molecules of similar mass (*e.g.* fragments) has not yet been demonstrated, so this approach is not generally considered feasible for medium-to-high-throughput fragment screening.

16.7.3 Looking at the Protein or the Ligand

Various mass spectrometric approaches for the quantification of protein–ligand interactions have been reported and their advantages and limitations discussed.[120] These strategies include methods to look at the protein targets by hydrogen/deuterium exchange and mass spectrometry such as stability of unpurified proteins from rates of H/D exchange (SUPREX)[121] and protein–ligand interactions in solution by mass spectrometry, titration, and H/D exchange (PLIMSTEX).[122] A different approach involves ligand detection by mass spectrometry using a competitive binding assay.[123]

SUPREX measures protein stability against chemical denaturation using MALDI mass spectrometry and H/D exchange. Changes in stability are detected in the presence of the ligand, and the extent of this stabilization is then correlated to the dissociation constant of the protein–ligand complex[121] (Figure 16.7A). In contrast, PLIMSTEX measures changes in H/D exchange of the protein at varying concentrations of ligand by ESI-MS. The increased protection of backbone amide protons upon formation of the protein–ligand complex is plotted against ligand concentration to yield the dissociation constant[122] (Figure 16.7B).

The competitive displacement of a known high-affinity ligand (marker) can be followed quantitatively by ESI-MS.[123] This offers an attractive alternative to

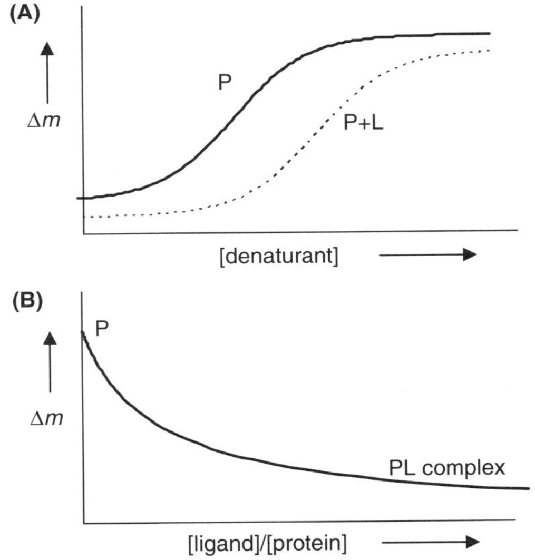

Figure 16.7 Schematic illustration of SUPREX (A) and PLIMSTEX (B) mass spectrometric methods.

radioactivity- and fluorescence-based assays for competition methods because 'native' markers can be used, *i.e.* there is no need to label the marker. The ease of automation and the possibility to use compound mixtures may allow this approach to be implemented for high-throughput ligand screening, with potential applications to fragment identification.

16.8 Thermal Shift

Ligand-induced conformational stabilization of proteins is a well-understood phenomenon in which the ligand provides enhanced stability to proteins on binding.[124] By measuring the compound-dependent difference in T_m at a series of compound concentrations, the dissociation constant (K_d) of that compound can be calculated. Traditionally, thermal-shift assays have been conducted using differential scanning calorimeters (DSCs) that monitor the change in heat capacity as proteins undergo temperature-induced melting transitions.[125] In addition, biophysical methods that employ temperature-dependent changes in absorbance, fluorescence, and circular dichroism have also been used to perform thermal shift assays.[126]

Ligand-induced conformational stabilization of proteins has been exploited in high-throughput assays.[124,127] These assays monitor the binding of a fluorescent dye to the denatured protein. Hits are identified as compounds that produce an increase in T_m at least three-fold greater than the standard deviation ($\sim 3 \times 0.2\,^{\circ}C$). The potential of this technique in drug discovery has been recognized,[125] and it has recently been applied to discover small-molecule antagonists of the HDM2-p53 protein–protein interaction.[128–130] Thermal-shift assays were also used to characterize slow-binding, high-affinity p38α MAP kinase inhibitors, in combination with DSC, CD, and ITC, by Kroe *et al.* at Boehringer Ingelheim Pharmaceuticals.[131] The shifts in T_m observed in the thermal denaturation method for a series of inhibitors were found to correlate well with binding affinities obtained with the other methods.[131]

One of the potential advantages of thermal-shift assays for fragment-based drug design is the ability to measure affinities that are out of reach of other techniques. Applications include the characterization and ranking of slow-binding, high-affinity leads developed from small fragments, which can be difficult with standard equilibrium methods.

16.9 Isothermal Titration Calorimetry

ITC measures the heat associated with a binding event directly and allows a full thermodynamic characterization of an interaction in a single experiment.[132] In an ITC experiment a solution of the ligand is titrated into a solution of the protein at constant temperature.[133] Changes in temperature due to the heat absorbed or released upon binding are detected as a function of the number of injections. When correctly designed, an ITC experiment provides values of both the dissociation constant (K_d) and the enthalpy of binding (ΔH), from which the entropy of binding (ΔS) can be calculated. Dissociation constants (K_d) in the range of 10^{-8} M < K_d < 10^{-3} M are directly accessible by ITC, although indirect methods extend this range for both weak ($K_d > 10^{-3}$ M) and tight ($K_d < 10^{-9}$ M) affinities.[134] Turnbull and Daranas have reported a theoretical treatment and practical considerations about the feasibility of using ITC to study very weak affinity systems.[135] This study is likely to extend the applicability of ITC to fragment-based drug discovery.

Changes in binding interactions are sometimes not detected as changes in affinity because of enthalpy–entropy compensation.[125,136] Therefore, knowledge of binding enthalpy and entropy, rather than just affinity, can help understand the effect of particular structural modifications for drug design, not only during lead optimization, but also in the early stages of fragment elaboration. The elucidation of complex binding mechanisms and the ability to perform calorimetric assays to monitor enzyme kinetics are other examples of the utility of ITC in the drug-discovery process.[137]

Recent papers discuss the application of ITC for factorizing the thermodynamics of inhibitor binding to target proteins, *e.g.* the serine proteases trypsin and thrombin.[138] ITC has also been applied to characterize the binding thermodynamics of hydroxamic and carboxylic acid inhibitors with stromelysin,[139] and several classes of inhibitors of HIV-1 protease[140] and cyclin-dependent kinase 2.[126]

Potential high-throughput applications of biocalorimetry in drug discovery have been discussed.[125,138] Amongst the limitations of ITC are the requirement for relatively large amounts (10^{-8}–10^{-7} mol) of protein and the long times (2–3 h) necessary to run experiments, which currently limit the use of ITC as a screening tool in drug discovery. However, this is likely to change as new automated instrumentation is capable of running up to 288 samples unattended. Furthermore, recent developments of nanocalorimetric microarrays[141] and experimental approaches (*e.g.* fast single-injection experiments[142]) may allow biocalorimetry to be used as an information-rich screening tool for drug discovery.

16.10 Surface Plasmon Resonance

SPR is an optical technique which allows real-time monitoring of binding of an analyte present in solution with a receptor immobilized on a surface and characterizing the kinetics of the interaction.[143,144] Advances in instrumentation and experimental design allow the analysis of protein interactions over a wide range of affinities (millimolar to picomolar) and kinetic rate constants.[145] The difficulties with measuring very strong interactions (K_d < 10 nM) are related to slow dissociation rate constants, and to diffusion limitations, *i.e.* the rate at which the analyte binds the receptor can exceed the rate at which it is delivered to the surface. In contrast, non-specific binding tends to be a problem when studying weak affinities.[143,146]

SPR-based biomolecular interaction analysis has been used to study small-molecule inhibitors of p38α MAP kinase[147] and millimolar affinity fragments binding to the phosphotyrosine pocket of Src homology 2 (SH2) domains.[148] The weak interactions (10^{-2}–10^{-5} M range) of small purine fragments (130–450 Da) with human liver glycogen phosphorylase was investigated using SPR.[149] This study demonstrated that SPR could be used to screen up to 200 compounds per day and characterize their interactions with a protein partner 500-fold greater in mass.[149] An SPR-based fragment screening of compounds immobilized in a chemical microarray platform to identify inhibitors of Factor VIIa has been reported.[150]

The greatest challenge of using SPR to characterize the interaction of fragments with the protein targets is the intrinsic small signal associated with the association of low MW compounds. However, recent improvements in the sensitivity of Biacore instruments have enabled detection of binding of analytes with MW as low as 100 Da.[151] Routine analysis of small-molecule interactions is now possible,[152] with potential applications in fragment-based drug discovery for fragment screening and identification and SARs.

16.11 Concluding Remarks

The first compounds developed using fragment-based approaches are now moving into clinical trials. LY517717 (Eli Lilly) targets Factor Xa and has currently completed successfully Phase II clinical trials for the treatment of venous thromboembolism.[153,154] AT7519 (Astex Therapeutics), a cell cycle inhibitor against cyclin-dependent kinases, is in Phase I clinical trials for the treatment of cancer.[155] Other fragment-derived compounds have entered pre-clinical toxicological studies and should soon enter clinical trials.[156] However, fragment-based drug design is still in its infancy and its success will be better assessed as compounds generated by these strategies come near commercialization. Nevertheless, the approach is very attractive and therefore the development of novel methods to detect weak binding interactions and their optimization and applications for fragment screening are the subject of intense research.

Acknowledgments

We thank Rob van Montfort and Steve Howard for help with Figure 16.4 and Marc O'Reilly for early discussions.

References

1. C. R. Beddell, P. J. Goodford, F. E. Norrington, S. Wilkinson and R. Wootton, *Br. J. Pharmacol.*, 1976, **57**, 201–209.
2. P. J. Goodford, J. Stlouis and R. Wootton, *Br. J. Pharmacol.*, 1980, **68**, 741–748.
3. T. L. Blundell, *Nature*, 1996, **384**, 23–26.
4. S. F. Campbell, *Clin. Sci.*, 2000, **99**, 255–260.

5. M. Miller, J. Schneider, B. K. Sathyanarayana, M. V. Toth, G. R. Marshall, L. Clawson, L. Selk, S. B. H. Kent and A. Wlodawer, *Science*, 1989, **246**, 1149–1152.
6. R. Lapatto, T. L. Blundell, A. Hemmings, J. Overington, A. Wilderspin, S. Wood, J. R. Merson, P. J. Whittle, D. E. Danley, K. F. Geoghegan, S. J. Hawrylik, S. E. Lee, K. G. Scheld and P. M. Hobart, *Nature*, 1989, **342**, 299–302.
7. J. N. Varghese, *Drug Dev. Res.*, 1999, **46**, 176–196.
8. L. W. Hardy and A. Malikayil, *Curr. Drug Discovery*, 2003, 15–20.
9. R. W. Spencer, *Biotechnol. Bioeng.*, 1998, **61**, 61–67.
10. P. Seneci and S. Miertus, *Mol. Diversity*, 2000, **5**, 75–89.
11. N. Bailey, A. W. J. Cooper, M. J. Deal, A. W. Dean, A. L. Gore, M. C. Hawes, D. B. Judd, A. T. Merritt, R. Storer, S. Travers and S. P. Watson, *Chimia*, 1997, **51**, 832–837.
12. R. E. Dolle, *J. Comb. Chem.*, 2004, **6**, 623–679.
13. A. L. Hopkins and C. R. Groom, *Nat. Rev. Drug Discovery*, 2002, **1**, 727–730.
14. R. D. Taylor, P. J. Jewsbury and J. W. Essex, *J. Comput. Aided Mol. Des.*, 2002, **16**, 151–166.
15. J. W. M. Nissink, C. Murray, M. Hartshorn, M. L. Verdonk, J. C. Cole and R. Taylor, *Proteins: Struct. Funct. Genet.*, 2002, **49**, 457–471.
16. R. A. Friesner, J. L. Banks, R. B. Murphy, T. A. Halgren, J. J. Klicic, D. T. Mainz, M. P. Repasky, E. H. Knoll, M. Shelley, J. K. Perry, D. E. Shaw, P. Francis and P. S. Shenkin, *J. Med. Chem.*, 2004, **47**, 1739–1749.
17. D. B. Kitchen, H. Decornez, J. R. Furr and J. Bajorath, *Nat. Rev. Drug Discovery*, 2004, **3**, 935–949.
18. D. C. Rees, M. S. Congreve, C. W. Murray and R. Carr, *Nat. Rev. Drug Discovery*, 2004, **3**, 660–672.
19. D. A. Erlanson, R. S. McDowell and T. O'Brien, *J. Med. Chem.*, 2004, **47**, 3463–3482.
20. R. Carr, M. S. Congreve, C. W. Murray and D. C. Rees, *Drug Discovery Today*, 2005, **10**, 987–992.
21. E. R. Zartler and M. J. Shapiro, *Curr. Opin. Chem. Biol.*, 2005, **9**, 366–370.
22. A. M. Davis, S. J. Teague and G. J. Kleywegt, *Angew. Chem., Int. Ed.*, 2003, **42**, 2718–2736.
23. V. Mountain, *Chem. Biol.*, 2003, **10**, 95–98.
24. G. L. Card, L. Blasdel, B. P. England, C. Zhang, Y. Suzuki, S. Gillette, D. Fong, P. N. Ibrahim, D. R. Artis, G. Bollag, M. V. Milburn, S. H. Kim, J. Schlessinger and K. Y. Zhang, *Nat. Biotechnol.*, 2005, **23**, 201–207.
25. H. J. Boehm, M. Boehringer, D. Bur, H. Gmuender, W. Huber, W. Klaus, D. Kostrewa, H. Kuehne, T. Luebbers, N. Meunier-Keller and F. Mueller, *J. Med. Chem.*, 2000, **43**, 2664–2674.

26. J. Fejzo, C. A. Lepre, J. W. Peng, G. W. Bemis, Ajay, M. A. Murcko and J. M. Moore, *Chem. Biol.*, 1999, **6**, 755–769.
27. A. Schuffenhauer, S. Ruedisser, A. L. Marzinzik, W. Jahnke, M. J. Blommers, P. Selzer and E. Jacoby, *Curr. Top. Med. Chem.*, 2005, **5**, 751–762.
28. C. A. Lipinski, F. Lombardo, B. W. Dominy and P. J. Feeney, *Adv. Drug Delivery Rev.*, 2001, **46**, 3–26.
29. M. S. Congreve, R. Carr, C. W. Murray and H. Jhoti, *Drug Discovery Today*, 2003, **8**, 876–877.
30. S. L. McGovern, B. T. Helfand, B. Feng and B. K. Shoichet, *J. Med. Chem.*, 2003, **46**, 4265–4272.
31. S. L. McGovern and B. K. Shoichet, *J. Med. Chem.*, 2003, **46**, 1478–1483.
32. C. W. Murray and M. L. Verdonk, *J. Comput. Aided Mol. Des.*, 2002, **16**, 741–753.
33. M. J. Hartshorn, C. W. Murray, A. Cleasby, M. Frederickson, I. J. Tickle and H. Jhoti, *J. Med. Chem.*, 2005, **48**, 403–413.
34. A. L. Hopkins, C. R. Groom and A. Alex, *Drug Discovery Today*, 2004, **9**, 430–431.
35. I. D. Kuntz, K. Chen, K. A. Sharp and P. A. Kollman, *Proc. Natl. Acad. Sci. U. S. A.*, 1999, **96**, 9997–10002.
36. T. J. Stout, C. R. Sage and R. M. Stroud, *Structure*, 1998, **6**, 839–848.
37. P. J. Hajduk, J. Dinges, J. M. Schkeryantz, D. Janowick, M. Kaminski, M. Tufano, D. J. Augeri, A. Petros, V. Nienaber, P. Zhong, R. Hammond, M. Coen, B. Beutel, L. Katz and S. W. Fesik, *J. Med. Chem.*, 1999, **42**, 3852–3859.
38. M. D. Wendt, T. W. Rockway, A. Geyer, W. McClellan, M. Weitzberg, X. M. Zhao, R. Mantei, V. L. Nienaber, K. Stewart, V. Klinghofer and V. L. Giranda, *J. Med. Chem.*, 2004, **47**, 303–324.
39. A. L. Gill, M. Frederickson, A. Cleasby, S. J. Woodhead, M. G. Carr, A. J. Woodhead, M. T. Walker, M. S. Congreve, L. A. Devine, D. Tisi, M. O'Reilly, L. C. A. Seavers, D. J. Davis, J. Curry, R. Anthony, A. Padova, C. W. Murray, R. A. E. Carr and H. Jhoti, *J. Med. Chem.*, 2005, **48**, 414–426.
40. A. L. Gill, A. Cleasby and H. Jhoti, *ChemBioChem*, 2005, **6**, 506–512.
41. W. P. Jencks, *Proc. Natl. Acad. Sci. U. S. A.*, 1981, **78**, 4046–4050.
42. M. I. Page and P. Jencks, *Proc. Natl. Acad. Sci. U. S. A.*, 1971, **68**, 1678–1683.
43. P. J. Hajduk, G. Sheppard, D. G. Nettesheim, E. T. Olejniczak, S. B. Shuker, R. P. Meadows, D. H. Steinman, G. M. Carrera, P. A. Marcotte, J. Severin, K. Walter, H. Smith, E. Gubbins, R. Simmer, T. F. Holzman, D. W. Morgan, S. K. Davidsen, J. B. Summers and S. W. Fesik, *J. Am. Chem. Soc.*, 1997, **119**, 5818–5827.
44. N. I. Howard, C. Abell, W. Blakemore, G. Chessari, M. S. Congreve, S. Howard, H. Jhoti, C. W. Murray, L. C. A. Seavers and R. L. M. van Montfort, *J. Med. Chem.*, 2006, **49**, 1346–1355.

45. I. Huc and J. M. Lehn, *Proc. Natl. Acad. Sci. U. S. A.*, 1997, **94**, 2106–2110.
46. J. M. Lehn, *Chem. Eur. J.*, 1999, **5**, 2455–2463.
47. J. M. Lehn and A. V. Eliseev, *Science*, 2001, **291**, 2331–2332.
48. O. Ramstrom and J. M. Lehn, *Nat. Rev. Drug Discovery*, 2002, **1**, 26–36.
49. S. Otto, R. L. E. Furlan and J. K. M. Sanders, *Drug Discovery Today*, 2002, **7**, 117–125.
50. W. G. Lewis, L. G. Green, F. Grynszpan, Z. Radic, P. R. Carlier, P. Taylor, M. G. Finn and K. B. Sharpless, *Angew. Chem., Int. Ed.*, 2002, **41**, 1053–1057.
51. M. Hochgurtel, H. Kroth, D. Piecha, M. W. Hofmann, C. Nicolau, S. Krause, O. Schaaf, G. Sonnenmoser and A. V. Eliseev, *Proc. Natl. Acad. Sci. U. S. A.*, 2002, **99**, 3382–3387.
52. M. S. Congreve, D. J. Davis, L. Devine, C. Granata, M. O'Reilly, P. G. Wyatt and H. Jhoti, *Angew. Chem., Int. Ed.*, 2003, **42**, 4479–4482.
53. H. F. Gilbert, *Methods Enzymol.*, 1995, 8–28.
54. D. A. Erlanson, J. W. Lam, C. Wiesmann, T. N. Luong, R. L. Simmons, W. L. Delano, I. C. Choong, M. T. Burdett, W. M. Flanagan, D. Lee, E. M. Gordon and T. O'Brien, *Nat. Biotechnol.*, 2003, **21**, 308–314.
55. R. Carr and H. Jhoti, *Drug Discovery Today*, 2002, **7**, 522–527.
56. M. M. Hann, A. R. Leach and G. Harper, *J. Chem. Inf. Comput. Sci.*, 2001, **41**, 856–864.
57. T. L. Blundell, H. Jhoti and C. Abell, *Nat. Rev. Drug Discovery*, 2002, **1**, 45–54.
58. P. Kuhn, K. Wilson, M. G. Patch and R. C. Stevens, *Curr. Opin. Chem. Biol.*, 2002, **6**, 704–710.
59. A. Sharff and H. Jhoti, *Curr. Opin. Chem. Biol.*, 2003, **7**, 340–345.
60. M. S. Congreve, C. W. Murray and T. L. Blundell, *Drug Discovery Today*, 2005, **10**, 895–907.
61. T. L. Blundell, C. Abell, A. Cleasby, M. J. Hartshorn, I. J. Tickle, E. Parasini and H. Jhoti, in *Drug Design: Cutting Edge Approaches*, ed. D. R. Flowers, Royal Society of Chemistry, Cambridge, 2002, pp. 53–59.
62. V. L. Nienaber, P. L. Richardson, V. Klighofer, J. J. Bouska, V. L. Giranda and J. Greer, *Nat. Biotechnol.*, 2000, **18**, 1105–1108.
63. S. W. Muchmore and P. J. Hajduk, *Curr. Opin. Drug Discovery Devel.*, 2003, **6**, 544–549.
64. K. N. Allen, C. R. Bellamacina, X. C. Ding, C. J. Jeffery, C. Mattos, G. A. Petsko and D. Ringe, *J. Phys. Chem.*, 1996, **100**, 2605–2611.
65. B. Meyer and T. Peters, *Angew. Chem., Int. Ed.*, 2003, **42**, 864–890.
66. M. Pellecchia, D. S. Sem and K. Wuthrich, *Nat. Rev. Drug Discovery*, 2002, **1**, 211–219.
67. B. J. Stockman and C. Dalvit, *Prog. Nucl. Magn. Reson. Spectrosc.*, 2002, **41**, 187–231.
68. C. A. Lepre, J. M. Moore and J. W. Peng, *Chem. Rev.*, 2004, **104**, 3641–3676.

69. P. J. Hajduk, T. Gerfin, J. M. Boehlen, M. Haberli, D. Marek and S. W. Fesik, *J. Med. Chem.*, 1999, **42**, 2315–2317.
70. A. Ross and H. Senn, *Drug Discovery Today*, 2001, **6**, 583–593.
71. A. Ross and H. Senn, *Curr. Top. Med. Chem.*, 2003, **3**, 55–67.
72. S. B. Shuker, P. J. Hajduk, R. P. Meadows and S. W. Fesik, *Science*, 1996, **274**, 1531–1534.
73. K. Pervushin, R. Riek, G. Wider and K. Wuthrich, *Proc. Natl. Acad. Sci. U. S. A.*, 1997, **94**, 12366–12371.
74. D. Henrichsen, B. Ernst, J. L. Magnani, W. T. Wang, B. Meyer and T. Peters, *Angew. Chem., Int. Ed.*, 1999, **38**, 98–102.
75. M. Mayer and B. Meyer, *J. Med. Chem.*, 2000, **43**, 2093–2099.
76. C. Dalvit, M. Flocco, S. Knapp, M. Mostardini, R. Perego, B. J. Stockman, M. Veronesi and M. Varasi, *J. Am. Chem. Soc.*, 2002, **124**, 7702–7709.
77. C. Dalvit, M. Fasolini, M. Flocco, S. Knapp, P. Pevarello and M. Veronesi, *J. Med. Chem.*, 2002, **45**, 2610–2614.
78. W. Jahnke, P. Floersheim, C. Ostermeier, X. L. Zhang, R. Hemmig, K. Hurth and D. P. Uzunov, *Angew. Chem., Int. Ed.*, 2002, **41**, 3420–3423.
79. A. H. Siriwardena, F. Tian, S. Noble and J. H. Prestegard, *Angew. Chem., Int. Ed.*, 2002, **41**, 3454–3457.
80. C. Dalvit, M. Flocco, M. Veronesi and B. J. Stockman, *Comb. Chem. High Throughput Screening*, 2002, **5**, 605–611.
81. C. Dalvit, P. E. Fagerness, D. T. A. Hadden, R. W. Sarver and B. J. Stockman, *J. Am. Chem. Soc.*, 2003, **125**, 7696–7703.
82. P. J. Hajduk, E. T. Olejniczak and S. W. Fesik, *J. Am. Chem. Soc.*, 1997, **119**, 12257–12261.
83. C. A. Lepre, J. Peng, J. Fejzo, N. Abdul-Manan, J. Pocas, M. Jacobs, X. L. Xie and J. M. Moore, *Comb. Chem. High Throughput Screening*, 2002, **5**, 583–590.
84. W. Jahnke, *ChemBioChem*, 2002, **3**, 167–173.
85. W. Jahnke, S. Rudisser and M. Zurini, *J. Am. Chem. Soc.*, 2001, **123**, 3149–3150.
86. W. Jahnke, L. B. Perez, C. G. Paris, A. Strauss, G. Fendrich and C. M. Nalin, *J. Am. Chem. Soc.*, 2000, **122**, 7394–7395.
87. W. Jahnke, A. Florsheimer, M. J. Blommers, C. G. Paris, J. Heim, C. M. Nalin and L. B. Perez, *Curr. Top. Med. Chem.*, 2003, **3**, 69–80.
88. W. Jahnke, M. J. Blommers, C. Fernandez, C. Zwingelstein and R. Amstutz, *ChemBioChem*, 2005, **6**, 1607–1610.
89. M. Mayer and B. Meyer, *Angew. Chem. Int. Ed.*, 1999, **38**, 1784–1788.
90. C. Dalvit, P. Pevarello, M. Tato, M. Veronesi, A. Vulpetti and M. Sundstrom, *J. Biomol. NMR*, 2000, **18**, 65–68.
91. C. Dalvit, G. Fogliatto, A. Stewart, M. Veronesi and B. J. Stockman, *J. Biomol. NMR*, 2001, **21**, 349–359.
92. A. Ciulli and C. Abell, *Biochem. Soc. Trans.*, 2005, **33**, 767–771.
93. M. Pellecchia, *Chem. Biol.*, 2005, **12**, 961–971.
94. S. Vanwetswinkel, R. J. Heetebrij, J. van Duynhoven, J. G. Hollander, D. V. Filippov, P. J. Hajduk and G. Siegal, *Chem. Biol.*, 2005, **12**, 207–216.

95. E. R. Zartler, J. Hanson, B. E. Jones, A. D. Kline, G. Martin, H. P. Mo, M. J. Shapiro, R. Wang, H. P. Wu and J. L. Yan, *J. Am. Chem. Soc.*, 2003, **125**, 10941–10946.
96. G. L. Glish and R. W. Vachet, *Nat. Rev. Drug Discovery*, 2003, **2**, 140–150.
97. D. B. Kassel, *Chem. Rev.*, 2001, **101**, 255–267.
98. S. M. Schermann, D. A. Simmons and L. Konermann, *Expert Rev. Proteomics*, 2005, **2**, 475–485.
99. D. A. Erlanson, J. A. Wells and A. C. Braisted, *Annu. Rev. Biophys. Biomol. Struct.*, 2004, **33**, 199–223.
100. D. A. Erlanson, A. C. Braisted, D. R. Raphael, M. Randal, R. M. Stroud, E. M. Gordon and J. A. Wells, *Proc. Natl. Acad. Sci. U. S. A.*, 2000, **97**, 9367–9372.
101. A. C. Braisted, J. D. Oslob, W. L. Delano, J. Hyde, R. S. McDowell, N. Waal, C. Yu, M. R. Arkin and B. C. Raimundo, *J. Am. Chem. Soc.*, 2003, **125**, 3714–3715.
102. M. R. Arkin and J. A. Wells, *Nat. Rev. Drug Discovery*, 2004, **3**, 301–317.
103. M. E. Webb, E. Stephens, A. G. Smith and C. Abell, *Chem. Comm.*, 2003, 2416–2417.
104. M. Karas, U. Bahr and T. Dulcks, *Fresenius J. Anal. Chem.*, 2000, **366**, 669–676.
105. A. Schmidt, U. Bahr and M. Karas, *Anal. Chem.*, 2001, **73**, 6040–6046.
106. A. J. R. Heck and R. H. H. van den Heuvel, *Mass Spectrom. Rev.*, 2004, **23**, 368–389.
107. F. Sobott, M. G. McCammon, H. Hernandez and C. V. Robinson, *Philos. Trans., Ser. A.*, 2005, **363**, 379–389.
108. F. Sobott and C. V. Robinson, *Curr. Opin. Struct. Biol.*, 2002, **12**, 729–734.
109. J. A. Loo, *Int. J. Mass Spectrom.*, 2000, **200**, 175–186.
110. E. W. Chung, D. A. Henriques, D. Renzoni, C. J. Morton, T. D. Mulhern, M. C. Pitkeathly, J. E. Ladbury and C. V. Robinson, *Protein Sci.*, 1999, **8**, 1962–1970.
111. F. Sobott, J. L. P. Benesch, E. Vierling and C. V. Robinson, *J. Biol. Chem.*, 2002, **277**, 38921–38929.
112. A. A. Rostom, P. Fucini, D. R. Benjamin, R. Juenemann, K. H. Nierhaus, F. U. Hartl, C. M. Dobson and C. V. Robinson, *Proc. Natl. Acad. Sci. U. S. A.*, 2000, **97**, 5185–5190.
113. M. A. Tito, K. Tars, K. Valegard, J. Hajdu and C. V. Robinson, *J. Am. Chem. Soc.*, 2000, **122**, 3550–3551.
114. B. N. Pramanik, P. L. Bartner, U. A. Mirza, Y. H. Liu and A. K. Ganguly, *J. Mass Spectrom.*, 1998, **33**, 911–920.
115. M. G. McCammon, H. Hernandez, F. Sobott and C. V. Robinson, *J. Am. Chem. Soc.*, 2004, **126**, 5950–5951.
116. M. G. McCammon, D. J. Scott, C. A. Keetch, L. H. Greene, H. E. Purkey, H. M. Petrassi, J. W. Kelly and C. V. Robinson, *Structure*, 2002, **10**, 851–863.

117. K. De Vriendt, K. Sandra, T. Desmet, W. Nerinckx, J. Van Beeumen and B. Devreese, *Rapid Commun. Mass Spectrom.*, 2004, **18**, 3061–3067.
118. K. Benkestock, C. K. Van Pelt, T. Akerud, A. Sterling, P. O. Edlund and J. Roeraade, *J. Biomol. Screen.*, 2003, **8**, 247–256.
119. C. A. Keetch, H. Hernanndez, A. Sterling, M. Baumert, M. H. Allen and C. V. Robinson, *Anal. Chem.*, 2003, **75**, 4937–4941.
120. K. Breuker, *Angew. Chem. Int. Ed.*, 2004, **43**, 22–25.
121. K. D. Powell, S. Ghaemmaghami, M. Z. Wang, L. Ma, T. G. Oas and M. C. Fitzgerald, *J. Am. Chem. Soc.*, 2002, **124**, 10256–10257.
122. M. M. Zhu, D. L. Rempel, Z. Du and M. L. Gross, *J. Am. Chem. Soc.*, 2003, **125**, 5252–5253.
123. G. Hofner and K. T. Wanner, *Angew. Chem. Int. Ed.*, 2003, **42**, 5235–5237.
124. M. W. Pantoliano, E. C. Petrella, J. D. Kwasnoski, V. S. Lobanov, J. Myslik, E. Graf, T. Carver, E. Asel, B. A. Springer, P. Lane and F. R. Salemme, *J. Biomol. Screen.*, 2001, **6**, 429–440.
125. G. A. Holdgate and W. H. Ward, *Drug Discovery Today*, 2005, **10**, 1543–1550.
126. T. W. Mayhood and W. T. Windsor, *Anal. Biochem.*, 2005, **345**, 187–197.
127. M. C. Lo, A. Aulabaugh, G. X. Jin, R. Cowling, J. Bard, M. Malamas and G. Ellestad, *Anal. Biochem.*, 2004, **332**, 153–159.
128. P. Raboisson, J. J. Marugan, C. Schubert, H. K. Koblish, T. B. Lu, S. Y. Zhao, M. R. Player, A. C. Maroney, R. L. Reed, N. D. Huebert, J. Lattanze, D. J. Parks and M. D. Cummings, *Bioorg. Med. Chem. Lett.*, 2005, **15**, 1857–1861.
129. B. L. Grasberger, T. B. Lu, C. Schubert, D. J. Parks, T. E. Carver, H. K. Koblish, M. D. Cummings, L. V. LaFrance, K. L. Milkiewicz, R. R. Calvo, D. Maguire, J. Lattanze, C. F. Franks, S. Y. Zhao, K. Ramachandren, G. R. Bylebyl, M. Zhang, C. L. Manthey, E. C. Petrella, M. W. Pantoliano, I. C. Deckman, J. C. Spurlino, A. C. Maroney, B. E. Tomczuk, C. J. Molloy and R. F. Bone, *J. Med. Chem.*, 2005, **48**, 909–912.
130. D. J. Parks, L. V. LaFrance, R. R. Calvo, K. L. Milkiewicz, V. Gupta, J. Lattanze, K. Ramachandren, T. E. Carver, E. C. Petrella, M. D. Cummings, D. Maguire, B. L. Grasberger and T. B. Lu, *Bioorg. Med. Chem. Lett.*, 2005, **15**, 765–770.
131. R. R. Kroe, J. Regan, A. Proto, G. W. Peet, T. Roy, L. D. Landro, N. G. Fuschetto, C. A. Pargellis and R. H. Ingraham, *J. Med. Chem.*, 2003, **46**, 4669–4675.
132. S. Leavitt and E. Freire, *Curr. Opin. Struct. Biol.*, 2001, **11**, 560–566.
133. T. Wiseman, S. Williston, J. F. Brandts and L. N. Lin, *Anal. Biochem.*, 1989, **179**, 131–137.
134. Y. L. Zhang and Z. Y. Zhang, *Anal. Biochem.*, 1998, **261**, 139–148.
135. W. B. Turnbull and A. H. Daranas, *J. Am. Chem. Soc.*, 2003, **125**, 14859–14866.
136. W. H. Ward and G. A. Holdgate, *Prog. Med. Chem.*, 2001, **38**, 309–376.

137. M. J. Todd and J. Gomez, *Anal. Biochem.*, 2001, **296**, 179–187.
138. F. Dullweber, M. T. Stubbs, D. Musil, J. Sturzebecher and G. Klebe, *J. Mol. Biol.*, 2001, **313**, 593–614.
139. M. H. Parker, E. A. Lunney, D. F. Ortwine, A. G. Pavlovsky, C. Humblet and C. G. Brouillette, *Biochemistry*, 1999, **38**, 13592–13601.
140. A. Velazquez-Campoy, Y. Kiso and E. Freire, *Arch. Biochem. Biophys.*, 2001, **390**, 169–175.
141. F. R. Salemme, *Nat. Biotechnol.*, 2004, **22**, 1100–1101.
142. N. Markova and D. Hallen, *Anal. Biochem.*, 2004, **331**, 77–88.
143. P. A. van der Merwe, in *Protein–Ligand Interactions: A Practical Approach*, ed. S. E. Harding and B. Z. Chowdhry, Oxford University Press, Oxford, 2000, pp. 137–170.
144. M. A. Cooper, *Nat. Rev. Drug Discovery*, 2002, **1**, 515–528.
145. T. A. Morton and D. G. Myszka, *Methods Enzymol.*, 1998, **295**, 268–294.
146. D. G. Myszka, M. D. Jonsen and B. J. Graves, *Anal. Biochem.*, 1998, **265**, 326–330.
147. D. Casper, M. Bukhtiyarova and E. B. Springman, *Anal. Biochem.*, 2004, **325**, 126–136.
148. G. Lange, D. Lesuisse, P. Deprez, B. Schoot, P. Loenze, D. Benard, J. P. Marquette, P. Broto, E. Sarubbi and E. Mandine, *J. Med. Chem.*, 2003, **46**, 5184–5195.
149. J. L. Ekstrom, T. A. Pauly, M. D. Carty, W. C. Soeller, J. Culp, D. E. Danley, D. J. Hoover, J. L. Treadway, E. M. Gibbs, R. J. Fletterick, Y. S. Day, D. G. Myszka and V. L. Rath, *Chem. Biol.*, 2002, **9**, 915–924.
150. S. Dickopf, M. Frank, H. D. Junker, S. Maier, G. Metz, H. Ottleben, H. Rau, N. Schellhaas, K. Schmidt, R. Sekul, C. Vanier, D. Vetter, J. Czech, M. Lorenz, H. Matter, M. Schudok, H. Schreuder, D. W. Will and H. P. Nestler, *Anal. Biochem.*, 2004, **335**, 50–57.
151. D. G. Myszka, *Anal. Biochem.*, 2004, **329**, 316–323.
152. M. J. Cannon, G. A. Papalia, I. Navratilova, R. J. Fisher, L. R. Roberts, K. M. Worthy, A. G. Stephen, G. R. Marchesini, E. J. Collins, D. Casper, H. Qiu, D. Satpaev, S. F. Liparoto, D. A. Rice, I. I. Gorshkova, R. J. Darling, D. B. Bennett, M. Sekar, E. Hommema, A. M. Liang, E. S. Day, J. Inman, S. M. Karlicek, S. J. Ullrich, D. Hodges, T. Chu, E. Sullivan, J. Simpson, A. Rafique, B. Luginbuhl, S. N. Westin, M. Bynum, P. Cachia, Y. J. Li, D. Kao, A. Neurauter, M. Wong, M. Swanson and D. G. Myszka, *Anal. Biochem.*, 2004, **330**, 98–113.
153. J. W. Liebeschuetz, S. D. Jones, P. J. Morgan, C. W. Murray, A. D. Rimmer, J. M. E. Roscoe, B. Waszkowycz, P. M. Welsh, W. A. Wylie, S. C. Young, H. Martin, J. Mahler, L. Brady and K. Wilkinson, *J. Med. Chem.*, 2002, **45**, 1221–1232.
154. G. Agnelli, S. K. Haas, K. A. Krueger, A. W. Bedding and J. T. Brandt, *Blood*, 2005, **106**, 278.
155. M. S. Squires, R. E. Feltell, D. Smith, J. E. Lewis, N. G. Gallagher, J. F. Lyons and P. G. Wyatt, *Clin. Cancer Res.*, 2005, **11**(Suppl. 1), B194.
156. S. Borman, *Chem. Eng. News*, 2005, **83**, 28–30.

CHAPTER 17

A Link Means a Lot: Disulfide Tethering in Structure-Based Drug Design

JEANNE A. HARDY

The University of Massachusetts Amherst, Department of Chemistry, 710 N. Pleasant St., Amherst, MA 01003, USA

17.1 Introduction: What is Disulfide Tethering?

Tethering is a disulfide-based drug-discovery technology that grew out of the era when combinatorial chemistry was a burgeoning new development. What separates Tethering from other combinatorial approaches is its site-directed nature. The disulfide link that is formed allows a wide variety of advantages and analysis not available with other drug methods, thus in Tethering, a link means a lot. Site-directedness makes it ideal for interrogating traditionally difficult drug targets, such as protein–protein interaction surfaces, and new binding sites. It has also led to candidates for linking in a combinatorial manner, new fragments for addition to existing scaffolds, and entirely new classes of pharmacophores for several different protein targets. At the very essence of tethering is the concept that structural information can and should inform drug design and discovery.

Tethering was conceptualized and developed by a team assembled and directed by James Wells at Sunesis Pharmaceuticals and first publicly reported by Erlanson and coworkers in 2000.[1] The progress of Tethering and its many applications has been extensively reviewed,[2–4] thus overlap in content between this chapter and those reviews is bound to exist. Nevertheless, this review focuses on the practical considerations during Tethering and on the strong role that structural information plays during Tethering interrogation.

The inaugural report of Tethering focused on the cysteine-containing active site of thymidylate synthase (TS), a protein that is essential to synthesis of deoxythymidine monophosphate (dTMP) from deoxyuridine monophosphate (dUMP), which has been identified as a cancer target.[1] To probe TS and develop Tethering, a library of 1200 disulfide-containing molecules was synthesized to allow proof of concept. Each molecule in the library consisted of two elements: a monophore, which is the unique chemical entity, and a linker region containing a thiol for disulfide exchange (Figure 17.1). Tethering screening was performed in pools of ten compounds per pool, where each compound had a unique molecular weight. Pools of compounds were incubated with TS under high reductant concentrations where disulfide exchange was rapid. Compounds that interacted even weakly with the surface, a pocket, or a binding site on the protein near a cysteine residue, and thus had an increased residence time, could promote the formation of a disulfide bond between the cysteine thiol and the compound thiol. Binding of the monophores to TS entropically stabilized the interactions and favored the disulfide-bonded protein.

One of the most fascinating aspects of screening for drug leads by Tethering is that it is a tunable activity. At moderate reductant concentrations, monophores with too weak an inherent activity are reduced off, so that only interactions driven by interactions of the monophore with the protein remain intact. Covalently modified protein can then be analyzed by mass spectrometry or a functional assay so that bound compounds can be identified. In the presence of

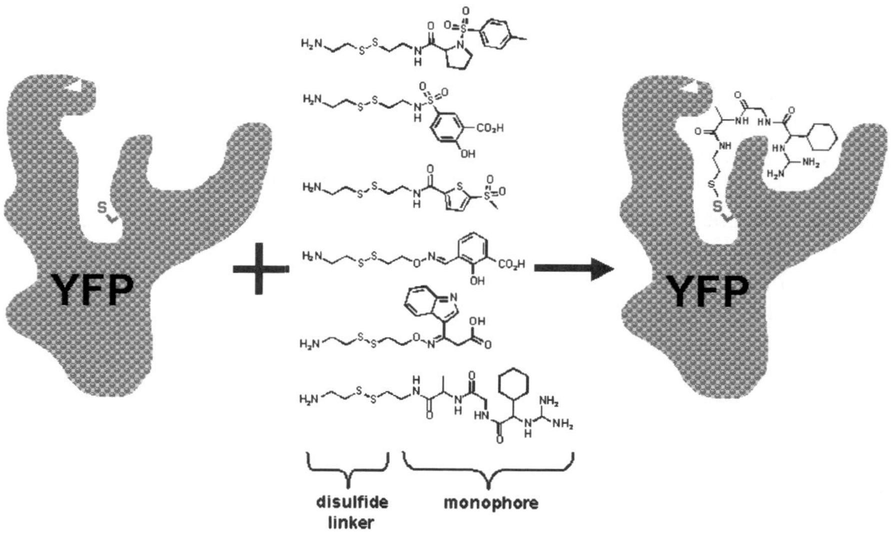

Figure 17.1 Tethering schematic. The general scheme of Tethering is shown for YFP (your favorite protein), which contains an exposed and free cysteine thiol. Disulfide-containing compounds are incubated with various concentrations of reductant and YFP. Monophores that specifically bind to surfaces or cavities on YFP facilitate the formation of a disulfide bond between the native or introduced cysteine residue. Non-specific interactors are reduced off by the moderate amount of reductant present. Captured fragments can be easily identified by mass spectrometry of the covalent protein–small molecule complex.

very high levels of reductant, any observed binding or inhibition is relieved. This aspect of the technique is very useful for showing that the mechanism of inhibition occurs through binding at the targeted site, via the disulfide, and not through some non-specific mechanism of inhibition such as micelle formation or protein denaturation. In the case of TS, the results were analyzed by mass spectrometry. The most strongly interacting compound was *N*-tosyl-D-proline (TP). The specificity and affinity of TP for TS was so significant that even when 100 compounds were included in the pool with TP, TS was still bound selectively by TP, showing that exquisite specificity is a result of the monophore only.

Although TS has five cysteines, the active site was thought to be the most reactive and was the only site at which compounds bound. Numerous subsequent studies have confirmed the finding that compound binding is specific for a single cysteine residue or a unique binding orientation. In addition, it has been generally observed that fragments that bind to one protein are specific for that protein and do not bind to other proteins. In the case of TS, the active site cysteine was mutated to serine (C146S) and TP binding was abolished. However, in that context, if a neighboring leucine residue was also mutated to cysteine (C146S/L143C) TP could still bind. Crystal structures of the wild-type TS and C146S/L143C show the TP to be in the same orientation (Figure 17.2). Introduction of a cysteine residue at a different adjacent position did not result in productive binding of TP. This suggests that there is some flexibility afforded by the linker portion of the disulfide-containing compounds. When the thiol was removed from TP to generate thiol-free-TP, that compound still bound to TS ($K_I = 1.1$ mM) and the conformation of the TP moiety in the crystallized complex was the same as in the covalently bound compound. Thus, the interaction is exquisitely driven by binding of the monophore region and is largely independent of the linker.

From the first success with TS, Tethering has been developed into a robust, rapid, and highly successful protocol for deriving new pharmacophores as pharmaceutical leads. During Tethering, fragments are equilibrated under conditions that promote disulfide exchange. Typically, these experiments are performed at concentrations of 0.1–2 mM β-mercaptoethanol (β-ME). There are two standard ways that the strength of the interaction between the small-molecule fragment and the protein are measured. The first method is to determine the β-ME_{50}, which measures the amount of β-ME needed to cause 50% labeling of the cysteine-containing protein. Another way to measure the strength of a Tethering interaction is by assessment of a dose–response-50 (DR_{50}) for a compound titration at a given level of reductant (usually 1–4 mM β-ME). Small molecule–protein pairs with very favorable interaction energies are less likely to dissociate. Typical β-ME_{50} values for strongly selected compounds range from 0.7 to 5 mM.

In developing Tethering into a successful commercial venture, one consideration was the composition of pools that would be ideal for decoding the binding molecule while still yielding tractable drug leads. During Tethering the average molecular mass in disulfide-containing molecules is 250 Da.

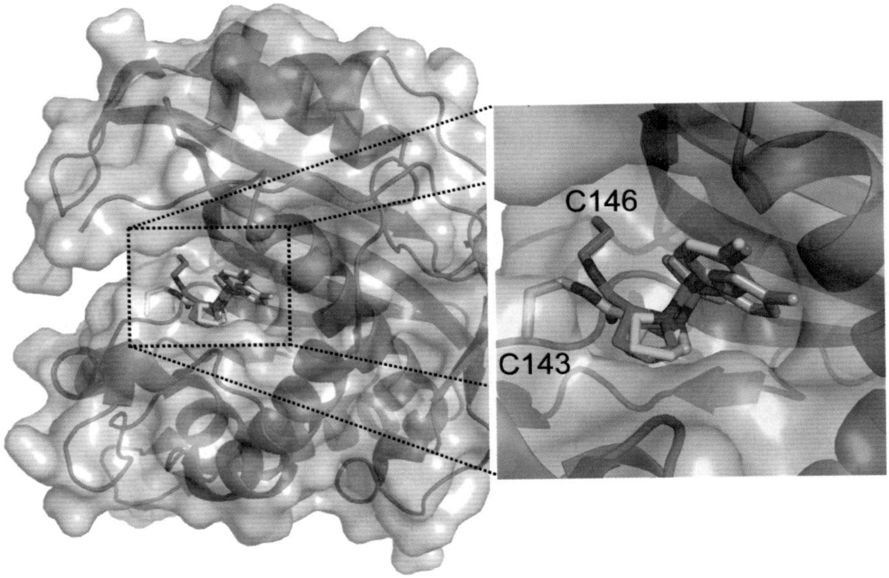

Figure 17.2 Tethering to thymidylate synthase. Thymidylate synthase (green ribbons and gray surface) is shown bound to *N*-tosyl-D-proline at active-site cysteine C146 (green sticks, PDB ID = 1F4C), or at introduced cysteine C143 (yellow sticks, PDB ID = 1F4D) or non-covalently associated with the active site (orange sticks, PDB ID = 1F4E). The overlap in the conformations of the *N*-tosyl-D-proline moieties shows that binding is determined by the monophore and is independent of the cysteine linker.

This size allows for fragment recombination and many avenues for future optimization. In development of the pools, attention was paid to the chemical nature of the compounds so that each compound exhibited drug-like properties as well as structural and chemical diversity. Thus, overall the frequency of hits is low. For screening 12 cysteine mutants in interleukin-2 (IL-2), the frequency of a hit was 0–1% overall, where the highest hit rate for an individual cysteine was 1.3% and four of the positions produced 0–0.08% hit rates.[5] In C5a the hit rate for the four cysteine residues studied ranged from 0 to 0.6%.[6] In screening the caspase-3 active-site, no hits out of a 10,000 compound library were found.[7] Screening by Tethering is a readily tunable activity, which is a hallmark of its adaptability, and thus varying the concentration of reductant dramatically changes the number of hits observed. The success of Tethering is monitored by stringent hit rate, biological sense of hit sites (either from other assay data or based on crystal structures), expectation that hits display sharp structure–activity relationships (SARs) – typically an enantiomer is expected to have at least ten-fold or more decreased activity as the originally identified compound – and the requirement that a hit must be reversible by addition of high concentrations of reductant.

Most screens based on Tethering have used covalent modification as measured by mass spectrometry as the readout. Two examples demonstrate that Tethering can also be monitored by functional assays. A functional screen for the inhibition of the intrinsic apoptotic pathway was measured as a function of caspase-3 activity in a cell-lysate based assay. One compound (FICA, see Section 17.7), which was subsequently shown to be an allosteric inhibitor of caspase-3 and -7, was discovered based on that screen.[7] Tethering has also been applied to the integral membrane protein, the C5a receptor. It was possible to adapt traditional cell-based assays for IP3 accumulation and C5a ligand-binding assays with membrane fractions containing C5a receptor to conditions where the level of reductant could be varied so that Tethering experiments could be performed.[6,8]

In developing a site-directed drug-discovery platform, other types of reversible bond-forming moieties could have been selected, but much of the success of Tethering probably lies in the selection of a disulfide as the covalent link, as it can form and break under mild conditions. An additional advantage is that, since Tethering is reversible, it is easy to rule out non-specific modes of inhibition, such as aggregation or denaturation. In the vast majority of cases where the strength of the interaction between the monophore and the protein is sufficient to bind in the absence of a disulfide, the crystal structure of the disulfide-bonded and disulfide-free versions of the compounds are virtually identical.

17.2 Success of Native Cysteine Tethering

Many of the targets probed using Tethering have been enzymes with active-site cysteine residues [caspase-1, caspase-3, caspase-7, protein tyrosine phosphatase 1B (PTP1B), and TS (Table 17.1)]. In these cases the site of disulfide interaction is often presumed to be the catalytically important cysteine. Using native cysteines of any ilk for Tethering, an important consideration is whether the selected cysteine will react more readily with the library of thiol-containing compounds than any other cysteine residue on the protein. This concern was directly tested in TS where four non-conserved buried cysteine residues are present. These residues appeared far less reactive to disulfide exchange than the active site cysteine. The active-site cysteine readily reacted with cystamine ($S–CH_2–CH_2–NH_2$), however the active-site mutant wherein cysteine was replaced with serine (C146S) did not react with cystamine, suggesting that the active-site cysteine was the only exposed and disulfide-exchange-reactive residue.[1]

The active-site cysteine is not necessarily the point of interaction in Tethering. If binding of the disulfide compounds was purely driven by electrophilic propensities, then the most activated cysteines should always preferentially be bound by the disulfide-containing compounds. Highly electrophilic moieties, such as methylthiolsulfonates, fluoro- and chloroketones, and nitrosylating

Table 17.1

Protein	*Tethering*	*Extended tethering*	*Breakaway Tethering*	*Native Cysteine*	*Engineered Cysteine*	*MS Screen*	*Functional Screen*	*Conformational change by Binding*
Thymidylate Synthase	+			+		+		
IL-2	+				+	+		+
caspase-1		+		+		+		+
caspase-3	+	+		+		+	+	
caspase-3 allosteric site	+			+		+	+	+
caspase-7 allosteric site	+			+		+		+
PTP-1B			+		+	+		
PTP-1B allosteric site	+			+		+		+
C5a	+			+	+		+	ND

compounds, such as *S*-nitroso-*N*-acetyl-D,L-penicillamine (SNAP), label exposed cysteines as a function of the pK_a and surface exposure of the cysteine residue. During Tethering experiments, binding of the disulfide-containing compounds is driven by interactions of the monophore with the protein rather than by the electrophilicity of the thiol, so the active site is not the only site that can be labeled. An example of this is in PTP1B where an introduced cysteine could be preferentially alkylated even in the presence of the selected site cysteine residue.[9]

Native cysteine residues have also been used productively for extended Tethering in caspase-3[10–12] and caspase-1[13,14] even in the presence of other exposed cysteine residues (see Section 17.5). One native-cysteine Tethering success occurred serendipitously at a non-active-site cysteine in a cavity distal from the active site. Tethering to this distal cysteine resulted in the discovery of a new allosteric site that is present in caspase-3 and caspase-7 (see Hardy *et al.*[7] and Section 17.7). Similarly, in probing PTP1B with a series of alkylating agents, one thiol-selective fluorogenic compound, 4-(aminosulfonyl)-7-fluoro-2,1,3-benzoxadiazole (ABDF), was discovered that selectively modified a single native cysteine (C121) outside the active site and allosterically inactivated the protein.[15]

17.3 Role of Structure in Engineered-cysteine Tethering

A major advantage that Tethering holds over other methods of drug discovery is the ability to specifically target one particular region of the protein that may be the active site, an allosteric site, a novel cavity, or a protein binding surface. If the region one hopes to target is not near a native cysteine, a cysteine residue must be engineered into the protein to allow Tethering to proceed. Thus, a big question in using engineered cysteines is where to engineer the cysteines. Very often visual inspection is used, and all of the exposed residues within the appropriate distance (4–8 Å) of the selected site are exhaustively or selectively mutated to cysteine.[16] Typically, mutation of one surface residue to cysteine does not have dramatic deleterious effects on activity, so it is possible to introduce cysteine residues at various locations around the site of interest.

An early case study in the introduction of new cysteine residues for Tethering was carried out in IL-2 (T-cell cytokine interleukin-2).[5,16–20] Arkin and co-workers selected a series of ten residues by visual inspection to mutate to cysteine for Tethering at an adaptable binding site that had been predicted to contact the IL-2 receptor (Figure 17.3).[5,17] None of the introduced cysteine residues disrupted production or folding of IL-2, however, the same residues that had been shown by previous mutagenesis studies to disrupt IL-2 binding to the IL-2 receptor also disrupted binding when mutated to cysteine. Not all of the introduced residues had the same hit rate in a Tethering assay, underscoring the general observation that some residues are intrinsically more fruitful for Tethering. This is likely because of the makeup of the library, the

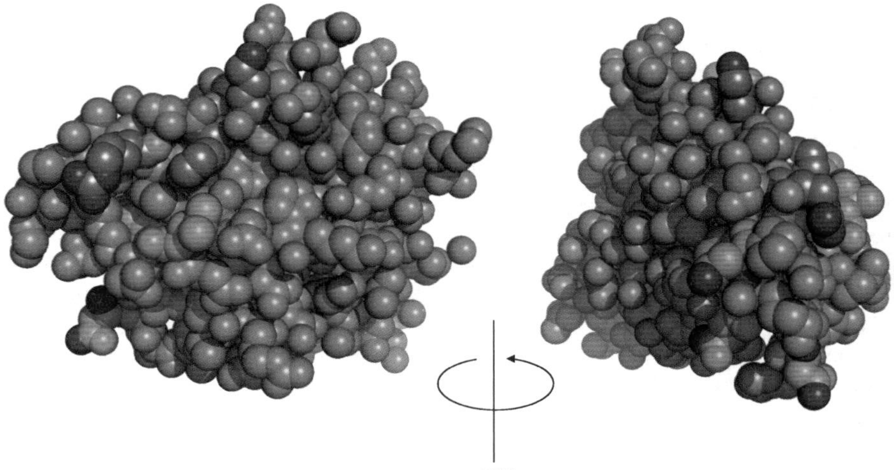

Figure 17.3 Residues mutated to cysteine for Tethering on IL-2. Residues that had lower reactivity (captured fewer than seven compounds per residue) were in the highly ordered region of the IL-2 binding site. These residues are blue and include, from left to right, in the panel on the left R38, F42, K43, and Y45. The residues that had higher reactivity (captured greater than 20 compounds per residue) were in the adaptable region of the IL-2 binding site. These residues are drawn in pink and include residues L72 and N77 in the panel on the left. In the panel on the right the pink residues, from top to bottom, are L72, N33, Y31, N30, and N77.

positioning of the residues, and the conformations that are accessible from that residue conjugated with small molecules. In the case of IL-2, residues that display the highest conformational flexibility also have the highest hit rate.[5]

Computational methods can also be used to suggest which residues might be best for cysteine introduction and subsequent Tethering interrogation. In these calculations, probes based on the linker regions present in the compounds in the disulfide library are attached to the candidate residues *in silico* and subjected to molecular dynamics simulations. The trajectories of the linkers toward or away from the region of interest can be used to determine which residues are the most likely to result in successful Tethering.

Depending on the position of the engineered cysteine and the flexibility of the linker, fragments discovered at one cysteine can also bind when an adjacent amino acid is substituted by cysteine. In the case of TS, the tosyl-proline fragment bound with similar affinity and in the same orientation to the active-site cysteine (C146) and to a cysteine introduced at position 143 (L143C), but not at other residues (e.g. H147C), presumably because the geometry was non-optimal[1] (Figure 17.2). For C5a, approximately 10% of the hits were reactive with more than one of the engineered cysteine residues.[6] In IL-2, no two introduced cysteine residues captured the same disulfide-containing fragments.[5]

This suggests, in general, that hits bind to a single cysteine with discriminating orientation.

Reactivity of the compounds from the disulfide library with other non-target cysteine residues is sometimes a consideration. Although Tethering is predominantly driven by interactions of the monophore with the surface of the protein, the reactivity of the compound library is, to some extent, governed by the pK_a of the cysteines on the surface of the protein. Activated or hyper-exposed cysteines can interfere with Tethering at the desired location. The general reactivity of non-candidate cysteine residues can often be monitored by reaction with small thiol-containing compounds such as β-ME, oxidized glutathione, or cystamine followed by peptide mapping with mass spectrometry. If additional non-candidate cysteine residues show strong reactivity, it often simplifies the analysis of the disulfide-library screening to "scrub" all other cysteine residues from the surface of the protein by mutation to alanine or serine. In screens on cysteine-free variants any covalent adducts can readily be attributed to the cysteine of interest rather than to other surface-exposed cysteine residues.

Removal of one surface cysteine facilitated the screening of caspase-3, a heterotetrameric enzyme composed of two small and two large subunits. The most reactive cysteine residue on the large subunit was the active-site cysteine (C163), however a second cysteine on the small subunits was fully exposed in the existing caspase-3 crystal structure and was also reactive. It was thus necessary to mutate that residue to serine (C184S) to make interpretation of the mass spectral data unambiguous (unpublished data).

Whereas sometimes overly reactive cysteine residues can confound or complicate a Tethering experiment, there are also examples of an introduced cysteine being successfully probed in the presence of an active-site cysteine. In PTP1B an R47C mutation allowed alkylation by an extender molecule preferentially, even though the catalytic cysteine residue (C215) was still present in the protein construct. This is likely because C215 sits at the bottom of a deep hole,[9] and can thus be protected from easy disulfide exchange. This recapitulates the results of a naïve screen of caspase-3, where no hits were found against the active-site cysteine residue. The caspase-3 active site is believed to be very mobile and mostly unstructured in the absence of substrate. Optimal caspase substrates contain a negatively charged aspartate moiety. Since the Tethering library was designed to have drug-like properties and charged entities are usually considered to be non-drug-like, it is not surprising that no hits against the caspase-3 active site were isolated. Together the examples of PTP1B and caspase-3 underscore the conclusion that not all active-site cysteines are structurally optimal for a Tethering reaction to occur. On the other hand, many introduced cysteines are perfectly competent for capturing hits via Tethering.

The vast majority of Tethering experiments have been on proteins of known structure, so that sites for Tethering could be chosen based on inspection of the binding pocket to be probed. Particularly when introducing cysteine residues, it is very useful, if not obligatory, to have structural information available. One question of direct interest is whether Tethering is also applicable to proteins where a high-resolution structure is not available. G-Protein coupled receptors (GPCRs) are of tremendous importance in drug discovery as up to half of marketed drugs target GPCRs,[21] however structures exist for just a handful of GPCRs. For example, there is no structure available for the C5a receptor. A homology model of the C5a receptor based on the rhodopsin structure has been published[22] and conclusive mutagenic studies have also pointed to the location of a ligand-binding pocket.[22–24]

With only indirect structural data on the binding surface of the C5a receptor in hand, Buck and Wells opened a new avenue for Tethering by asking whether this was sufficient structural detail to mount a successful Tethering campaign.[8] They first sought to determine that the location implicated by mutagenesis was, indeed, the ligand-binding site. Based on the homology model, they selected four sites at which cysteine residues were introduced.[8] Previous data had also indicated that six amino acid peptides from the C-terminal sequence of the C5a ligand were themselves sufficient to agonize or antagonize the receptor, dependant on their precise sequence.[22] Buck and Wells used the three amino acid sequence of the agonist or antagonist peptides and added an N-terminal cysteine residue. The homologous three amino acid peptides themselves were too short to either angonize or antagonize C5a like the six amino acid peptides had. The addition of the cysteine allowed for Tethering of these peptides to the cysteine-containing receptors. The four amino acid cysteine-containing peptides maintained the agonist or antagonist properties of the parent peptides, and also confirmed the location of the ligand-binding site to be near residues P113, G262, and L117. Thus, they concluded that these residues were appropriate residues for a screen for small molecules by Tethering.

In the case of C5a, four mutant cysteine positions were each screened against a library of 10,000 compounds. The three sites that were most appropriate for binding of the cysteine-containing peptides – P113C with 65 inhibitors, G262C with 36 inhibitors, and K117C with 24 inhibitors identified – were also successful for identifying small-molecule agonists and antagonists of C5a activity. An interesting feature of the C5a receptor is that structurally similar peptides can function as either agonists or antagonists. Of the small-molecule inhibitors identified by Tethering nearly equal amounts functioned as agonists and antagonists of C5a ligand binding. This is particularly interesting given that the small-molecule inhibitors are only 1/4 to 1/2 of the size of the minimally active peptide inhibitors.

17.4 Cooperative Tethering

Cooperative Tethering utilizes engineered cysteine residues on proximal regions to interrogate binding fragments that interact cooperatively with one another or to interrogate cooperative interactions between known binding elements and newly discovered disulfide-containing monophores. Many binding sites are very adaptive, which, stated differently, means that the energy barriers between various conformations are very small so a relatively small change in the system (*i.e.* binding of a Tethering monophore) can favor another conformational state. In these adaptive sites, it is therefore not surprising that there is a great deal of inherent cooperativity that can be accessed during binding at adjacent sites.

Cooperative Tethering has been most successfully applied in the case of IL-2.[16] IL-2 has historically been classified as a difficult target for drug discovery because drug molecules would have to bind to the interface of IL-2 that interacts with the IL-2 receptor. Compared to enzyme active sites that are often designed for binding and recognizing small molecules and that often contain rich functionalities in the catalytic residues, protein surfaces that promote contact with other proteins are comparably featureless and flat (Figure 17.4). Methods for probing such surfaces are few. Tethering is one of the few direct ways that a protein–protein interface can be specifically probed. Use of cooperative Tethering led to the development of a series of compounds that bound to IL-2 and strongly antagonized IL-2 receptor binding.

The interrogated IL-2 binding site had previously been identified as being involved in the IL-2:IL-2–receptor interface. This binding site was in an adaptive region, the conformation of which changed upon binding to non-peptidic small molecules such as **1** (Figure 17.4).[5] Hyde and coworkers observed that the adaptive region extended beyond the residues that were occupied by early compounds, and selected several adjacent residues for cysteine substitutions (N30C, Y31C, and N33C) to probe a larger portion of the adaptive region with Tethering.[16] These residues were selected from the crystal structure and none had any effect on the structure or function of IL-2.

Screening IL-2 by Tethering at these residues yielded 132 hits in this extended adaptive region. To probe the likelihood that these fragments could be coupled with **1** to generate a stronger binding inhibitor, binding of the identified disulfide-containing small molecules was tested in the presence of **1**. Fully 33% of the disulfide compounds showed an increase in conjugation efficiency (positive cooperativity), 44% showed a decrease in conjugation efficiency (negative cooperativity), and 23% showed no change in conjugation efficiency (no cooperativity). This cooperativity was confirmed using two-site Tethering with a disulfide-containing version of **1** that could be covalently coupled to the K43C mutation in IL-2. This allowed direct assessment of the fractional binding of **1** and hits from Tethering. Surface plasmon resonance (SPR) also confirmed the interaction between **1** and disulfide-free versions of the compound identified by Tethering and demonstrated that the cooperativity was reciprocal between **1** and hits derived from Tethering. Moreover, these non-covalent results emphasized that the binding of **1** did not simply change the accessibility of the disulfide required for Tethering, but induced important changes in the binding surface that contributed to binding of the monophore. The cooperativity between **1** and hits from Tethering was significant and contributed –2 kcal mol^{-1} $\Delta\Delta G$ to synergistic binding.

Ultimately, several classes of compounds were developed by combining known fragments, such as **1**, and fragments discovered by Tethering.[5,16–18,20] The most potent of these displayed a 60 nM IC_{50}.[17] The crystal structure of this

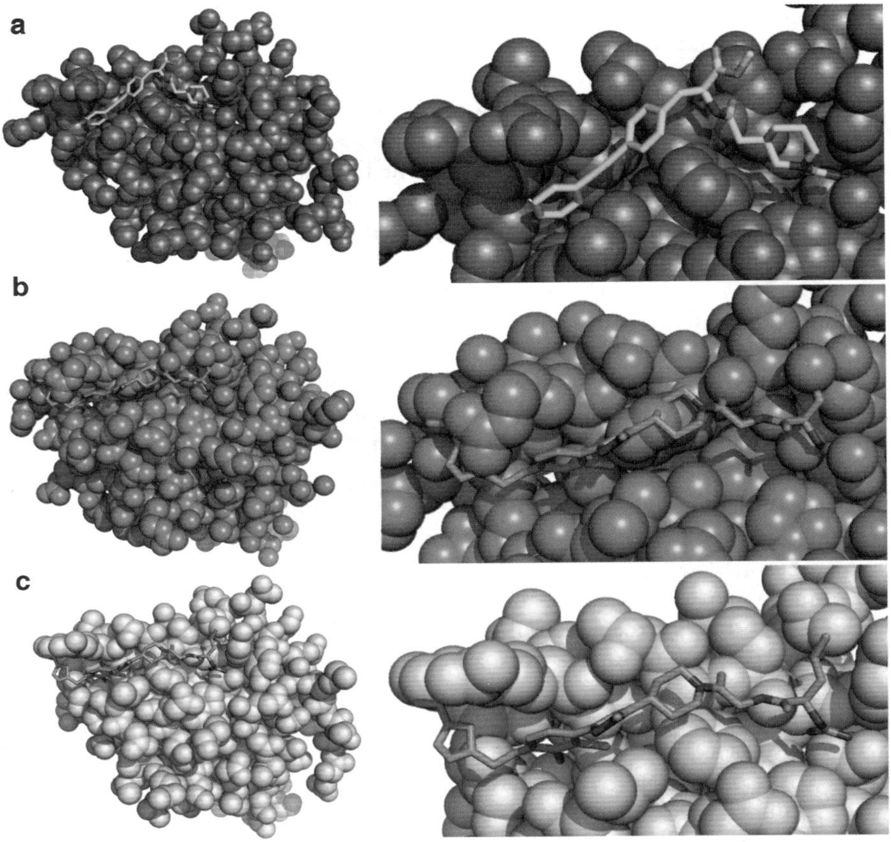

Figure 17.4 The adaptability of the IL-2 surface. (a) The structure of IL-2 (red) bound to **1** (yellow sticks) shows conformational changes in the adaptive site, which is occupied by the hydrophobic biaryl moiety (PDB ID 1M48). (b) The steric clashes that occur when a 60 nM inhibitor (pink sticks) is superimposed on the unliganded IL-2 structure (PDB ID 1M47, green). (c) Many conformational changes are wrought by the binding of the 60 nM inhibitor derived from tethering (PDB ID 1PY2, pink sticks on yellow protein surface), which carves out its own binding interface.

compound with IL-2 revealed that the compound induced massive structural rearrangement to promote binding to the adaptive region,[19] emphasizing why such strong cooperativity was observed. This work emphatically demonstrates that cooperative Tethering is a useful way to develop potent inhibitors, especially on dynamic regions of a protein surface.

17.5 Extended Tethering

The idea of Tethering with extenders (also called extended Tethering in some publications) is that a disulfide-containing small-molecule anchor can be used in the active site or binding site to probe adjacent regions for chemical moieties that bind a little further afield. This anchor is often an irreversible alkylator that reacts specifically with a single cysteine (native or engineered) on the protein of interest.

Extended Tethering has most widely been used against the caspases where the active-site cysteine is prime for alkylation by an "extender" molecule. Caspases have high specificity for cleaving after aspartate residues so extenders in which a thiol-reactive alkylating agent is placed adjacent to an aspartic acid moiety readily and specifically modify the active site. Although the 13 human caspases may all prove to be valuable drug targets, extended Tethering has, to date, only been performed with caspase-3 (an apoptotic executioner caspase)[12] and on caspase-1 (an inflammatory caspase).[14] Because of a family-wide specificity for binding to aspartic acid elements, the same extender could be used for both caspase-1 and caspase-3. These extenders were based on the structural and biochemical knowledge that all caspases need an acidic moiety to mimic the natural aspartate substrate that binds in the S1 pocket. Whether from a substrate or from some other source, an acidic moiety is necessary to nucleate formation of a correctly positioned active site and catalytic residues. Small, charged molecules, such as malonate, are capable of nucleating this conformation.[25] The extender used for both caspase-1 and caspase-3 mimics the natural substrates. Because the methods and results are similar, we focus here on the work on caspase-1, which is more recent. Nevertheless, the dramatic success of extended Tethering on caspase-1 relies heavily on the successful precedent set by extended Tethering in caspase-3, in which a salicylic acid discovered from an extended Tethering screen combined with medicinal chemistry optimization resulted in a series of inhibitors with K_is of 20 nM.[10–12]

Tethering with extenders was essential in the development of potent inhibitors of caspase-1.[13,14] After modification of the extender, the crystal structure of caspase-1 demonstrated that the extender specifically alkylated the active-site cysteine, and not any of the other four cysteines in the same subunit. The structure also showed the extender pointing, as predicted, toward the outer edge of the binding cavity (this region is termed the S3 and S4 binding pockets because the third and fourth amino acids of the peptide typically bind here, Figure 17.5). Screening by Tethering identified ten fragments that bound with high affinity. When these fragments were converted to non-covalent analogs, however, the affinity was poor, generally greater than 100 μM, demonstrating that their affinity was dependent on the interactions between the extender and the S1 pocket in the active site. The extender itself exhibited a binding affinity of only 110 μM.

One of the strengths of extended Tethering is that it suggests a defined way that the extender plus the hit can be converted to a useful inhibitor. The alkylating cyclooxy-methylketone on the extender can be replaced by an aldehyde, which is reversible although covalent. The disulfide region can be converted atom by atom for replacement by methylene units. This procedure resulted in an initial lead compound with 150 nM affinity for caspase-1 and 57-fold selectivity over caspase-5.

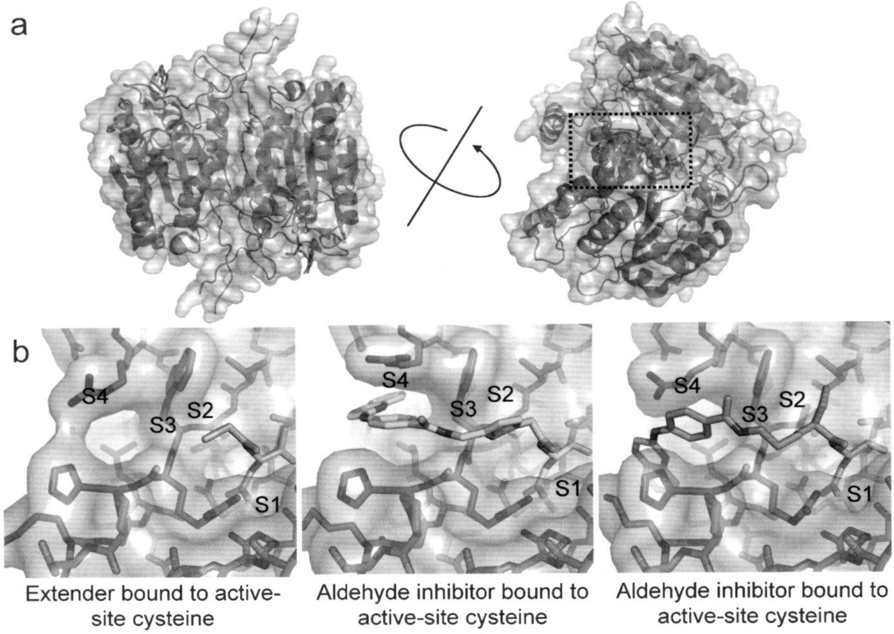

Figure 17.5 Extended Tethering in caspase-1. (a) A typical view of caspase-1 (ribbons) with an aldehyde inhibitor bound in the two active sites. Rotation around the indicated axis orients caspase-1 to look into one active site. The zoom region shown in (b) is marked by the dashed square. (b) The extender bound to caspase-1 occupies the S1 pocket (PDB ID 1RWK, left panel). S1–S4 binding pockets are marked. A compound derived from tethering generated by atom-for-atom replacement and addition of an ethyl moiety to bind in the S2 pocket results in a 150 nM inhibitor that orients the terminal tricyclic moiety in a downward orientation in the S4 pocket (PDB ID 1RWN, right panel). On the other hand, replacement of the disulfide with a bulkier thiophene group that does not enter the S2 pocket results in a 340 nM inhibitor that orients the terminal tricyclic moiety in a upward orientation in the S4 pocket (PDB ID 1RWM, central panel).

Guided by work in caspase-3 extended Tethering, the caspase-1 investigators knew that additional affinity could be added by modification of the region that bound to the S2 pocket. Based on the specificity of caspase-1, hydrophobic groups were introduced in this region. These additions resulted in sets of related compounds that contained the same constituents for binding to the S3 and S4 pockets, but which could bind to two different cavities outside of the S4 pocket that had not been previously exploited for drug discovery. Viewing the SARs of the compounds, it became clear that modification of the region binding to S2 was critical in determining the conformation of the distal end of the inhibitors. This suggests a cooperative coupling between the S2 and S4 regions of the caspase-1 binding site that had not previously been recognized and that does not seem to be present in caspase-3. The success of extended Tethering in this situation is that the tricyclic fragment discovered by Tethering that led to the highest affinity inhibitors had modest-enough affinity on its own that the authors concluded it could not have been derived from any traditional functional screen. In this case Tethering was seminal in the identification of this fragment.

In the caspases, a known binding moiety was used as the extender, but another variation on Tethering with extenders would utilize a high-affinity hit, derived from naïve Tethering, as the extender to probe distal regions of the binding pocket. In principal this could be repeated more than once, generating longer and longer extender molecules. The presumptive reason this has not been attempted is that third-generation extended tethers would be too long and have too high a molecular weight to generate a fruitful drug lead. Another future development that will be exciting to witness is the application of Tethering with extenders to an *in situ* situation, a course that is ripe with possibilities.

In early studies on caspase-3, naïve Tethering against the active-site cysteine did not result in any hits. Why, then, was Tethering with extenders so successful? A lack of hits from naïve Tethering is likely both because the length of molecules present in the compound collection was insufficiently long to bypass the P1 pocket. Because the library had been culled to contain fragments with drug-like properties (charged molecules are generally considered to be non-drug like), the library contained very few fragments composed of aspartate mimics. For both caspase-3 and caspase-1 it is clear that a charged fragment that mimics the aspartate that is present in all caspase substrates is necessary to enhance binding affinity.

17.6 Breakaway Tethering

In proteins with cysteines at the active site, native-cysteine Tethering is of great utility. Tethering with extenders has been useful for caspases, which contain active-site cysteines, largely because the available chemistries for modifying the active-site cysteines are robust and specificity elements for substrate binding are near the catalytic residue. One concern with engineering cysteine residues in active sites or binding sites is that the residue that would ideally be mutated for fragment discovery may be important for binding or catalysis. To circumvent these complications, Erlanson and co-workers developed a strategy they termed breakaway Tethering,[9] which is useful for probing sites that are narrow, deep, fragile, or significantly and negatively impacted by the introduction of a cysteine residue for Tethering. In short, breakaway Tethering is ideal for any protein where modification within the active site is not desirable.

PTP1B, a negative regulator of insulin-receptor phosphorylation and signaling, a pharmaceutical target for type-2 diabetes, is such a protein. The active site of PTP1B is deep, and because it binds to doubly charged phosphotyrosine residues, could be severely impacted by introduced cysteines in the pocket, so it was a good candidate for the development of breakaway Tethering.[9] To avoid disruption of the binding pocket, a cysteine residue was introduced outside of the binding pocket (R47C), which was roughly 10 Å from the active site and which was predicted to point toward the active site when mutatated (Figure 17.6). This introduced cysteine could be alkylated by a breakaway extender. The breakaway extender was designed as a derivative of a known active-site binder oxalic acid, a phosphotyrosine mimetic. The oxalic acid was linked to the extender via a thioester so that it could be cleaved under mild conditions by exposure to hydroxylamine, leaving the extender conjugated to the distal cysteine.[9] It was impressive that this approach worked despite the fact that

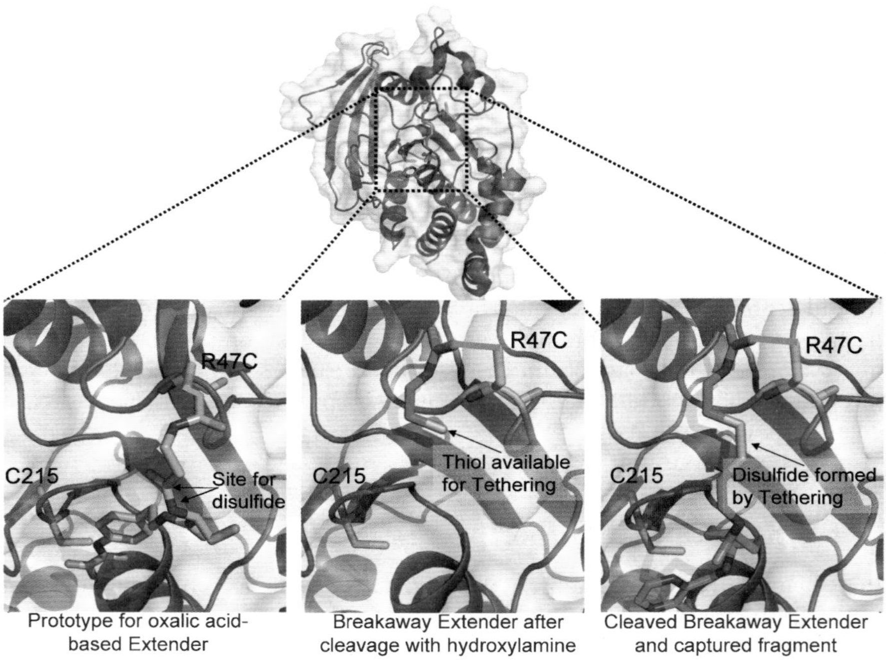

Figure 17.6 Breakaway Tethering in PTP1B. PTP1B (green ribbons, PDB ID=1NWE) is shown modified with a prototype oxalic acid-containing extender molecule attached to a cysteine residue that has been specifically engineered for Tethering (R47C, lower left panel). The prototype extender was the inspiration for the breakaway extender, which contained a disulfide bond at the positions equivalent to the carbons marked with arrows at the site for disulfide. Unfortunately, no structure of the oxalic acid breakaway extender is available. Treatment of the breakaway-extender-modified PTP1B with hydroxylamine exposes the free thiol which can then be interrogated with a library of disulfide-containing compounds (lower center panel). Compounds that bind to the active-site pocket, which is in a deep and fragile cavity, are able to form a disulfide bond with the cleaved extender, as was the case for the fragment captured in the PTP1B active site in the lower right panel (PDB ID 1NWL).

PTP1B contains an active-site cysteine. Alkylation occurred preferentially at the introduced cysteine, probably because the active-site cysteine was protected from alkylation by the oxalic acid. The charges on the oxalic acid mimic phosphotyrosine so that binding occurs preferentially in the direction that orients the extender toward the external cysteine at position 47. This again underscores the importance of non-covalent binding of the monophore driving the formation of the covalent linkage.

Once the oxalic acid was released, the extender-modified PTP1B was used to screen the library of disulfide-containing compounds. Novel phosphotyrosine mimics, of different classes than had been previously discovered through medicinal chemistry efforts or from traditional high-throughput screening (HTS), were ultimately developed based on the breakaway Tethering effort. In the crystal structures of the hits observed from breakaway Tethering, the monophores all sat in the deep active-site pocket as expected. The non-covalent monophores competitively inhibited PTP1B with a K_i of 4.1 mM, which is a notable improvement over extant phosphotyrosine mimetics with K_is of greater than 10 mM and comparable to phosphotyrosine, the native substrate, which binds PTP1B with a K_m of 4.9 mM. This work established breakaway Tethering as yet one more useful adaptation of a clearly powerful technology, enabling discovery efforts in one more category of target – those with fragile active sites.

17.7 Discovery of Novel Allosteric Sites with Tethering

Within the past several years, several prominent drugs that have entered the marketplace have been structurally shown to act at allosteric sites [*e.g.* Gleevec (Glivec) which binds to c-Abl[26] and the human immunodeficiency virus (HIV) non-nucleoside reverse transcriptase inhibitors (NNRTIs)[27]]. These developments have dramatically increased interest in exploiting allosteric sites in other proteins. Tethering is an ideal technology for both discovery and exploitation of novel allosteric sites. Tethering has been applied to the apoptotic executioner caspases, caspase-3 and -7. Caspases derive their name from their properties as cysteine aspartate proteases because they contain an active-site cysteine and cleave substrates after aspartate residues. An initial screen of caspase-3 against the disulfide-containing compound library using mass spectrometry as a read out was expected to elucidate compounds that bound the caspase-3 active-site cysteine in the large subunit. (Active caspases are heterotetramers composed of two large and two small subunits. The catalytic histidine–cysteine dyad is in the large subunit.) Surprisingly, no compounds were found that bound to the large subunit. One compound, DICA, was identified that bound strongly to a native non-active-site cysteine (C264) in the small subunit. In a separate functional screen for inhibitors of the intrinsic apoptotic pathway (see Section 17.1) another small molecule (FICA) was identified that bound to C264. This small-subunit cysteine exists in the bottom of a deep cavity at the dimer interface (Figure 17.7a). Both of these compounds acted as covalent stoichiometric inhibitors of caspase-3 and caspase-7 activity.

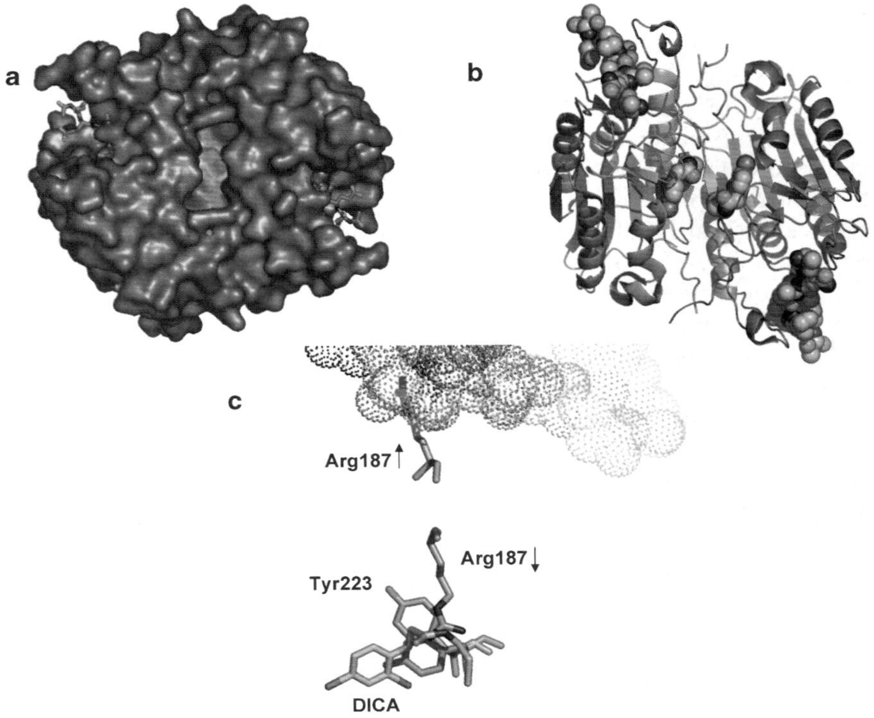

Figure 17.7 Allosteric Tethering in caspase-3 and -7. (a) The structure of caspase-7 (PDB ID 1F1J) bound at the active-site cysteine (green surface patch) to peptide-based inhibitor DEVD-FMK (green sticks) shows a large cavity (pink surface patch) at the dimer interface. Two native cysteine residues (yellow surface patches at C290 in caspase-7) are present in the cavity and were productive for tethering in the mass spectrometry-based and functional assays. (b) The new caspase-7 allosteric binding site (yellow spheres for compounds in the allosteric site) is at the dimer interface of the two caspase monomers (blue and green) and is spatially distant from the substrate-binding groove (peptide inhibitor in orange spheres). (c) Allosteric inhibitor DICA (yellow sticks) sterically clashes with the active conformation of caspase-7 (green sticks) and forces caspase-7 into an inactive conformation (purple sticks). This is accomplished when DICA contacts Tyr223 (green sticks), forcing Tyr223 (purple sticks) into the up position. This conformation of Tyr223 is in steric contact with Arg187 (green sticks), which subsequently causes Arg187 (purple sticks) to the up position. The up position of Arg187 (purple sticks) observed in the presence of the allosteric inhibitor is incompatible with binding of peptide (orange dots) in the active site.

The crystal structure of caspase-7 with the allosteric inhibitors revealed the mechanism of inhibition (Figures 17.7b and 17c). The stochiometric nature of inhibition was apparent when the crystal structure of these compounds was compared with the structure of caspase-7 in the presence of a substrate mimic (the covalent inhibitor z-DEVD–FMK). When DEVD binds to cleaved caspase-7, the presence of peptide orders the loops that together compose the substrate-binding cleft. Part of that ordering includes the movement of the L2 loop (the N-terminal end of a loop that is cleaved to convert caspase from the inactive zymogen to the cleaved and active caspase) toward the core of the protein, burying an arginine residue (R187, immediately adjacent to the active-site residue C186) in the core of caspase-7. When R187 is in this downward conformation, it sterically constrains the position of tyrosine Y223 into a downward conformation. This active conformation is incompatible with binding of FICA or DICA at the dimer interface cavity. When FICA or DICA bind to caspase-7 they constrain Y223 to adopt only the up conformation. This Y223 conformation is incompatible with binding of substrate because there is not room for R187 to bury in the protein core.

An additional layer of inhibition is also suggested by the FICA and DICA crystal structures. The binding of FICA and DICA appears to be driven nearly entirely by hydrophobic (entropically driven) interactions, partitioning the small molecules into the solvent-protected cavity. The hydrophobic nature of the compounds is certainly influential in their further protection from solvent by the conformation of one of the active-site loops (L2′, the C-terminal side of the loop cleaved upon conversion of the zymogen to the active caspase). In addition to interactions of FICA and DICA with the dimer cavity, there are strong hydrophobic interactions of the residues on the L2′ loop with the compounds. Either of these mechanisms appears to be sufficient to inactivate caspases: caspases that are incapable of binding substrate because of indirect competition between the L2 loop residue 187 burial and the allosteric site are inactive; caspases with the L2′ loop locked over the allosteric site instead of in the position that can allow substrate binding are also inactive.

An allosteric site in PTP1B, a negative regulator of the insulin receptor, was also discovered using Tethering, but in a somewhat less traditional format. In the context of searching for alkylating agents that could specifically label the active-site cysteine residue in PTP1B for use in Tethering with extenders, one cysteine reactive compound, ABDF, was discovered.[15] Modification of PTP1B by ABDF was rapid, reversible, and quantitative. Many of the other alkylating agents tested were equally reactive with three or more of the cysteine residues in PTP1B, presumably two solvent-exposed cysteines (C32 and C92) and the active-site cysteine (C215). ABDF was unique and noteworthy in that it selectively modified just one residue (C121, Figure 17.8), initially assumed to be the active-site cysteine, since the active-site cysteine has a pK_a of ~5.

Binding of ABDF quantitatively to PTP1B resulted in a 7.4-fold decrease in V_{max}, with no significant change in K_m. ABDF binding also did not result in full catalytic inhibition, even at quantitative labeling conditions, leading Hansen and co-workers to suspect that ABDF was not binding to the active site.[15] Electrospray ionization (ESI) mass spectrometry and peptide mapping confirmed that ABDF was selectively modifying the non-active-site residue C121.

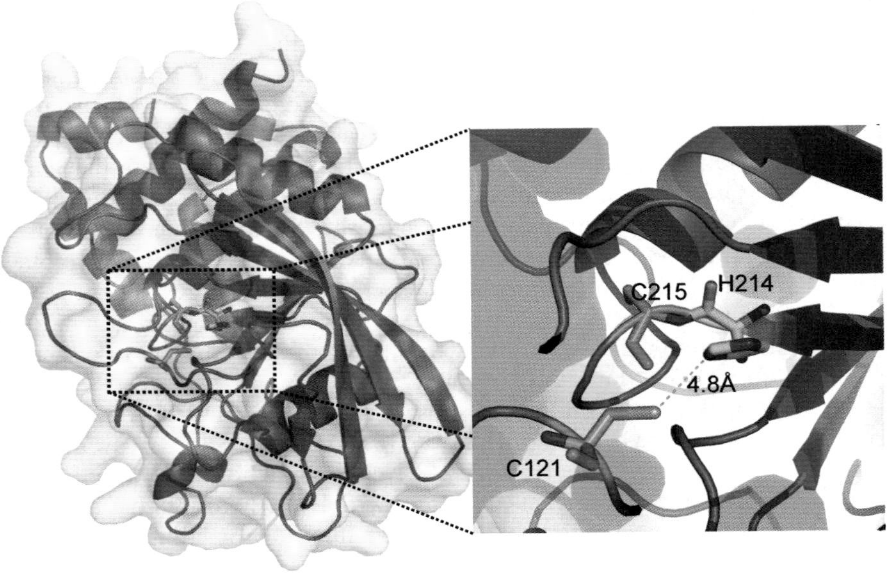

Figure 17.8 PTP1B ABDF-binding site. The serendipitous allosteric site discovered using ABDF at C121 is depicted as orange sticks. In this unliganded structure, C121 is pointed into the core of the protein. It is likely to change conformation upon binding to ABDF. The active-site cysteine (C215 pink sticks) and adjacent histidine (214 yellow sticks) are 4.8 Å from the cysteine that is modified.

C121 is conserved in most related phosphatases, including the phosphatases LAR and TCPTP. ABDF also inhibits LAR and TCPTP in a time- and dose-dependant manner, as would be expected for a rapidly binding, covalent compound. ABDF does not inactivate CD45, suggesting that there is some selectivity of binding and inhibition of ABDF. Although PTP1B readily crystallizes in the absence of ABDF or in the presence of other allosteric inhibitors which lock the WPD loop in an inactive conformation,[28] crystals of PTP1B in complex with ABDF could not be obtained, so the detailed mechanism of inhibition is not clear. Nevertheless, looking at the structure of PTP1B it is clear that the cysteine side-chain of C121 points toward the core, making hydrophobic interactions with Y124, which is in a hydrogen bonding network with H214, the residue adjacent to the active site. The inward direction of C121 in the crystal structures would not have been predicted to be ripe for tethering. Thus, conformational flexibility of PTP1B must allow the C121 thiol to become solvent accessible. It seems likely that binding of ABDF causes conformational changes in this region which, in Rube Goldberg fashion, is similar to the situation with FICA and DICA bound to caspase-7, where the conformation of the active site is disturbed. In contrast with the caspase-7 allosteric mechanism, any changes conferred to the active site do not have any effect on substrate binding. Discovery of allosteric sites involving cysteine residues is intriguing because it immediately suggests a mechanism by which native proteins could be targeted for drug discovery.

From the point of view of the discovery of new allosteric sites, it is interesting to note that examination of the structure of caspase-3 or -7 could have suggested that this site would be useful for small-molecule binding given its size, concavity, and through-protein distance to the active site (13 Å). It is likely that the new allosteric site in caspase-3 and -7 could have been selected based on geometric properties that were apparent in structure of caspase-7 with an active-site inhibitor or with no ligands bound.[29,30] The ideal future development of Tethering at allosteric sites is the ability to identify previously unexplored sites based on geometrical and proximity considerations and then exploit them as serendipitous allosteric sites.

Traditional HTS is poorly suited to probe allosteric sites such as the caspase and PTP1B allosteric sites because there is no mechanism to either target compounds toward any particular site, nor is there a ready means to determine the site of interaction. Tethering is by far the drug-discovery tool best suited to probe these sites. Tethering has the unique ability first to validate the ability of these sites to propagate an allosteric signal to the active site and second to identify fragments that bind to and modulate those sites.

17.8 Tethering as a Validation Tool

Small-molecule hits derived from HTS that do not act as quantitative competitive inhibitors are often discarded because they have a higher-than-acceptable rate of non-drug-like or artifactual inhibition (*e.g.* McGovern *et al.*[31,32]). This means that many legitimate and potentially useful allosteric inhibitors have likely been discarded as well. Because Tethering can probe protein interaction surfaces in a site-specific manner, it is the perfect tool for studying the location and mechanism of inhibition or activation of a small molecule that does not work as a competitive active-site inhibitor. In principle any small molecule could be converted to a disulfide-containing molecule and used with a series of cysteine mutants to determine the site of interaction. In the case of IL-2, a crystal structure of the complex between **1** and IL-2 had been determined so it was possible to engineer a disulfide-containing version of **1** with a three-methylene spacer that was competent to bind to a K43C mutant of IL-2 and antagonize binding in the same way that compound **1** behaved. It is likely that this same conversion would be successful in many other protein–small-molecule pairs.[16] In practice, if the site of interaction is completely unknown it would be a colossal task to find the site of interaction *ab initio*. Nevertheless, when some evidence suggests a site of interaction, Tethering is a useful technique for defining the site of interaction. Perhaps the best example of Tethering being used as a validation tool is with C5a receptor.

As previously noted, GPCRs are the most frequently targeted class of proteins for pharmaceutical control. Unfortunately, this class of proteins has been recalcitrant to structural studies due to difficulties of crystallization of integral membrane proteins. Tethering facilitates the use of disulfide capture as a means of determining the location of binding sites in the absence of other concrete structural information on C5a[8] (see Section 17.3). Tethering is particularly well-suited for determination of the site of binding when determination of a crystal structure of the complex is impossible to attain. The work on C5a validated the proposed site of peptide interaction on the C5a receptor and underscores the strength of Tethering to determine the site of action in difficult classes like GPCRs. For proteins interacting with peptides, this approach is also readily accessible.

Tethering was also useful in the validation of two sub-sites within the interface region of IL-2 that contacts the IL-2 receptor. Based on the number and type of hits that were observed across 12 engineered cysteine sites, the adaptive region was dissected into a rigid sub-site and an adaptive sub-site. This kind of information could be useful in directing lead optimization work in predicting what types of fragments are more likely to be successful.

17.9 Tethering *vs.* Traditional Medicinal Chemistry

A hallmarks of Tethering is that it can probe regions of protein space that are unapproachable for all practical purposes by traditional HTS combined with medicinal chemistry. Tethering can also rapidly identify binding fragments that would not be suggested by other methods because of the increased dynamic ranges of interaction energies accessible through Tethering. For example, a team applying Tethering to PTP1B was able to develop a new aryl-oxalamic acid pharmacophore in a way that was not obvious from traditional medicinal chemistry, and was much more rapid than would have been possible without Tethering.[9]

A previously addressed strength of Tethering is the predictive ability in how to recombine discovered fragments from screening. In studies on caspase-1, direct atom-for-atom replacement of atoms in an extended Tethering scheme gave the highest inhibition constant for caspase-1,[13] demonstrating the exquisite orientation specificity of the compounds selected by Tethering. However, when a series of linkers was introduced to replace the linker, the compound with the highest affinity was a benzenoid linker that had previously been exploited by Cytovia and Vertex in their development of caspase-1 inhibitors. Thus, even after using Tethering, a traditional medicinal chemistry approach is often required to improve the affinity of the monophore element. Is Tethering categorically better than traditional medicinal chemistry? Tethering certainly enjoys some distinct advantages over traditional medicinal chemistry, specifically in predicting recombination orientations.

Is Tethering faster? Tethering, particularly extended Tethering or breakaway Tethering, often provides a more direct means of performing SAR studies because it does not require synthesizing each compound by hand. In the case of IL-2, the results of Tethering suggested a focused set of just 20 compounds from which resulted an improvement in binding affinity from 3 μM to 60 nM.[17] In contrast, when those same compounds were interrogated with the standard medicinal chemistry approach, the binding affinity was significantly reduced. Even after extensive medicinal chemistry interrogation in various regions of the compound, the only approach that yielded a dramatic boost in affinity was the addition of a furanoic acid fragment that was discovered by Tethering at an adjacent residue.[20]

One arena where Tethering again offers a distinct advantage over other methods is that of non-active site binders. In these situations it is frequently difficult to conclude convincingly that the binding mode and mechanism are consistent with the inhibition or activation observed. When compounds are covalently liked via a Tethering interaction, much of the ambiguity is diminished. In addition, it is possible to derive structural information much earlier in the move from hit to lead with compounds that bind weakly. Because of disulfide stabilization, it is much more likely for the crystal structure of a low-affinity compound derived from Tethering to be determined than it is for low-affinity compound derived from a traditional high-throughput screen to be determined.

17.10 Tethering in Structural Determination

Determining the X-ray crystal or nuclear magnetic resonance (NMR) structure of small molecules in complex with the proteins to which they bind is often challenging. The concentrations of the complex required for crystallization or NMR spectroscopy are often prohibitively high and small-molecule inhibitors are often not soluble at concentrations required for crystallization or NMR. Even when compound solubility is not an issue, incomplete occupancy of the binding sites in all proteins in the sample can thwart structure determination.

The presence of small molecules can also affect crystal formation. In some cases, protein crystals will grow in the absence of a small-molecule inhibitor (or activator) but not in the presence of one, so the small molecule must be soaked into the crystalline lattice. Frequently this soaking disrupts and damages the crystal, so this is not a foolproof method for complex crystallization.

Crystallization of covalent complexes of compounds derived from Tethering offers several advantages. Because a covalent complex is formed stoichiometrically, one can be assured of full occupancy of all binding sites before the protein is crystallized or the solution structure determination proceeds. This confirmed occupancy dramatically increases the success of crystallization of compounds derived from Tethering. Additionally, because the investigator knows the site of modification the search for the location of the bound compound is simplified, which is especially helpful if the compound does not bind to the active site. Spurious binding of non-covalent small molecules to weak secondary sites sometimes occurs, but with disulfide-bound molecules this is not much of a risk, because small amounts of reducing agent can be included in the crystallization conditions to prevent any spurious binding.

17.11 The Challenge of Covalency

The advantages of Tethering as a tool for drug discovery in site determination, site-directedness, and the ability to combine monophores are clear. The most formidable challenge of the Tethering method is that of converting covalent fragments identified by Tethering to non-covalent molecules with drug-like properties. This necessary conversion requires empirical determination of a useful substituent to replace the disulfide moiety. This process usually requires the work of a team of medicinal chemists, so that hits derived from Tethering face some of the same challenges as hits derived from other HTS processes in terms of the effort required to produce a lead compound, which can be used in clinical trials, from a validated hit.

The success of converting a covalent disulfide-containing compound to a useful non-covalent compound depends exquisitely on the interaction energy of the monophore with the binding site on the protein. If the interaction energy of the monophore with the binding site is very high, then binding of the compound to the binding pocket is relatively free of the requirement for disulfide-bond formation to drive complex formation. For these compounds conversion to a non-covalent compound is very straightforward. Synthesis of an analog free of the disulfide bond is sufficient for conversion to a non-covalent analog. Although hits derived from the adaptive region of the IL-2 binding surface that were converted to non-covalent analogs by substitution of a methylene cap were competent to bind to IL-2 as assessed by SPR,[16] in general, examples of this simplistic type of conversion are rare. By definition and empirical observation through screening at various reductant concentrations, hits derived from Tethering have sufficient interaction energy between the monophore and the binding pocket that their resident time at the binding pocket promotes formation of a covalent disulfide bond between the compound and the protein. Nevertheless, for most Tethering hits, the interaction energy of the monophore with the binding pocket is not sufficient to promote meaningful binding on the time scale of biological assays in the absence of the covalent tether. In these situations when disulfide-free analogs are synthesized, the biological activity (*e.g.* inhibition of caspase activity) is no longer observed. In these cases, the monophore from Tethering can be used as an anchor for extended Tethering, or as a starting point for traditional synthetic medicinal chemistry efforts to generate analogs with improved binding affinity.

A related challenge of covalency is the ability to rank interaction energies of covalent hits in a way that predicts the interaction energies of their non-covalent counterparts. A standard method is via measurement of a β-ME$_{50}$ value (the β-ME concentration at which 50% of the protein in the sample is conjugated by a disulfide-containing compound, at a fixed compound concentration) or via measurement of a DR$_{50}$ (the disulfide-containing compound concentration at which 50% of the protein sample is covalently conjugated by the compound, at a fixed reductant concentration). Both of these values are straightforward to measure using standard mass spectrometry, however neither of these measures has been shown to have a strict relationship to the energetics of binding of the disulfide-free monophore. This is likely because the hydrophobic nature of the linker region contributes differently to the overall binding energetics for various compounds and their associated binding modes. This may be particularly true for introduced cysteine residues that can access more than one region of a binding site (*e.g.* caspase-7 FICA *vs.* DICA[7] or IL-2.[16])

Using extended Tethering, an irreversible warhead such as a chloro- or fluoro-methylketone attacks the active-site cysteine. This necessitates a modest medicinal chemistry effort to convert hits from extended Tethering to either non-covalent or reversibly covalent molecules. The caspase-1 team found that hits discovered from extended Tethering could be routinely converted from covalent to non-covalent molecules by simply performing atom-for-atom replacement of the atoms in the linker, and by replacement of the disulfide with an aldehyde moiety.[14] Aldehydes have also been widely used as covalent but reversible electrophiles against cysteine proteases. This replacement approach, while straightforward in terms of maintaining potency against the caspases, is not completely foolproof. Converted non-covalent monophores from Tethering with extenders on caspase-1 revealed two distinct binding modes when the linker was converted with a rigid thiophene linker rather than with an ethylene unit. This region of the compound sits in the hydrophobic S2 pocket and appears to determine the orientation of the molecule in the S4 binding pocket. Thus the binding mode is not uniquely dependent on the monophore, but is also influenced by the linker portion of the molecule.[14]

17.12 Hydrophobic Binders

The successes of Tethering are impressive and numerous. One challenge of Tethering is that hydrophobic pharmacophores are often selected. A general rule seems to be that hydrophobicity drives binding affinity while hydrogen bonds and overall shape complementarity drive specificity. Of the published molecules discovered from Tethering, a high proportion are relatively hydrophobic in nature, or derive their binding affinity largely through hydrophobic interactions. It is not surprising that a large number of small-molecule–protein interactions are driven largely by hydrophobic interactions.

Specific cases include C5a receptor for which the strongest compound was biaryl pyrrolidine. Though the exact mechanism of binding is not yet known, inspection of this molecule suggest that it would make predominantly hydrophobic interactions. The remaining reported hits for C5a were also hydrophobic in nature. When the adaptive region of IL-2 was probed by Tethering, the majority of the compounds selected were hydrophobic in nature.[5] Similarly, the best compound identified for allosterically inhibiting caspase-3 and -7 was a hydrophobic dicholorophenyl moiety. In the structures of caspase-7 with both FICA and DICA no direct hydrogen bonds between the small molecule and any protein atoms were observed. (The crystal structures were not of sufficiently high resolution to model water molecules accurately, so it is possible that water-mediated hydrogen bonds were formed.) Nevertheless, shape complementarity to the binding cavity and hydrophobic interactions appear to be the major driving force in this interaction.

Taking the published results of Tethering in sum, it seems fair to conclude that hydrophobic constituents are also favored to bind to the more adaptable regions. This was certainly the case for caspase-1 extended Tethering, where the best fragment discovered was a hydrophobic tricyclic moiety. Most of the binding affinity seems to arise from hydrophobic interactions with the residues that make up the binding pocket.[14] Is this apparent bias toward hydrophobic compounds a function of the compound collection or is it the inherent selectivity of the types of cavities that are being probed with Tethering?

Using cooperative Tethering on IL-2, fragments identified against Y31C and L72C were overlapping, meaning that fragments that were effective at Y31C were also effective when bound to L72C. Compound **1** itself is a hydrophobic biaryl moiety that caused a modest rearrangement of the surface relative to the unliganded structure (Figure 17.4). These fragments were predicted computationally to occupy a deep hydrophobic cavity on the surface of IL-2 that has been termed the "adaptive region".[17] The highest affinity compound produced from these studies (60 nM affinity) did not bind as computationally predicted, but caused a relatively dramatic rearrangement of this surface of IL-2 binding in a groove that had not previously been observed in any crystal structures (Figure 17.4).[19] This is an example of a small molecule carving out a new binding site based on physical, in this case hydrophobic, interactions that are addressable with small changes in the energy of the system. Thus, as with other methods of drug discovery, structure determination is the only unambiguous means to determine the binding mode of discovered fragments, particularly since hydrophobic constituents, which tend to partition to the core of the protein, are prevalent.

17.13 Conclusions: The Future of Tethering

Tethering has already been successfully applied to a wide variety of protein surfaces (Table 17.1) – somewhat featureless, but adaptable protein surfaces (IL-2), deep and fragile active sites (PTP1B), novel allosteric sites (caspase-3, caspase-7, PTP1B), cysteine-containing active sites (TS, caspases, PTP1B), proteins of known structure (most) and of unknown structure (C5a). Indeed, this might be a case where "more of the same" would be a great advance and we can expect that both natural cysteines and cysteines introduced all over the surfaces of a great variety of proteins will lead profitably to both our biological and pharmacological understanding. Applying the wealth of Tethering technologies to more protein targets promises a rich and fruitful road ahead.

A biological maxim might be that it is more probable to find inhibitory compounds than to find activating compounds. This is probably because there are nearly infinite ways to disrupt protein function (*e.g.* blocking ligand binding either directly or indirectly, disrupting proper protein folding, disrupting dynamics and conformational changes necessary to catalysis or binding) and any disruption can lead to loss of function. On the other hand, there are relatively fewer ways for a small molecule to activate a protein (*e.g.* bind and shift protein to a catalytically- or binding-competent form, stabilize the binding and transition states of the protein). Is it possible to identify activators using Tethering? Certainly it is possible due to the site-directed nature of Tethering and its unique ability to target and probe protein structure and conformation site-specifically, Tethering is perhaps the means of drug discovery with the best chance of developing activators for biologically important process. In fact, both agonist and antagonist small molecules have been derived from Tethering against the C5a receptor.[6] An exciting future frontier of Tethering will certainly be its application to the discovery of activating compounds.

To date, Tethering using disulfides has only been applied in a large-scale way at Sunesis Pharmaceuticals. Compared with traditional high-throughput compound screening and medicinal chemistry efforts, which have been performed on a large number of protein targets at a huge number of pharmaceutical companies and an increasing number of academic facilities, the application of Tethering has been modest in terms of both numbers of compounds and numbers of protein targets screened [currently publications exist on eight protein targets (see Table 17.1)]. In most publications, the library size reported was 10 000–30 000 compounds, however the fraction of mostly hydrophobic compounds in the compound collection has not been discussed. A larger and more diverse library of disulfide-containing compounds might improve the already marked success of Tethering and could influence the nature of the hits derived from Tethering. Nevertheless, the original intent of developing Tethering was that by directing drug discovery toward the pocket of interest it would be possible to find or generate (through mechanisms such as extended Tethering) high-affinity leads by searching a limited space, and it has been clearly demonstrated on a number of protein targets that this can be done with 10 000–30 000 compounds. Overall, Tethering can be measured as a resounding success both in the development of a robust new strategy for the discovery of protein surfaces that are good drug targets and for small molecules that bind to those sites.

References

1. D. A. Erlanson, A. C. Braisted, D. R. Raphael, M. Randal, R. M. Stroud, E. M. Gordon and J. A. Wells, Site-directed ligand discovery, *Proc. Natl. Acad. Sci. USA.*, 2000, **97**(17), 9367–9372.
2. D. A. Erlanson and S. K. Hansen, Making drugs on proteins: site-directed ligand discovery for fragment-based lead assembly, *Curr. Opin. Chem. Biol.*, 2004, **8**(4), 399–406.
3. D. A. Erlanson, R. S. McDowell and T. O'Brien, Fragment-based drug discovery, *J. Med. Chem.*, 2004, **47**(14), 3463–382.

4. D. A. Erlanson, J. A. Wells and A. C. Braisted, Tethering: fragment-based drug discovery, *Annu. Rev. Biophys. Biomol. Struct.*, 2004, **33**, 199–223.
5. M. R. Arkin, M. Randal, W. L. DeLano, J. Hyde, T. N. Luong, J. D. Oslob, D. R. Raphael, L. Taylor, J. Wang, R. S. McDowell, J. A. Wells and A. C. Braisted, Binding of small molecules to an adaptive protein–protein interface, *Proc. Natl. Acad. Sci. U. S. A.*, 2003, **100**(4), 1603–1608.
6. E. Buck and J. A. Wells, Disulfide trapping to localize small-molecule agonists and antagonists for a G protein-coupled receptor, *Proc. Natl. Acad. Sci. U. S. A.*, 2005, **102**(8), 2719–2724.
7. J. A. Hardy, J. Lam, J. T. Nguyen, T. O'Brien and J. A. Wells, Discovery of an allosteric site in the caspases, *Proc. Natl. Acad. Sci. U. S. A.*, 2004, **101**(34), 12461–12466.
8. E. Buck, H. Bourne and J. A. Wells, Site-specific disulfide capture of agonist and antagonist peptides on the C5a receptor, *J. Biol. Chem.*, 2005, **280**(6), 4009–4012.
9. D. A. Erlanson, R. S. McDowell, M. M. He, M. Randal, R. L. Simmons, J. Kung, A. Waight and S. K. Hansen, Discovery of a new phosphotyrosine mimetic for PTP1B using breakaway tethering, *J. Am. Chem. Soc.*, 2003, **125**(19), 5602–5603.
10. D. A. Allen, P. Pham, I. C. Choong, B. Fahr, M. T. Burdett, W. Lew, W. L. DeLano, E. M. Gordon, J. W. Lam, T. O'Brien and D. Lee, Identification of potent and novel small-molecule inhibitors of caspase-3, *Bioorg. Med. Chem. Lett.*, 2003, **13**(21), 3651–3655.
11. I. C. Choong, W. Lew, D. Lee, P. Pham, M. T. Burdett, J. W. Lam, C. Wiesmann, T. N. Luong, B. Fahr, W. L. DeLano, R. S. McDowell, D. A. Allen, D. A. Erlanson, E. M. Gordon and T. O'Brien, Identification of potent and selective small-molecule inhibitors of caspase-3 through the use of extended tethering and structure-based drug design, *J. Med. Chem.* 2002, **45**(23), 5005–5022.
12. D. A. Erlanson, J. W. Lam, C. Wiesmann, T. N. Luong, R. L. Simmons, W. L. DeLano, I. C. Choong, M. T. Burdett, W. M. Flanagan, D. Lee, E. M. Gordon and T. O'Brien, *In situ* assembly of enzyme inhibitors using extended tethering, *Nat. Biotechnol.*, 2003, **21**(3), 308–314.
13. B. T. Fahr, T. O'Brien, P. Pham, N. D. Waal, S. N. Baskaran, B. C. Raimundo, J. W. Lam, M. M. Sopko, H. E. Purkey and M. J. Romanowski, Tethering identifies fragment that yields potent inhibitors of human caspase-1, *Bioorg. Med. Chem. Lett.*, 2006, **16**(3), 559–562.
14. T. O'Brien, T. Fahr, M. M. Sopko, J. W. Lam, N. D. Waal, B. C. Raimundo, H. E. Purkey, P. Pham and M. J. Romanowski, Structural analysis of caspase-1 inhibitors derived from Tethering, *Acta Crystallogr., Sect. F: Struct. Biol Crystallogr. Commun.*, 2005, **61**(5), 451–458.
15. S. K. Hansen, M. T. Cancilla, T. P. Shiau, J. Kung, T. Chen and D. A. Erlanson, Allosteric inhibition of PTP1B activity by selective modification

of a non-active site cysteine residue, *Biochemistry*, 2005, **44**(21), 7704–77012.
16. J. Hyde, A. Braisted, M. Randal and M. R. Arkin, Discovery and characterization of cooperative ligand binding in the adaptive region of interleukin-2, *Biochemistry*, 2003, **42**(21), 6475–64783.
17. A. C. Braisted, J. D. Oslob, W. L. Delano, J. Hyde, R. S. McDowell, N. Waal, C. Yu, M. R. Arkin and B. C. Raimundo, Discovery of a potent small molecule IL-2 inhibitor through fragment assembly, *J. Am. Chem. Soc.*, 2003, **125**(13), 3714–3715.
18. B. C. Raimundo, J. D. Oslob, A. C. Braisted, J. Hyde, R. S. McDowell, M. Randal, N. D. Waal, J. Wilkinson, C. H. Yu and M. R. Arkin, Integrating fragment assembly and biophysical methods in the chemical advancement of small-molecule antagonists of IL-2: an approach for inhibiting protein–protein interactions, *J. Med. Chem.*, 2004, **47**(12), 3111–3130.
19. C. D. Thanos, M. Randal and J. A. Wells, Potent small-molecule binding to a dynamic hot spot on IL-2, *J. Am. Chem. Soc.*, 2003, **125**(50), 15280–15281.
20. N. D. Waal, W. Yang, J. D. Oslob, M. R. Arkin, J. Hyde, W. Lu, R. S. McDowell, C. H. Yu and B. C. Raimundo, Identification of nonpeptidic small-molecule inhibitors of interleukin-2, *Bioorg. Med. Chem. Lett.*, 2005, **15**(4), 983–987.
21. M. Gurrath, Peptide-binding G protein-coupled receptors: new opportunities for drug design, *Curr. Med. Chem.*, 2001, **8**(13), 1605–1648.
22. B. O. Gerber, E. C. Meng, V. Dotsch, T. J. Baranski and H. R. Bourne, An activation switch in the ligand binding pocket of the C5a receptor, *J. Biol. Chem.*, 2001, **276**(5), 3394–3400.
23. T. J. Baranski, P. Herzmark, O. Lichtarge, B. O. Gerber, J. Trueheart, E. C. Meng, T. Iiri, S. P. Sheikh and H. R. Bourne, C5a receptor activation. Genetic identification of critical residues in four transmembrane helices, *J. Biol. Chem.*, 1999, **274**(22), 15757–15765.
24. J. A. DeMartino, Z. D. Konteatis, S. J. Siciliano, G. Van Riper, D. J. Underwood, P. A. Fischer and M. S. Springer, Arginine 206 of the C5a receptor is critical for ligand recognition and receptor activation by C-terminal hexapeptide analogs, *J. Biol. Chem.*, 1995, **270**(27), 15966–1599.
25. M. J. Romanowski, J. M. Scheer, T. O'Brien and R. S. McDowell, Crystal structures of a ligand-free and malonate-bound human caspase-1: implications for the mechanism of substrate binding, *Structure*, 2004, **12**(8), 1361–1371.
26. C. Pargellis, L. Tong, L. Churchill, P. F. Cirillo, T. Gilmore, A. G. Graham, P. M. Grob, E. R. Hickey, N. Moss, S. Pav and J. Regan, Inhibition of p38 MAP kinase by utilizing a novel allosteric binding site, *Nat. Struct. Biol.*, 2002, **9**(4), 268–272.
27. R. Esnouf, J. Ren, C. Ross, Y. Jones, D. Stammers and D. Stuart, Mechanism of inhibition of HIV-1 reverse transcriptase by non-nucleoside inhibitors, *Nat. Struct. Biol.*, 1995, **2**(4), 303–308.

28. C. Wiesmann, K. J. Barr, J. Kung, J. Zhu, D. A. Erlanson, W. Shen, B. J. Fahr, M. Zhong, L. Taylor, M. Randal, R. S. McDowell and S. K. Hansen, Allosteric inhibition of protein tyrosine phosphatase 1B, *Nat. Struct. Mol. Biol.*, 2004, **11**(8), 730–737.
29. J. Chai, E. Shiozaki, S. M. Srinivasa, Qi Wu, P. Dataa, E. S. Alnemri and Y. Shi, Structural basis of caspase-7 inhibition by XIAP, *Cell*, 2001, **104**(5), 769–780.
30. Y. Wei, T. Fox T, S. P. Chambers, J. Sintchak, J. T. Coll, J. M. Golec, L. Swenson, K. P. Wilson and P. S. Charifson, The structures of caspases-1, -3, -7 and -8 reveal the basis for substrate and inhibitor selectivity, *Chem. Biol.*, 2000, **7**(6), 423–432.
31. S. L. McGovern, B. T. Helfand, B. Feng and B. K. Shoichet, A specific mechanism of nonspecific inhibition, *J. Med. Chem.*, 2003, **46**(20), 4265–4272.
32. S. L. McGovern, E. Caselli, N. Grigorieff and B. K. Shoichet, A common mechanism underlying promiscuous inhibitors from virtual and high-throughput screening, *J. Med. Chem.*, 2002, **45**(8), 1712–1722.

CHAPTER 18

The Impact of Protein Kinase Structures on Drug Discovery

CHAO ZHANG[a] AND SUNG-HOU KIM[b]

[a]Plexxikon Inc., Berkeley CA, USA
[b]Department of Chemistry, University of California, Berkeley CA, USA

18.1 Introduction

Protein phosphorylation regulates many aspects of cellular functions. Abnormal phosphorylation is implicated in a myriad of human diseases. Eukaryotic protein kinases make up an unusually large family of proteins related by a homologous catalytic domain (known as the protein kinase domain). The sequenced human genome revealed 478 distinct kinase genes,[1] offering a rich collection of targets for developing novel therapeutics for human diseases.

Protein kinases have become the second most explored family of proteins (after G-protein coupled receptors) as drug targets, accounting for more than 20% of drug-discovery projects at many pharmaceutical companies.[2] Small molecule inhibitors are being developed for kinases across all major subfamilies (for detailed phylogenetic classification of the kinome, see Manning *et al.*[1] and Hanks and Hunter[3]). The endeavor led in 2001 to the commercialization of Gleevec (Glivec, imatinib mesylate, STI-571, Novartis),[4] the first protein-kinase inhibitor used as a drug for human disease. This was followed by the Food and Drug Administration's (FDA's) approval of two epidermal growth factor receptor (EGFR) kinase inhibitors, Iressa (gefitinib, ZD-1839, AstraZeneca)[5] and Tarceva (erlotinib, OSI-774, OSI Pharmaceuticals/Roche/Genentech),[6] both for non-small cell lung cancer treatment. Other small molecules that target specific kinases are currently in various stages of clinical testing. Significant developments in this field are expected in the next decade, including breakthroughs for treating some of the most devastating human diseases.[7,8]

One of the important catalysts for the discovery of kinase inhibitors is the availability of the three-dimensional structures of kinases and their complexes with inhibitors. The early 1990s witnessed the publication of the first kinase structure (cAMP-dependent protein kinase or PKA),[9] which unveiled a characteristic bilobular fold with a conserved ATP-binding site sandwiched by the two lobes (Figure 18.1A). This was followed by the structures of CDK2,[10] ERK2,[11] twitchin kinase,[12] and the kinase domain of insulin receptor,[13] and then by a series of structures of kinase-inhibitor complexes.[14,15] The structures of more than 60 protein kinases (Figure 18.1B) have now been solved [Figure 18.1C shows the number of kinase structures deposited in the Protein Data Bank (PDB) every year from 1995 to 2004]. Many of these are inhibitor complexes containing small molecules targeting the structurally conserved ATP-binding sites. The diverse biological functions of kinases, not only as enzymes, but also as signaling molecules, are reflected at molecular level in the numerous ways by which the activities of kinases are regulated. An increasing

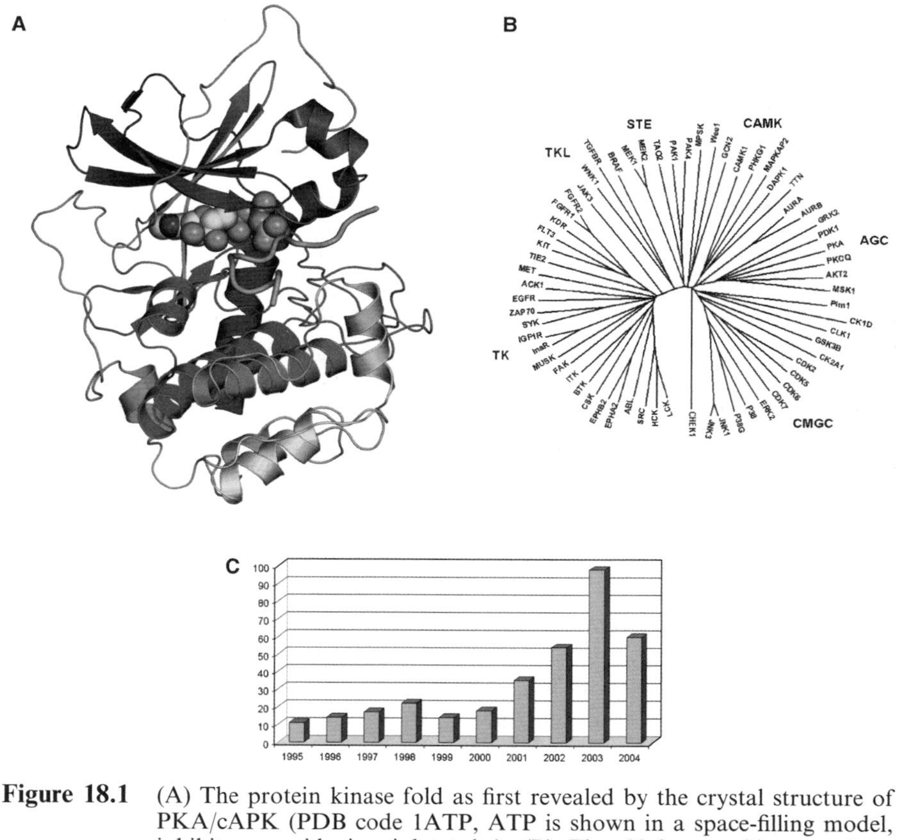

Figure 18.1 (A) The protein kinase fold as first revealed by the crystal structure of PKA/cAPK (PDB code 1ATP, ATP is shown in a space-filling model, inhibitor peptide in violet tube). (B) The 64 human kinases whose structures have been solved (group names: TK, tyrosine kinase; TKL, tyrosine kinase-like; STE, homologs of yeast Sterile 7, Sterile 11, Sterile 20 kinases; CAMK, calcium-/calmodulin-dependent protein kinase; AGC, containing PKA, PKG, PKC families; CMGC, containing CDK, MAPK, GSK3, CLK families). (C) Number of kinase structures deposited each year from 1995 to 2004.

number of such regulations are now captured by X-ray crystallography, revealing the remarkable structural plasticity of the protein kinase domain[16] and novel opportunities for structure-guided design of a new generation of kinase inhibitors.

18.2 The Hinge Region and the Concept of Kinase Inhibitor Scaffold

All kinase structures determined to date exhibit a conserved bilobular fold; the hinge linking the smaller N-terminal lobe to the larger C-terminal lobe plays an important role in the specific binding of ATP, the common cofactor of kinases. The ATP-binding pocket consists of a number of conserved structural elements that interact with the various functional groups of ATP (Figure 18.2). For example, the binding of the phosphates involves conserved catalytic site residues, including the aspartate from the activation loop (A-loop), asparagine from the catalytic loop, lysine from the third β-strand (β3) of the N-terminal lobe, and the glycine-rich phosphate-binding fragment (also called P-loop). However, it is the adenine-binding site that has been at the center of attention for kinase-inhibitor discovery. The hydrophobic cleft formed by conserved residues from the main β-sheet of the N-terminal lobe and conserved residues

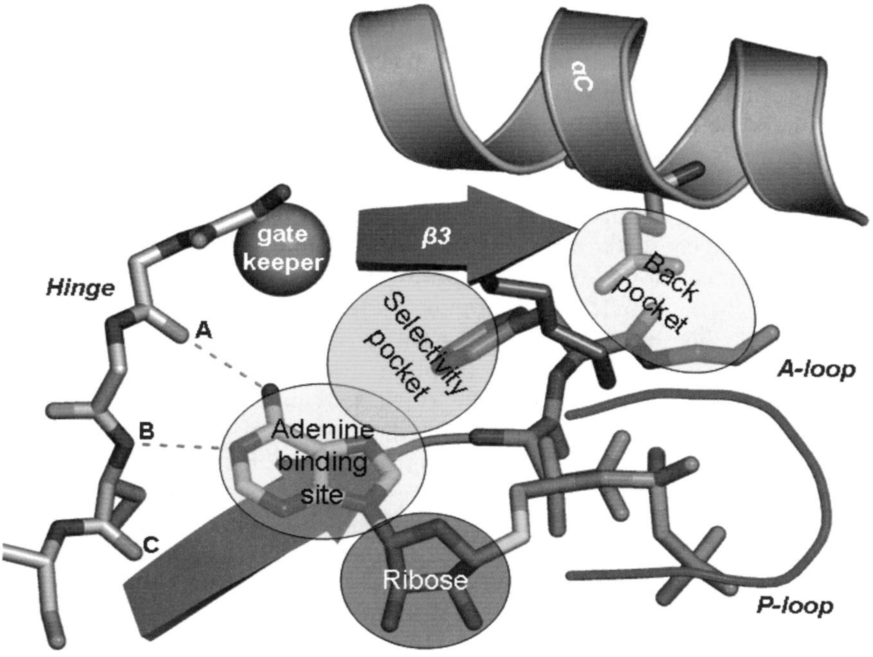

Figure 18.2 A regional map showing the ATP-binding site and other potential inhibitor-binding pockets in a protein kinase.

from a small β-sheet of the C-terminal lobe shows an overall preference for a ring-like structure, whereas the backbone amine and carbonyl groups of the hinge dictates specific recognition of adenine and its competitors.

The hinge lines the ATP-binding pocket with two hydrogen bond acceptors [in Figure 18.2, the main chain carbonyl of the first hinge residue (Hinge-1) is designated as 'A', the main chain carbonyl group of the third hinge residue (Hinge-3) is designated as 'C'), and a hydrogen bond donor (the main chain nitrogen of Hinge-3) is denoted 'B'). The nearly linear arrangement of three groups forming hydrogen bonds represents a distinct feature of the adenine-binding pocket of the kinases. A survey of the published structures of protein kinase–small molecule complexes suggests that a vast majority of the inhibitors contain polar groups near the hinge to make at least one hydrogen bond. The hydrogen bond that is almost always formed involves the hydrogen bond donor B of the hinge. In addition to this interaction, many inhibitors also form a second hydrogen bond with either of the two hydrogen bond acceptors (A or C). For example, the structures of the "dual cosubstrate" protein kinase CK2 complexed with ATP and GTP demonstrate that the two substrates form distinct hydrogen-bond interactions with the hinge: the A–B pair is used in ATP binding whereas the B–C pair is used in GTP binding.[17] Compounds that form a triplet of hydrogen bonds with the hinge (*i.e.* making use of all three hinge hydrogen-bonding groups A, B, and C) have also been observed. Figure 18.3 provides a catalog of structures showing how kinase inhibitors of different chemical classes bind to their respective targets through interactions with the hinge regions. An interesting deviation from the A, B, and C hydrogen-bond triad was observed in p38α MAP kinase where a peptide flip between Hinge-3 and Hinge-4 (Met-109 and Gly-110 in p38α numbering)[18] alters the hydrogen-bond polarity of C (from an acceptor to a donor). This change is crucial for the binding of a group of highly selective p38 inhibitors.

Because the binding of ATP is essential for kinase activities, agents targeting the structurally conserved ATP-binding sites account for the majority of the kinase inhibitors known today. The discovery of these inhibitors has largely been guided by the search for compounds containing the "privileged" chemical moieties that meet the interaction requirements of the adenine-binding site and the subsequent optimization of the compounds for potent and selective binding to individual kinases. Because these privileged moieties are kept constant throughout chemical exploration of other parts of the molecules (the R-groups), they are also called scaffolds. The key interactions that a scaffold uses to bind a target protein often involve residues or structural features that are conserved in the target family (*e.g.* the adenine-binding site in the kinase); therefore, the scaffolds discovered based on one family member (*e.g.* CDK2 or p38) can sometimes provide a good starting point for developing inhibitors for other members of the family. Thus, scaffold-based lead discovery is a powerful and efficient strategy for rapid generation of novel drug candidates with high potency and specificity by exploiting the similarities and differences among a set of related targets.

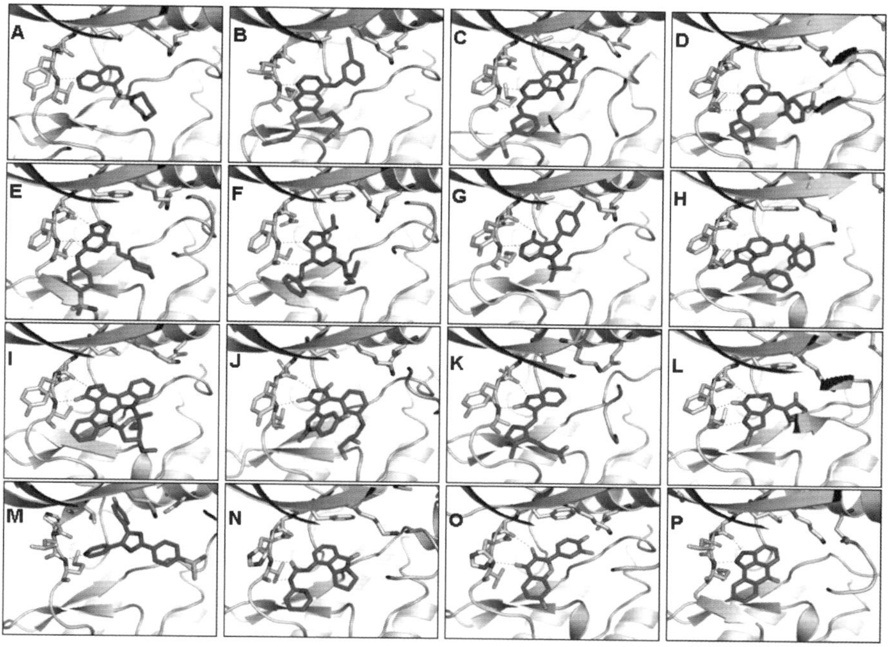

Figure 18.3 Selected co-crystal structures illustrating distinct kinase-inhibitor scaffolds. (A) PKA with Fasudil (1Q8W); (B) EGFR kinase with erlotinib/GW572016 (1M17); (C) ABL kinase with PD173955 (1M52); (D) CDK2 with a 4,6-bis-amino-pyrimidine CDK4 inhibitor (1V1K); (E) CDK2 with NU6102 (1H1S); (F) CDK5 with roscovitine (1UNL); (G) LCK with PP2 (1QPE); (H) CDK2 with an amino-imidazopyridine inhibitor (1PYE); (I) CHEK1 with staurosporine (1NVR); (J) GSK3β with an indolyl-maleimide inhibitor (1R0E); (K) FGFR1 kinase with SU5402 (1FGI); (L) CDK2 with hymenialdisine (1DM2); (M) p38 MPAK with SB203580 (1A9U); (N) CDK4 mimic CDK2 with a pyrazolyl-urea inhibitor (1GII); (O) CDK6 with Fisetin (1XO2); and (P) JNK3 with SP600125 (1PMV). The inhibitors are shown in sticks, as are the side chains of the three hinge residues, the conserved lysine from β3, and the gatekeeper residue from β5.

18.3 High-throughput Crystallography for the Discovery of Novel Scaffolds

18.3.1 High Potency-High Specificity-High Molecular (H3) Weight Screening

A majority of the known kinase inhibitors are derived from a dozen or so widely used scaffolds.[19] Additional novel scaffolds that enable the generation of leads with distinct biological profiles and pharmacological properties are needed to fulfill the great therapeutic potential of the kinase family. High-throughput screening (HTS), which has been applied to many target families to identify leads, has yielded relatively few new kinase-inhibitor scaffolds. The lack of success of HTS in novel scaffold discovery can be attributed to the large size and complexity of compounds screened, as well as to the screening criteria of high potency and high specificity. Chemical moieties that have scaffold-like properties are frequently missed because the remaining, less desirable parts of the molecules prevent the compounds from attaining the activity threshold required for detection under such criteria. Even in cases when the scaffold-containing compounds present themselves as HTS hits, the identification of the scaffold components is not trivial and usually requires extensive chemistry efforts and structure–activity relationship analyses.

18.3.2 Low Potency-Low Specificity-Low Molecular Weight (L3) Screening

To break away from the high potency, high specificity, and high molecule weight HTS paradigm, a new approach to scaffold discovery emphasizing screening of low potency, low specificity, and low molecular weight compounds has emerged. Low molecular weight (250–350Da) compounds provide a rich, untapped source of potential new scaffolds for drug discovery. Such molecules are large enough to exhibit weak but detectable biological activity; but at the same time they still have room to accommodate additional substitutions for designing-in potency and specificity while retaining favorable pharmacological properties. With the advance in the speed of co-crystallography, a structure-guided approach can be used to mine compounds of low molecular weight. To identify the most promising candidates, one starts with compounds that meet a minimal inhibitory activity criterion and then uses co-crystallography to evaluate the quality of compounds as scaffolds based on their interactions with key features of the protein and their potential for modification to increase potency and selectivity. This approach leads to the rapid discovery of novel scaffolds with exceptional binding efficiency. These scaffolds can be developed into multiple lead series that maintain low complexity and molecular weight, enhancing the possibility of successful drug development.

Recently, scientists at Plexxikon Inc., published their work in using co-crystallography of a serine–threonine kinase (PIM1) with low molecular weight molecules as a means to the discovery of novel chemical scaffolds.[20] In this experiment, over 200 low molecular weight (< 350 Da) compounds with weak inhibitory activity ($> 30\%$ inhibition of PIM1 kinase activity at 200 μM compound concentration) in a biochemical assay were identified. After a high-throughput co-crystallography campaign, over 150 crystallographic data sets were collected and their structures determined and 70 co-crystal structures with compounds bound in the ATP binding site were obtained. As shown in Figure 18.4, the majority of the co-crystallized ligands are found between the hinge and the conserved Lys-67–Glu-89 salt bridge. PIM1 is an unusual kinase in that it

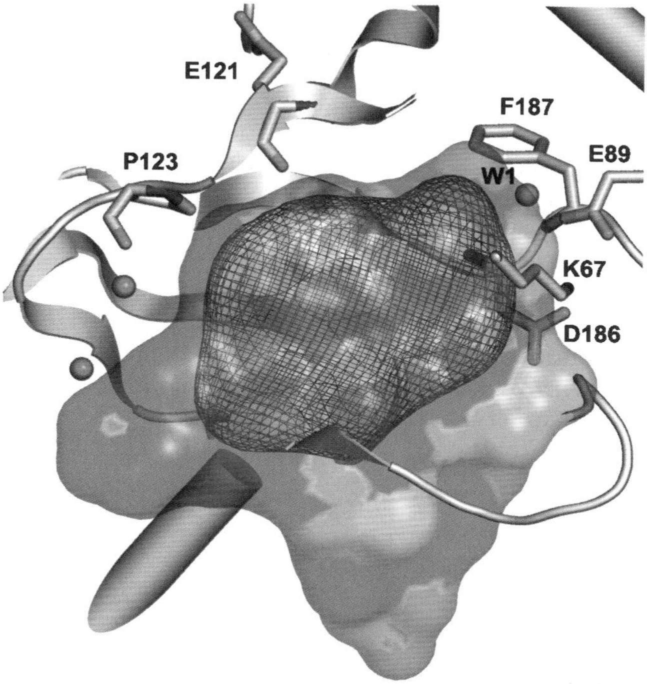

Figure 18.4 Molecular surface of the 70 low molecular weight compounds that have been co-crystallized with PIM1. The standard atom-type coloring scheme is used to render the ligand surface (carbons in green, oxygen red, nitrogen blue, and halogen purple). The mesh inside the molecular surface encloses a region where a large number of compounds overlap. For clarity, only protein residues and the three water molecules that were observed in a majority of the 70 co-crystal structures are shown.

possesses a proline at Hinge-3 (Pro-123 in PIM1 numbering) where the corresponding residues in other kinases provide the canonical hydrogen-bonding donor B (Figure 18.2). Because residue Pro-123 in PIM1 is incapable of making a hydrogen bond due to its lack of the amide hydrogen, the atoms in the compounds that are in closest proximity to Pro-123 are hydrophobic in nature, rather than hydrogen-bond accepting as is found routinely in other kinase-inhibitor complexes. This ensemble of scaffold candidates reveals how diverse compounds can satisfy the binding requirement of a kinase and provides the foundation for optimization of selected scaffolds.

18.4 The Gatekeeper Residue and the Selectivity Pocket

Once scaffolds have been identified, evolving scaffolds into selective drug leads requires a detailed understanding of the specific structural features of individual targets. Most of the first-generation kinase inhibitors occupy the same space as ATP, raising the concern that target-specific inhibitors may not be possible for kinases. The work of Tong *et al.*[21] in revealing an interior pocket adjacent to the ATP-binding site of p38 MAP kinase represents a key breakthrough. The SB202190 and SB203580 series are the first ATP-competitive inhibitors that show relative specificity for p38. The structure of p38 in complex with SB203580 or close analogs[21,22] shows that only a portion of the compound overlaps with ATP, whereas the 4-fluorophenyl moiety binds in a small hydrophobic groove between Lys-53 and Thr-106 (Figure 18.5A). The presence of a small side chain at position 106 is critical, as its mutation to a bulkier hydrophobic residue makes p38 insensitive to B203580. In CDK2, one of the most exploited kinases by structure-based inhibitor design,[14,23] the residue at the equivalent position is Phe-80, and the interior pocket is not accessible (Figure 18.5B). The residues in kinases that correspond to Thr-106 of p38 or Phe-80 of CDK2 thus serve as gatekeepers to the interior pocket and control its accessibility to inhibitors.

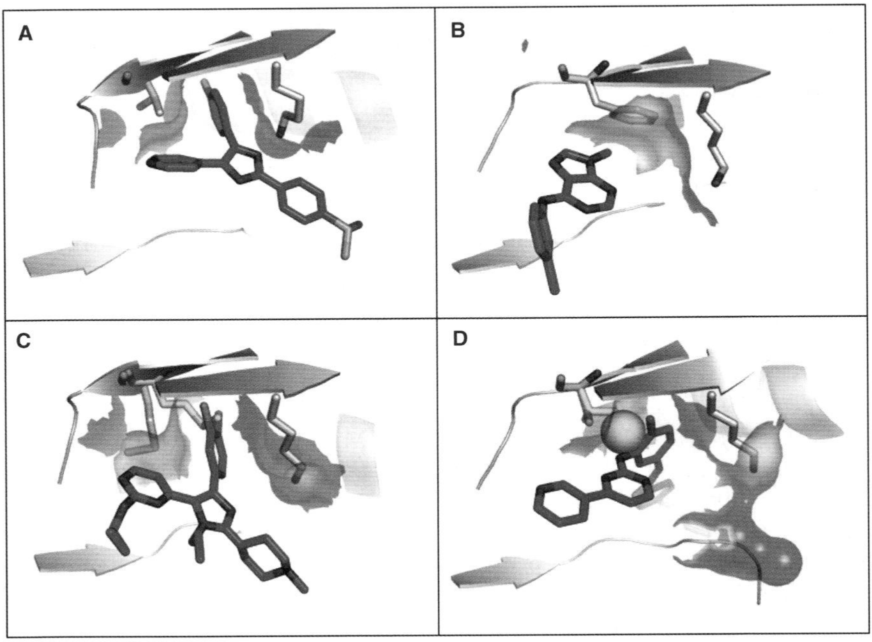

Figure 18.5 The gatekeeper residue. (A) The fluorophenyl group of p38-selective inhibitor SB203580 binds to a hydrophobic cleft between gatekeeper Thr-106 and Lys-53 (1A9U). (B) The Phe-80 in CDK2 closes down the interior pocket (illustrated here is the CDK2-Purvalanol B co-structure, 1CKP). (C) Met-146 of JNK3 moved down by 3 Å to accommodate the dichlorophenyl side chain of an imidazolepyrimidine inhibitor (Met-146 in the bound and unbound states are colored in gray and yellow, respectively, based on 1PMN/1PMV). (D) T315I substation interferes with imatinib binding to the ABL kinase (1IEP).

Exploiting the differences in the gatekeeper residue and the interior pocket (often referred to as the selectivity pocket) is an effective strategy to develop highly specific kinase inhibitors.[24] Based on the size of the gatekeeper residue, human kinases can be separated into two groups. About a quarter of the kinases in humans have a small gatekeeper residue (threonine, valine, serine, or cysteine) and, for this group of kinases, selective inhibitors can be designed by exploiting the unique shape and chemical features of the selectivity pocket. The remaining kinases possess a bulkier side chain (phenylalanine, leucine, isoleucine, methionine, or glutamine) at the gatekeeper position, and the access to the interior pocket from the ATP-binding site is highly restricted. As a result, inhibitors for these kinases reside completely in the ATP-binding pocket. An interesting exception is JNK3 where the gatekeeper Met-146 side chain moves away from the Lys-93 upon binding of the imidazolepyrimidine-based inhibitors to create room for the hydrophobic substituents of the inhibitors (Figure 18.5C).[25] It remains to be seen whether similar induced-fit mechanisms can be employed by inhibitors of other kinases that also possess a large yet flexible residue as the gatekeeper.

The importance of the gatekeeper residue is demonstrated in drug-resistant mutation, a well-recognized problem that could beset the benefit of target-based therapy.[26] The therapeutic inactivation of an essential protein creates selective pressures for diseased cells to evolve mechanisms of resistance. Recent clinical data indicate that tumor cells can produce drug-resistant variants of the targeted protein. For example, resistance to the ABL kinase inhibitor imatinib frequently results from point mutations within the BCR–ABL kinase domain that interfere with drug binding, one of which is the substitution of the gatekeeper threonine by isoleucine (T315I, Figure 18.5D.[27]) Although introducing the bulkier isoleucine side chain diminishes imatinib binding, it does not affect ATP binding, which explains why the imatinib-refractory T315I allele retains the tumor-promoting function.

18.5 The Conformational States of the DFG Motif and the Opening of the Back Pocket

The activation loop controls the catalytic activity in most kinases by switching between different states. The beginning portion of the activation loop, named the DFG motif after the conserved aspartate–phenylalanine–glycine triad in the region, is particularly relevant for small-molecule binding. The DFG motif in a fully active kinase adopts a highly conserved 'open' conformation (see Figure 18.6A for an example).[28] The aspartate of the DFG triad points into the ATP pocket (ready to bind the ATP cofactor), whereas the phenylalanine is held at the base of the selectivity pocket, contributing to the largely hydrophobic surface of the pocket (other lipophilic side chains from the β-strands 3, 4, and 5 and αC helix also participate). The salt bridge formed between the conserved β3 lysine and αC glutamate, along with the first two main-chain nitrogens of the DFG triad, constitutes a polar patch on one side of the pocket. In kinases with

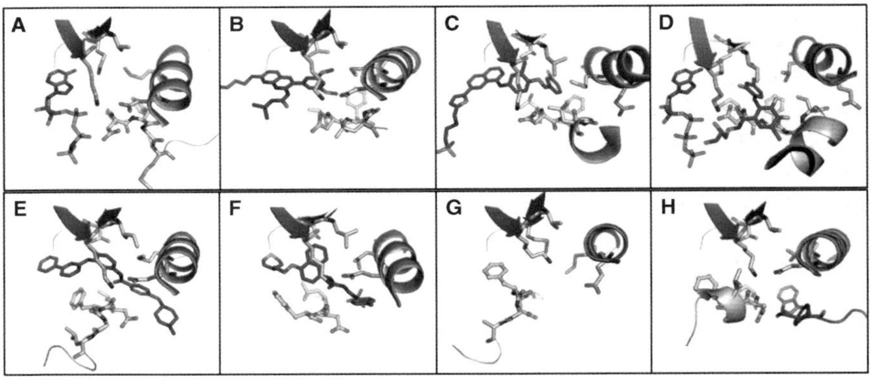

Figure 18.6 Representative crystal structures showing distinct DFG conformations and their effects on the shape and properties of the selectivity pocket. (A) Insulin receptor kinase in complex with ATP (1IR3); (B) FGFR1 kinase with PD173074 (2FGI); (C) EGFR kinase with GW572016 (1XKK); (D) MEK1 with PD318088, an analog of PD184352 (1S9J); (E) ABL kinase with imatinib (1IEP); (F) p38 MAPK with BIRB796 (1KV2); (G) inactive insulin receptor kinase (1IRK); (H) auto-inhibited state of c-Kit kinase (1T45; Trp-557 from the N-terminal juxtamembrane domain is shown in blue).

a small gatekeeper residue, the selectivity pocket can accommodate an aromatic ring with small substituents (the FGFR1 kinase-PD173074[29] complex is given as an example in Figure 18.6B).

The DFG motif can deviate from the open conformation and adopt a more compact structure (*e.g.* one α-helical turn or a β-turn) that folds in between the β3-lysine and the αC-helix. This causes a concerted shift in the αC-helix and the separation of the lysine–glutamate salt bridge. These structural changes have been observed in a number of kinases and, in most cases, they result in an expansion of the selectivity pocket for larger substituents. For example, in the structure of the EGFR kinase–GW572016 complex[30] (Figure 18.6C), the 3-fluorobenzyloxy group of GW572016 occupies the space deep in the back of the pocket and such a binding is responsible for the unusually low off-rate of the inhibitor. The structure of MEK1 bound with a PD184352 analogue (PD318088)[31] represents an even more dramatic example where the whole backside between the β3-lysine and the activation loop is opened up. The creation of the new pocket is coupled to the formation of a stable two-turn α-helical structure by residues adjacent to the DFG motif in the activation loop (Figure 18.6D). Note that the new pocket merges with the ATP pocket near the phosphate binding site rather than the adenine binding site (MEK1 has a large gatekeeper residue Met-143). Here we name this pocket the MEK-like back pocket to distinguish it from the ABL kinase-like back pocket revealed earlier by the ABL kinase–imatinib complex structure (see below). The MEK-like back pocket exists independent of the gatekeeper residue.

The structure of imatinib [Gleevec (Glivec)] in complex with the ABL kinase (Figure 18.6E)[32] revealed, for the first time, that an inhibitor can displace the DFG motif and project into a large opening between the αC helix and the activation loop. Because the new pocket is also located on the back side of the ATP pocket, we name it the ABL kinase-like back pocket. Of all the structural changes required for imatinib binding, the distortion at the DFG motif is the most noticeable. An approximately 120° rotation about the ϕ main chain torsion angle of Asp-381 brings Phe-382 out to the ATP pocket, and turns Asp-381 to face the back pocket. This conformational change creates an inactive state of the kinase because the flipped-out phenylalanine obstructs ATP binding. Several other kinase inhibitors, including BIRB796 (p38 MAPK inhibitor, Figure 18.6F[33]), BAY93-9006 (B-RAF inhibitor[34]), AAL993 (KDR inhibitor[35]), have subsequently been shown to bind in the same region in space as imatinib.

The very first crystal structure of the insulin receptor tyrosine kinase domain[13] exhibits the same DFG conformation as the imatinib-bound ABL kinase structure (Figure 18.6G). Subsequently published structures of CDK6 bound to tumor suppressors INK4a and INK4d[36] and the structure of unphosphorylated AKT2 kinase also show the same inactive-state DFG conformation.[37] Therefore, the opening of the ABL kinase-like back pocket can occur in the absence of small-molecule inhibitors. Recently, it has been shown that the ABL kinase-like back pocket is the direct target of kinase autoinhibition by the juxtamembrane domain in both Flt-3 kinase and c-Kit kinase[38,39] (Figure 18.6H). Compounds that bind solely to the back pocket may mimic the role of the autoinhibitory element and prevent kinases from attaining the active conformation.

18.6 Allosteric Inhibitors, Non-ATP Competitive Inhibitors, and Irreversible Inhibitors

Structural identification of the back pockets in a number of kinases has intensified the effort to discover inhibitors targeting regions that are less conserved and spatially distinct from the ATP-binding pocket. These inhibitors are expected to be more selective than the inhibitors that use the ATP pocket. The ABL kinase-like back pocket is an allosteric binding site because compounds binding to this pocket stabilize a conformation of the DFG motif that is incompatible with ATP binding. Inhibitors such as imatinib, BIRB796, and BAY43-9006 represent the first generation of compounds that utilize the allosteric binding site. However, these compounds still contain structural overlaps with ATP. The first example of an inhibitor that binds exclusively in the allosteric back pocket is a *N*,*N*′-diaryl urea-based compound (a precursor of BIRB796) in complex with p38 MAP kinase,[33] although the compound still takes advantage of the hydrophobic cleft made available by the small gatekeeper residue (Thr-106) in p38. Given that kinases with large gatekeeper residues (*e.g.* insulin receptor kinase domain, AKT2, CDK6, *etc.*) can adopt the

ABL-like conformation in the inactive states, the discovery of allosteric inhibitors for kinases with a large gatekeeper may also be possible.

The MEK1/MEK2 inhibitor PD98059 is one of the few synthetic protein-kinase inhibitors that are not ATP competitive. It has been shown that both PD98059 and the subsequently developed more potent compound PD184352 prevent the activation of MEK1 much more potently than they inhibit MEK1 kinase activity.[40] The crystal structures of MEK1 and MEK2 in complexes with ATP and PD184352-like inhibitors provided the first structural illustrations of noncompetitive inhibition of protein kinases by small-molecule inhibitors.[31] The binding of the noncompetitive inhibitors to the back pocket of MEK1 and MEK2 does not require significant conformational change in the DFG motif, and thus does not preclude ATP binding. Understanding the structural basis of noncompetitive inhibition makes it possible to design other novel noncompetitive kinase inhibitors.

CI-1033 (PD183805) is the first irreversible protein kinase inhibitor that has advanced into clinical testing.[41] CI-1033 is a pan-erbB receptor tyrosine kinase inhibitor that effectively blocks signal transduction through all four members of the EGFR kinase family. CI-1033 contains a mild Michael acceptor on the 6-position of the quinazoline scaffold, which allows very specific alkylation of Cys-773 in the ATP pocket of the EGFR kinases. A survey of the human kinome shows that cysteines can be present at several key residue positions inside the ATP-binding pocket. Covalently attacking these cysteines by a mild Michael acceptor can be used as a general strategy to discover selective, irreversible kinase inhibitors.[42] Because the cysteines are located inside the ATP pocket, the irreversible inhibitor still needs to bind potently into the ATP pocket to place the Michael acceptor in close proximity with the intended cysteine. Therefore, it is possible to rationally design molecules that are active only when bound to the target kinase. For diseases such as cancer, the irreversible nature of these compounds may offer potential advantages, including prolonged target suppression for maximum anti-tumor activity and minimal requirement for *in vivo* pharmacokinetics.

18.7 Discovering Kinase Inhibitors in a 500-Dimensional Space

Kinase inhibitors represent a new generation of drugs that act only on specific cellular targets that play a major role in particular diseases. While the discovery of highly selective compounds allows for target-specific intervention, complex diseases such as cancer call for drugs that can target multiple pathways involved in the disease process. Clinical experience with several kinase inhibitors demonstrated that inhibitors with potencies against a combination of targets might be valuable in achieving useful activity for treating complex diseases.[43] Technological advances in bioassay now make it possible to screen an extended number of compounds against the majority of the ~500 human kinases and to determine the full target-inhibition profile.[44] The inhibition profile data can be analyzed together with the phenotypes elicited by these compounds to distinguish combinations of kinases whose inhibition can have a synergistically beneficial effect for a particular indication from those whose inhibition leads to adverse effects.

In light of this paradigm shift, what are the best strategies for identifying small-molecule kinase inhibitors that have the desired biological effects? Inhibitor specificity is not dictated by the primary intended target, but by the entire spectrum of molecular recognition patterns exhibited by the kinome. Because all protein kinases share a conserved bilobular fold, a three-dimensional model can be built for every kinase with sufficient quality to identify most of the residues involved in inhibitor binding. Bioinformatics strategies are playing an important role in identifying variations in the protein sequences that may serve as specificity determinants. In addition to the experimentally determined structures of 64 kinases (Figure 18.1B), reliable homology models can now be constructed for more than half of the human kinome (using the X-ray structures as templates). The unprecedented level of structural understanding in combination with the large amount of data that are being produced by kinome-wide inhibitor-binding assays provides a framework to critically evaluate the specificity hypotheses generated by bioinformatics analyses and create strategies for the rapid identification of candidate inhibitors with the desired target-inhibition profiles. By using a scaffold-based approach, these strategies can be carried out by systematically adding molecular features that satisfy the shared requirement of intended targets while minimizing liabilities towards undesired off-targets.

Acknowledgement

The authors thank their colleagues at Plexxikon, Inc., for stimulating discussions. CZ thanks Drs. Rick Artis and Peter Hirth for suggestions and support.

References

1. G. Manning, D. B. Whyte, R. Martinez, T. Hunter and S. Sudarsanam, The protein kinase complement of the human genome, *Science*, 2002, **298**, 1912–1934.
2. P. Cohen, Protein kinases – the major drug targets of the twenty-first century?, *Nat. Rev. Drug Discovery*, 2002, **1**, 309–315.
3. S. K. Hanks and T. Hunter T, Protein kinases 6. The eukaryotic protein kinase superfamily: kinase (catalytic) domain structure and classification, *FASEB J.*, 1995, **9**, 576–596.
4. B. J. Druker, M. Talpaz, D. J. Resta, B. Peng, E. Buchdunger, J. M. Ford, N. B. Lydon, H. Kantarjian, R. Capdeville, S. Ohno-Jones and C. L. Sawyers, Efficacy and safety of a specific inhibitor of the BCR–ABL tyrosine kinase in chronic myeloid leukemia, *N. Engl. J. Med.*, 2001, **344**, 1031–1037.

5. A. J. Barker, K. H. Gibson, W. Grundy, A. A. Godfrey, J. J. Barlow, M. P. Healy, J. R. Woodburn, S. E. Ashton, B. J. Curry, L. Scarlett, L. Henthorn and L. Richards, Studies leading to the identification of ZD1839 (IRESSA): an orally active, selective epidermal growth factor receptor tyrosine kinase inhibitor targeted to the treatment of cancer, *Bioorg. Med. Chem. Lett.*, 2001, **11**, 1911–1914.
6. D. Soulieres, N. N. Senzer, E. E. Vokes, M. Hidalgo, S. S. Agarwala and L. L. Siu, Multicenter phase II study of erlotinib, an oral epidermal growth factor receptor tyrosine kinase inhibitor, in patients with recurrent or metastatic squamous cell cancer of the head and neck, *J. Clin. Oncol.*, 2004, **22**, 77–85.
7. T. Force, K. Kuida, M. Namchuk, K. Parang and J. M. Kyriakis, Inhibitors of protein kinase signaling pathways: emerging therapies for cardiovascular disease, *Circulation*, 2004, **109**, 1196–1205.
8. D. S. Krause and R. A. Van Etten, Tyrosine kinases as targets for cancer therapy, *N. Engl. J. Med.*, 2005, **353**, 172–187.
9. D. R. Knighton, J. H. Zheng, L. F. Ten Eyck, V. A. Ashford, N. H. Xuong, S. S. Taylor and J. M. Sowadski, Crystal structure of the catalytic subunit of cyclic adenosine monophosphate-dependent protein kinase, *Science*, 1991, **253**, 407–414.
10. H. L. De Bondt, J. Rosenblatt, J. Jancarik, H. D. Jones, D. O. Morgan and S. H. Kim, Crystal structure of cyclin-dependent kinase 2, *Nature*, 1993, **363**, 595–602.
11. F. Zhang, A. Strand, D. Robbins, M. H. Cobb and E. J. Goldsmith, Atomic structure of the MAP kinase ERK2 at 2.3 Å resolution, *Nature*, 1994, **367**, 704–711.
12. S. H. Hu, M. W. Parker, J. Y. Lei, M. C. Wilce, G. M. Benian and B. E. Kemp, Insights into autoregulation from the crystal structure of twitchin kinase, *Nature*, 1994, **369**, 581–584.
13. S. R. Hubbard, L. Wei, L. Ellis and W. A. Hendrickson, Crystal structure of the tyrosine kinase domain of the human insulin receptor, *Nature*, 1994, **372**, 746–754.
14. S. H. Kim, U. Schulze-Gahmen, J. Brandsen and W. F. de Azevedo Junior, Structural basis for chemical inhibition of CDK2, *Prog. Cell Cycle Res.*, 1996, **2**, 137–145.
15. R. A. Engh, A. Girod, V. Kinzel, R. Huber and D. Bossemeyer, Crystal structures of catalytic subunit of cAMP-dependent protein kinase in complex with isoquinolinesulfonyl protein kinase inhibitors H7, H8, and H89. Structural implications for selectivity, *J. Biol. Chem.*, 1996, **271**, 26157–26164.
16. M. Huse and J. Kuriyan, The conformational plasticity of protein kinases, *Cell*, 2002, **109**, 275–282.
17. K. Niefind, M. Putter, B. Guerra, O. G. Issinger and D. Schomburg, GTP plus water mimic ATP in the active site of protein kinase CK2, *Nat. Struct. Biol.*, 1999, **6**, 1100–1103.

18. C. E. Fitzgerald, S. B. Patel, J. W. Becker, P. M. Cameron, D. Zaller, V. B. Pikounis, S. J. O'Keefe and G. Scapin, Structural basis for p38alpha MAP kinase quinazolinone and pyridol-pyrimidine inhibitor specificity, *Nat. Struct. Biol.*, 2003, **10**, 764–769.
19. J. Dumas, Protein kinase inhibitors: emerging pharmacophores 1997–2000, *Expert Opin. Ther. Patents*, 2001, **11**, 405–429.
20. A. Kumar, V. Mandiyan, Y. Suzuki, C. Zhang, J. Rice, J. Tsai, D. R. Artis, P. Ibrahim and R. Bremer, Crystal structures of proto-oncogene kinase Pim1: a target of aberrant somatic hypermutations in diffuse large cell lymphoma, *J. Mol. Biol.*, 2005, **348**, 183–193.
21. L. Tong, S. Pav, D. M. White, S. Rogers, K. M. Crane, C. L. Cywin, M. L. Brown and C. A. Pargellis, A highly specific inhibitor of human p38 MAP kinase binds in the ATP pocket, *Nat. Struct. Biol.*, 1997, **4**, 311–316.
22. Z. Wang, B. J. Canagarajah, J. C. Boehm, S. Kassisa, M. H. Cobb, P. R. Young, S. Abdel-Meguid, J. L. Adams and E. J. Goldsmith, Structural basis of inhibitor selectivity in MAP kinases, *Structure*, 1998, **6**, 1117–1128.
23. N. S. Gray, L. Wodicka, A. M. Thunnissen, T. C. Norman, S. Kwon, F. H. Espinoza, D. O. Morgan, G. Barnes, S. LeClerc, L. Meijer, S. H. Kim, D. J. Lockhart and P. G. Schultz, Exploiting chemical libraries, structure, and genomics in the search for kinase inhibitors, *Science*, 1998, **281**, 533–538.
24. S. Blencke, B. Zech, O. Engkvist, Z. Greff, L. Orfi, Z. Horvath, G. Keri, A. Ullrich and H. Daub, Characterization of a conserved structural determinant controlling protein kinase sensitivity to selective inhibitors, *Chem. Biol.*, 2004, **11**, 691–701.
25. G. Scapin, S. B. Patel, J. Lisnock, J. W. Becker and P. V. LoGrasso, The structure of JNK3 in complex with small molecule inhibitors: structural basis for potency and selectivity, *Chem. Biol.*, 2003, **10**, 705–712.
26. H. Daub, K. Specht and A. Ullrich, Strategies to overcome resistance to targeted protein kinase inhibitors, *Nat. Rev. Drug Discovery.*, 2004, **3**, 1001–1010.
27. N. P. Shah, J. M. Nicoll, B. Nagar, M. E. Gorre, R. L. Paquette, J. Kuriyan and C. L. Sawyers, Multiple BCR–ABL kinase domain mutations confer polyclonal resistance to the tyrosine kinase inhibitor imatinib (STI571) in chronic phase and blast crisis chronic myeloid leukemia, *Cancer Cell*, 2002, **2**, 117–125.
28. S. R. Hubbard, Crystal structure of the activated insulin receptor tyrosine kinase in complex with peptide substrate and ATP analog, *EMBO J.*, 1997, **16**, 5572–5581.
29. M. Mohammadi, G. McMahon, L. Sun, C. Tang, P. Hirth, B. K. Yeh, S. R. Hubbard and J. Schlessinger, Structures of the tyrosine kinase domain of fibroblast growth factor receptor in complex with inhibitors, *Science*, 1997, **276**, 955–960.
30. E. R. Wood, A. T. Truesdale, O. B. McDonald, D. Yuan, A. Hassell, S. H. Dickerson, B. Ellis, C. Pennisi, E. Horne, K. Lackey, K. J. Alligood, D. W.

Rusnak, T. M. Gilmer and L. Shewchuk, A unique structure for epidermal growth factor receptor bound to GW572016 (Lapatinib): relationships among protein conformation, inhibitor off-rate, and receptor activity in tumor cells, *Cancer Res.*, 2004, **64**, 6652–6659.

31. J. F. Ohren, H. Chen, A. Pavlovsky, C. Whitehead, E. Zhang, P. Kuffa, C. Yan, P. McConnell, C. Spessard, C. Banotai, W. T. Mueller, A. Delaney, C. Omer, J. Sebolt-Leopold, D. T. Dudley, I. K. Leung, C. Flamme, J. Warmus, M. Kaufman, S. Barrett, H. Tecle and C. A. Hasemann, Structures of human MAP kinase kinase 1 (MEK1) and MEK2 describe novel noncompetitive kinase inhibition, *Nat. Struct. Mol. Biol.*, 2004, **11**, 1192–1197.
32. T. Schindler, W. Bornmann, P. Pellicena, W. T. Miller, B. Clarkson and J. Kuriyan, Structural mechanism for STI-571 inhibition of abelson tyrosine kinase, *Science*, 2000, **289**, 1938–1942.
33. C. Pargellis, L. Tong, L. Churchill, P. F. Cirillo, T. Gilmore, A. G. Graham, P. M. Grob, E. R. Hickey, N. Moss, S. Pav and J. Regan, Inhibition of p38 MAP kinase by utilizing a novel allosteric binding site, *Nat. Struct. Biol.*, 2002, **9**, 268–272.
34. P. T. Wan, M. J. Garnett, S. M. Roe, S. Lee, D. Niculescu-Duvaz, V. M. Good, C. M. Jones, C. J. Marshall, C. J. Springer, D. Barford and R. Marais, Mechanism of activation of the RAF-ERK signaling pathway by oncogenic mutations of B-RAF, *Cell*, 2004, **116**, 855–867.
35. P. W. Manley, G. Bold, J. Bruggen, G. Fendrich, P. Furet, J. Mestan, C. Schnell, B. Stolz, T. Meyer, B. Meyhack, W. Stark, A. Strauss and J. Wood, Advances in the structural biology, design and clinical development of VEGF-R kinase inhibitors for the treatment of angiogenesis, *Biochim. Biophys. Acta.*, 2004, **1697**, 17–27.
36. A. A. Russo, L. Tong, J. O. Lee, P. D. Jeffrey and N. P. Pavletich, Structural basis for inhibition of the cyclin-dependent kinase Cdk6 by the tumour suppressor p16INK4a, *Nature*, 1998, **395**, 237–243.
37. X. Huang, M. Begley, K. A. Morgenstern, Y. Gu, P. Rose, H. Zhao and X. Zhu, Crystal structure of an inactive Akt2 kinase domain, *Structure*, 2003, **11**, 21–30.
38. J. Griffith, J. Black, C. Faerman, L. Swenson, M. Wynn, F. Lu, J. Lippke and K. Saxena, The structural basis for autoinhibition of FLT3 by the juxtamembrane domain, *Mol. Cell*, 2004, **13**, 169–178.
39. C. D. Mol, D. R. Dougan, T. R. Schneider, R. J. Skene, M. L. Kraus, D. N. Scheibe, G. P. Snell, H. Zou, B. C. Sang and K. P. Wilson, Structural basis for the autoinhibition and STI-571 inhibition of c-Kit tyrosine kinase, *J. Biol. Chem.*, 2004, **279**, 31655–31663.
40. J. S. Sebolt-Leopold, D. T. Dudley, R. Herrera, K. Van Becelaere, A. Wiland, R. C. Gowan, H. Tecle, S. D. Barrett, A. Bridges, S. Przybranowski, W. R. Leopold and A. R. Saltiel, Blockade of the MAP kinase pathway suppresses growth of colon tumors *in vivo*, *Nat. Med.*, 1999, **5**, 810–816.

41. S. Campos, O. Hamid, M. V. Seiden, A. Oza, M. Plante, R. K. Potkul, P. F. Lenehan, E. P. Kaldjian, M. L. Varterasian, C. Jordan, C. Charbonneau and H. Hirte, Multicenter, randomized phase II trial of oral CI-1033 for previously treated advanced ovarian cancer, *J. Clin. Oncol.*, 2005, **23**, 5597–5604.
42. M. S. Cohen, C. Zhang, K. M. Shokat and J. Taunton, Structural bioinformatics-based design of selective, irreversible kinase inhibitors, *Science*, 2005, **308**, 1318–1321.
43. H. J. Broxterman and N. H. Georgopapadakou, Anticancer therapeutics: 'addictive' targets, multi-targeted drugs and new drug combinations, *Drug Resist. Updat.*, 2005, **8**, 183–197.
44. M. A. Fabian, W. H. Biggs 3rd, D. K. Treiber, C. E. Atteridge, M. D. Azimioara, M. G. Benedetti, T. A. Carter, P. Ciceri, P. T. Edeen, M. Floyd, J. M. Ford, M. Galvin, J. L. Gerlach, R. M. Grotzfeld, S. Herrgard, D. E. Insko, M. A. Insko, A. G. Lai, J. M. Lelias, S. A. Mehta, Z. V. Milanov, A. M. Velasco, L. M. Wodicka, H. K. Patel, P. P. Zarrinkar and D. J. Lockhart, A small molecule-kinase interaction map for clinical kinase inhibitors, *Nat. Biotechnol.*, 2005, **23**, 329–336.

Subject Index